Stephen Creagh

Recent advances in technology have made it possible to fabricate structures whose dimensions are much smaller than the mean free path of an electron. This is the first text-book to give a thorough account of the theory of electronic transport in such mesoscopic systems. Important concepts are illustrated by reference to relevant experimental results.

The book begins with a chapter summarizing the necessary background material. The next chapter introduces the 'transmission formalism' which is widely used in describing mesoscopic transport. The applicability of this formalism to different transport regimes is examined and practical methods for evaluating the transmission function are discussed. This formalism is then used to describe three key topics in mesoscopic physics: quantum Hall effect, localization, and double-barrier tunneling. Optical analogies to mesoscopic phenomena are discussed briefly. The book closes with a simple intuitive description of the non-equilibrium Green's function formalism and its relation to the transmission formalism.

Emphasizing basic concepts and techniques throughout, and complete with problems and solutions, the book will be of great interest to graduate students as well as to established researchers interested in mesoscopic physics and nanoelectronics.

Cambridge Studies in Semiconductor Physics and
Microelectronic Engineering: 3

EDITED BY
Haroon Ahmed
Cavendish Laboratory, University of Cambridge
Michael Pepper
Cavendish Laboratory, University of Cambridge
Alec Broers
Department of Engineering, University of Cambridge

ELECTRONIC TRANSPORT IN MESOSCOPIC SYSTEMS

TITLES IN THIS SERIES

ELECTRONIC TRANSPORT IN MESOSCOPIC SYSTEMS

SUPRIYO DATTA

Professor of Electrical Engineering
Purdue University

CAMBRIDGE
UNIVERSITY PRESS

Published by the Press Syndicate of the University of Cambridge
The Pitt Building, Trumpington Street, Cambridge CB2 1RP
40 West 20th Street, New York, NY 10011–4211, USA
10 Stamford Road, Oakleigh, Melbourne 3166, Australia

First published 1995

Printed in Great Britain at the University Press, Cambridge

A catalogue record for this book is available from the British Library

Library of Congress cataloguing in publication data

Datta, Supriyo, 1954–
Electronic transport in mesoscopic systems / Supriyo Datta.
p.   cm. – (Cambridge studies in semiconductor physics and
microelectronic engineering: 3)
ISBN 0 521 41604 3
1. Electron transport. 2. Semiconductors. 3. Green's functions.
4. Mesoscopic phenomena (Physics) I. Title. II. Series.
QC611.6.E45D38 1995
537.6′225–dc20   94-40111 CIP

ISBN 0 521 41604 3 hardback

WS

To Anuradha, Manoshi and Malika

Contents

Acknowledgements

I am grateful to Phil Bagwell, Markus Büttiker, Marc Cahay, Nick Giordano, Selman Hershfield, Rolf Landauer, Craig Lent and Mark Lundstrom for taking time off their busy schedules to look over earlier versions of the manuscript and provide me with useful feedback. Also, this book is based on a graduate course I developed at Purdue and I am indebted to the many students and ex-students who have contributed to my understanding in different ways.

A few common symbols

Symbol	Description	Units/Value
A	spectral function	$(\text{eV})^{-1}$
$\mathbf{A}$	vector potential	V s/m
B	magnetic field	10^4 G = 1 T = 1 V s/m^2
C	capacitance	F
D	diffusion coefficient	cm^2/s
e	electronic charge	-1.6×10^{-19} C
E	electric field	V/cm
E_f	equilibrium Fermi energy	eV
E_c	band-edge energy (bulk)	eV
E_s	band-edge energy (2-D)	eV
$f_0(E)$	equilibrium Fermi function	dimensionless
F_n	quasi-Fermi energy	eV
g	normalized conductance	dimensionless
G	conductance	Ω^{-1}
G^n	electron correlation function	
G^p	hole correlation function	$(\text{eV})^{-1}$ in a discrete
G^A	advanced Green's function	representation
G^R	retarded Green's function	
h	Planck's constant	6.63×10^{-34} J s
$\hbar$	$= h/2\pi$	
$i(E)$	current per unit energy	A/eV
I	current	A
J	current density	A/cm^2 (3-D)
		A/cm (2-D)
k_B	Boltzmann constant	0.087 meV/K
k_f	Fermi wavenumber	cm^{-1}

L	length	cm
L_m	mean free path	cm
L_φ	phase-relaxation length	cm
m	effective mass	for GaAs the standard value is $0.067\,m_0$. We will generally use $0.07\,m_0$ in our examples
m_0	free electron mass	9.1×10^{-31} kg
M	number of transverse modes	dimensionless
n_s	(areal) electron density	/cm^2
N_s	(2-D) density of states	$= m/\pi\hbar^2$ $\sim 2.9 \times 10^{10}$/cm^2 meV for GaAs
r	reflection amplitude	dimensionless
R	reflection probability	dimensionless
	resistance	Ω
S	area	cm^2
t	time	s
	transmission amplitude	dimensionless
T	temperature	K
	transmission probability	dimensionless
$\bar{T}$	transmission function	= (number of modes) × (transmission probability/mode)
U	potential energy	eV
v	velocity	cm/s
v_d	drift velocity	cm/s
v_f	Fermi velocity	cm/s
V	electrostatic potential	V
W	width	cm
Γ	energy broadening	eV
ε_n	cutoff energy	eV
ϑ	unit step function	$\vartheta(E) = 1$ if $E > 0$ $= 0$ if $E < 0$
λ_f	Fermi wavelength	cm
μ	mobility	cm^2/V s
	electrochemical potential	eV
ν	attempt frequency	s^{-1}
	linear density of scatterers	cm^{-1}
ρ	(2-D) resistivity	Ω

σ	(2-D) conductivity	Ω^{-1}
Σ^A	advanced self-energy	
Σ^R	retarded self-energy	(eV) in a discrete
Σ^{in}	inscattering function	representation
Σ^{out}	outscattering function	
τ_m	momentum-relaxation time	s
τ_φ	phase-relaxation time	s
$\omega_c/2\pi$	cyclotron frequency	s^{-1}

Please note that we have often used the terms 'electrochemical potential' (μ) and 'quasi-Fermi energy' (F_n) interchangeably.

Introductory remarks

It is well-known that the conductance (G) of a rectangular two-dimensional conductor is directly proportional to its width (W) and inversely proportional to its length (L); that is,

$$G = \sigma W/L$$

where the conductivity σ is a material property of the sample independent of its dimensions. How small can we make the dimensions (W and/or L) before this ohmic behavior breaks down? This question has intrigued scientists for a long time. During the 1980s it became possible to fabricate small conductors and explore this question experimentally, leading to significant progress in our understanding of the meaning of resistance at the microscopic level. What emerged in the process is a conceptual framework for describing current flow on length scales shorter than a mean free path. We believe that these concepts should be useful to a broad spectrum of scientists and engineers. This book represents an attempt to present these developments in a form accessible to graduate students and to non-specialists.

Small conductors whose dimensions are intermediate between the microscopic and the macroscopic are called mesoscopic. They are much larger than microscopic objects like atoms, but not large enough to be 'ohmic'. A conductor usually shows ohmic behavior if its dimensions are much larger than each of three characteristic length scales: (1) the de Broglie wavelength, which is related to the kinetic energy of the electrons, (2) the mean free path, which is the distance that an electron travels before its initial momentum is destroyed and (3) the phase-relaxation length, which is the distance that an electron travels before its initial phase is destroyed. These length scales vary widely from one material to another and are also strongly affected by temperature,

1

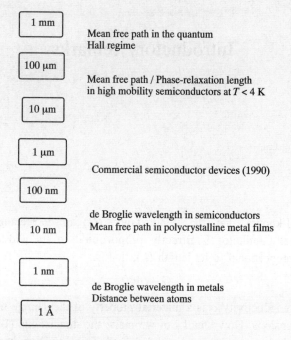

Fig. 0.1. A few relevant length scales. Note that
1 μm = 10^{-6} m = 10^{-4} cm
1 nm = 10^{-9} m = 10 angstroms (Å).

magnetic field etc. (Fig. 0.1). For this reason, mesoscopic transport phenomena have been observed in conductors having a wide range of dimensions from a few nanometers to hundreds of microns (that is, micrometers).

Mesoscopic conductors are usually fabricated by patterning a planar conductor that has one very small dimension to start with. For example, Fig. 0.2 shows a ring-shaped conductor having dimensions ~ 100 nm, patterned out of a polycrystalline gold film ~ 40 nm thick. This is the structure that was used for one of the landmark experiments in mesoscopic physics: the resistance of this ring was shown to oscillate as the magnetic field through it was changed because the magnetic field modified the interference between the electron waves traversing the two arms of the ring.

Although some of the pioneering experiments in this field were performed using metallic conductors, most of the recent work has been based on the gallium arsenide (GaAs)–aluminum gallium arsenide (AlGaAs) material system. Figure 0.3 shows a Hall bridge patterned out of a conducting layer ~ 10 nm thick formed at a GaAs–AlGaAs inter-

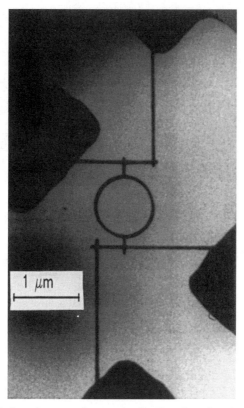

Fig. 0.2. Transmission electron micrograph of a ring-shaped resistor made from a 38 nm film of polycrystalline gold. The diameter of the ring is 820 nm and the thickness of the wires is 40 nm. Reproduced with permission from S. Washburn and R. A. Webb (1986). *Adv. Phys.* **35**, 375. The structures were fabricated by C. Umbach of IBM.

face. Four-terminal resistance measurements on such narrow conductors revealed many surprises, including negative resistance, that defied common sense (at least, common sense as extrapolated from macroscopic conductors).

In this book we will generally emphasize the work on semiconductors. However, the issues we will discuss have very little to do with device concepts or applications. There is an obvious difference in perspective between those interested in basic transport physics and those interested in device applications. From the point of view of basic physics the objective is to identify paradigms that help lay a clear conceptual framework for the subject. Low temperature low bias measurements are well suited for this purpose, because under these conditions the current is carried only by

Fig. 0.3. Scanning electron micrograph of a long wire 75 nm wide patterned from a GaAs–AlGaAs heterojunction. Four-terminal Hall measurements are made using voltage probes placed along the wire ~ 2 μm. apart. Reproduced with permission from M. L. Roukes, A. Scherer, S. J. Allen, H. G. Craighead, R. M. Ruthen, E. D. Beebe and J. P. Harbison (1987), *Phys. Rev. Lett.* **59**, 3011.

electrons at the Fermi energy. This is analogous to doing optical experiments with a monochromatic source. On the other hand, the commercial market for 'low temperature low bias devices' is severely limited. Interesting devices usually operate at room temperature under high bias such that transport occurs over a broad band of energies. Moreover, scattering processes are much stronger at higher temperatures. Most of the phenomena we will discuss would be washed out under these operating conditions. The only exception to this is resonant tunneling (see Chapter 6); useful device characteristics based on this phenomenon have been demonstrated at room temperature. But even with resonant tunneling we discuss only the basic physics and do not do justice to the device-related issues. We do not discuss the different material systems in which resonant tunneling has been observed or the different types of resonant tunneling phenomena (such as Γ–X tunneling or interband tunneling) that have been observed. Instead we focus on conceptual issues like the difference between coherent and sequential tunneling or between resonant tunneling and single-electron tunneling.

We have tried to emphasize basic concepts and techniques at the expense of details, as is appropriate for a text book. The few references at the end of each chapter are largely review articles and tutorial discussions. For a thorough review of this field we refer the reader to the article by C. W. J. Beenakker and H. van Houten entitled 'Quantum transport in semiconductor nanostructures' appearing in *Solid State Physics*, vol. 44, eds. H. Ehrenreich and D. Turnbull (New York, Academic Press, 1991). This article also provides a fairly comprehensive list of references until 1990.

Since we expect our readers to come from different backgrounds, we will start in Chapter 1 with a brief review of some basic concepts.

In Chapter 2 we discuss a concept that has proved very useful in understanding mesoscopic transport, namely that the current flow through a conductor is proportional to a 'transmission function' describing the ease with which electrons can transmit through it. The applicability of this concept to different regimes of transport is critically examined. In Chapter 3 we describe methods for calculating the transmission function. At the same time we relate the transmission formalism to other formalisms that are widely used in the literature.

Chapters 4, 5 and 6 use the transmission formalism to describe three major paradigms of mesoscopic physics namely the quantum Hall effect, localization and double-barrier tunneling. In Chapter 7 we briefly explore the similarities and differences between electron waves and electromagnetic waves pointing out optical analogies to different mesoscopic phenomena. The discussion is qualitative and can be read at any stage without much reference to the rest of the book.

Finally in Chapter 8 we describe the non-equilibrium Green's function formalism which is conceptually more complicated than the transmission formalism, but is more generally applicable. Indeed it provides a general framework for quantum transport in weakly interacting systems similar to that provided by the Boltzmann formalism for semiclassical transport.

All our discussions are based on a simple one-particle picture. We have avoided the use of advanced concepts like second quantized operators, even when discussing advanced topics like the Green's function formalism. Section 5.5 and Chapter 8 are relatively more difficult and could be skipped depending on the background of the students. But we believe that, with a little effort, even these sections can be understood by students who have taken a graduate level course in quantum mechanics.

Each chapter starts with an introduction and ends with a brief summary. It is followed by a set of detailed exercises designed to complement the material covered in the chapter. The solutions to these exercises are provided at the end of the book.

1

Preliminary concepts

We start this chapter with a brief review of some basic concepts. First in Section 1.1 we introduce the gallium arsenide (GaAs)/aluminum gallium arsenide (AlGaAs) material system which provides a very high quality two-dimensional conduction channel and has been widely used in mesoscopic experiments. Section 1.2 summarizes the free electron model that is commonly used to describe conduction electrons in metals and semiconductors. Next we discuss different characteristic lengths like the de Broglie wavelength, mean free path and the phase-relaxation length which determine the length scale at which mesoscopic effects appear (Section 1.3). The variation of resistance in the presence of a magnetic field is widely used to characterize conducting films. Both the low-field properties (Section 1.4) and the high-field properties (Section 1.5) yield valuable information regarding the electron density and mobility.

In Section 1.6 we introduce the concept of transverse modes which plays a prominent role in the theory of mesoscopic conductors and will appear repeatedly in this book. Finally in Section 1.7 we address an important conceptual issue that arises in the description of degenerate conductors, that is, conductors with a Fermi energy that is much greater than $k_B T$. Normally we view the current as being carried by all the conduction electrons which drift along slowly. However, in degenerate conductors it is more appropriate to view the current as being carried by a few electrons close to the Fermi energy which move much faster. One consequence of this is that the conductance of degenerate conductors is determined by the properties of electrons near the Fermi energy rather than the entire sea of electrons.

1.1 Two-dimensional electron gas (2-DEG)

Recent work on mesoscopic conductors has largely been based on GaAs–AlGaAs heterojunctions where a thin two-dimensional conducting layer is formed at the interface between GaAs and AlGaAs. To understand why this layer is formed consider the conduction and valence band line-up in the z-direction when we first bring the layers in contact (Fig. 1.1.1a). The Fermi energy E_f in the widegap AlGaAs layer is higher than that in the narrowgap GaAs layer. Consequently electrons spill over from the

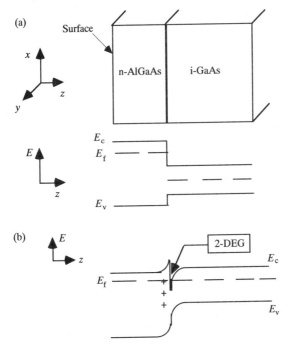

Fig. 1.1.1. Conduction and valence band line-up at a junction between an n-type AlGaAs and intrinsic GaAs, (*a*) before and (*b*) after charge transfer has taken place. Note that this is a cross-sectional view. Patterning (as shown in Fig. 0.3) is done on the surface (*x–y* plane) using lithographic techniques.

n-AlGaAs leaving behind positively charged donors. This space charge gives rise to an electrostatic potential that causes the bands to bend as shown. At equilibrium the Fermi energy is constant everywhere. The electron density is sharply peaked near the GaAs–AlGaAs interface (where the Fermi energy is inside the conduction band) forming a thin conducting layer which is usually referred to as the two-dimensional electron gas (2-DEG in short). The carrier concentration in a 2-DEG typically ranges from $2 \times 10^{11}/cm^2$ to $2 \times 10^{12}/cm^2$ and can be depleted by applying a negative voltage to a metallic gate deposited on the surface. The practical importance of this structure lies in its use as a field effect transistor [1.2, 1.3] which goes under a variety of names such as MODFET (MOdulation Doped Field Effect Transistor) or HEMT (High Electron Mobility Transistor).

Note that this structure is similar to standard silicon MOSFETs, where the 2-DEG is formed in silicon instead of GaAs. The role of the wide-gap AlGaAs is played by a thermally grown oxide layer (SiO_x). Indeed much of the pioneering work on the properties of two-dimensional conductors was performed using silicon MOSFETs [1.4].

Mobility

What makes the 2-DEG in GaAs very special is the extremely low scattering rates that have been achieved. The mobility (at low temperatures) provides a direct measure of the momentum relaxation time as limited by impurities and defects. Let us first briefly explain the meaning of mobility. In equilibrium the conduction electrons move around randomly not producing any current in any direction. An applied electric field **E** gives them a drift velocity $\mathbf{v}_d$ in the direction of the force $e\mathbf{E}$ as shown in Fig. 1.1.2. To relate the drift velocity to the electric field we note that, at

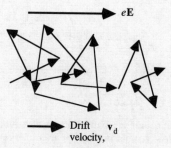

Fig. 1.1.2. In the presence of an electric field the electrons acquire a drift velocity superposed on their random motion.

steady-state, the rate at which the electrons receive momentum from the external field is exactly equal to the rate at which they lose momentum (**p**) due to scattering forces:

$$\left[\frac{d\mathbf{p}}{dt}\right]_{\text{scattering}} = \left[\frac{d\mathbf{p}}{dt}\right]_{\text{field}}$$

Hence, (τ_m: momentum relaxation time)

$$\frac{m\mathbf{v}_d}{\tau_m} = e\mathbf{E} \quad \Rightarrow \quad \mathbf{v}_d = \frac{e\tau_m}{m}\mathbf{E}$$

The mobility is defined as the ratio of the drift velocity to the electric field:

$$\mu \equiv \left|\frac{v_d}{E}\right| = \frac{|e|\tau_m}{m} \tag{1.1.1}$$

Mobility measurement using the Hall effect (see Section 1.5) is a basic characterization tool for semiconducting films. Once the mobility is known, the momentum relaxation time is readily deduced from Eq.(1.1.1).

In bulk semiconductors as we go down from room temperature, the momentum relaxation time increases at first due to the suppression of phonon scattering. But it does not increase any further once the phonon scattering is small enough that impurity scattering becomes the dominant mechanism (see Fig. 1.1.3). With a donor concentration of 10^{17}/cm^3 the highest mobility is less than 10^4 cm^2/V s. Higher mobilities can be obtained with undoped samples but this is not very useful since there are very few conduction electrons.

In a 2-DEG, on the other hand, carrier concentrations of 10^{12}/cm^2 in a layer of thickness ~100 Å (equivalent bulk concentration of 10^{18}/cm^3) have been obtained with mobilities in excess of 10^6 cm^2/V s (the current record is almost an order of magnitude larger than what is shown in Fig. 1.1.3). The reason is the spatial separation between the donor atoms in the AlGaAs layer and the conduction electrons in the GaAs layer. This reduces the scattering cross-section due to the impurities, leading to weaker scattering. Often an extra buffer layer of undoped AlGaAs is introduced between the GaAs and the n-AlGaAs in order to increase the separation between the 2-DEG in the GaAs and the ionized donors in the AlGaAs. This reduces the scattering but it also reduces the carrier concentration.

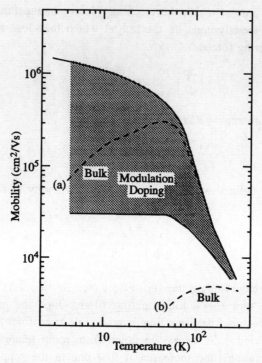

Fig. 1.1.3. Mobility vs. temperature in modulation-doped structures. Higher mobility (but lower carrier concentration) is obtained with thicker buffer layers. Also shown for comparison is the mobility in (a) high purity bulk GaAs and in (b) doped GaAs for use in FETs. Adapted with permission from Fig. 9 of T. J. Drummond, W. T. Masselink and H. Morkoc (1986). *Proc. IEEE*, **74**, 779. © 1986 IEEE

1.2 Effective mass, density of states etc.

Effective mass equation

Electronic conduction in semiconductors can take place either through electrons in the conduction band or through holes in the valence band. However, most experiments on mesoscopic conductors involve the flow of electrons in the conduction band and in this book we will assume that this is the case. The dynamics of electrons in the conduction band can be described by an equation of the form

$$\left[E_c + \frac{\left(i\hbar\nabla + e\mathbf{A} \right)^2}{2m} + U(\mathbf{r}) \right] \Psi(\mathbf{r}) = E\Psi(\mathbf{r}) \qquad (1.2.1)$$

where $U(\mathbf{r})$ is the potential energy due to space-charge etc., $\mathbf{A}$ is the vector potential and m is the effective mass. Although Eq.(1.2.1) looks just like the Schrödinger equation it is really what is called a single-band effective mass equation. The lattice potential, which is periodic on an atomic scale, does not appear explicitly in Eq.(1.2.1); its effect is incorporated through the effective mass m which we will assume to be spatially constant. Any band discontinuity ΔE_c at heterojunctions is incorporated by letting E_c be position-dependent.

It should be noted that the wavefunctions that we calculate from Eq.(1.2.1) are not the true wavefunctions but are smoothed out versions that do not show any rapid variations on the atomic scale. This can be seen easily by considering a homogeneous semiconductor with $U(\mathbf{r}) = 0$, $\mathbf{A} = 0$ and $E_c =$ constant. The wavefunctions satisfying Eq.(1.2.1) have the form of plane waves

$$\Psi(\mathbf{r}) = \exp[i\mathbf{k}.\mathbf{r}]$$

and not that of Bloch waves

$$\Psi(\mathbf{r}) = u_k(\mathbf{r})\exp[i\mathbf{k}.\mathbf{r}]$$

since the lattice potential is not included. The simplified description based on the single-band effective mass equation, Eq.(1.2.1), is usually adequate for conduction band electrons at low fields in GaAs–AlGaAs heterostructures and we will adopt it throughout this book. A review of the techniques used for detailed calculations of electronic states in heterostructures that go beyond the single-band effective mass equation can be found in the article by G. Bastard, J. A. Brum and R. Ferreira (1991), entitled 'Electronic states in semiconductor heterostructures', in *Solid State Physics*, vol. 44, eds. H. Ehrenreich and D. Turnbull, (New York, Academic Press).

Subbands

Consider the 2-DEG shown in Fig. 1.1.1b. The electrons are free to propagate in the x–y plane but are confined by some potential $U(z)$ in the z-direction. The electronic wavefunctions in such a structure (with $\mathbf{A} = 0$, assuming zero magnetic field) can be written in the form

$$\Psi(\mathbf{r}) = \phi_n(z)\exp(ik_x x)\exp(ik_y y)$$

with the dispersion relation:

$$E = E_c + \varepsilon_n + \frac{\hbar^2}{2m}\left(k_x^2 + k_y^2\right)$$

The index n numbers the different subbands each having a different wave-function $\phi_n(z)$ in the z-direction and a cut-off energy ε_n. Usually at low temperatures with low carrier densities only the lowest subband with $n = 1$ is occupied and the higher subbands do not play any significant role. We can then ignore the z-dimension altogether and simply treat the conductor as a two-dimensional system in the x–y plane. Specifically, instead of Eq.(1.2.1) we will use the following equation:

$$\left[E_s + \frac{(i\hbar\nabla + e\mathbf{A})^2}{2m} + U(x,y)\right]\Psi(x,y) = E\Psi(x,y) \tag{1.2.2}$$

where $E_s = E_c + \varepsilon_1$. We will generally adopt this simplification throughout the book when discussing semiconductors. With metallic films, however, the electron density is so high that even a film of thickness 10 nm has several tens of occupied subbands and it is more accurate to treat it as a three-dimensional conductor.

Band diagrams

For a free electron gas in the absence of magnetic fields, the eigenfunctions are obtained from Eq.(1.2.2) by setting $U = 0$ and $\mathbf{A} = 0$. The eigenfunctions normalized to an area S have the form

$$\Psi(x,y) = \frac{1}{\sqrt{S}}\exp(ik_x x)\exp(ik_y y) \tag{1.2.3}$$

with eigenenergies given by

$$E = E_s + \frac{\hbar^2}{2m}\left(k_x^2 + k_y^2\right) \tag{1.2.4}$$

The dispersion relation (Eq.(1.2.4)) is sketched in Fig. 1.2.1a. It is also common to draw band diagrams in real space (see Fig. 1.2.1b) showing only the bottom of the band (corresponding to $k_x = k_y = 0$) and the Fermi energy.

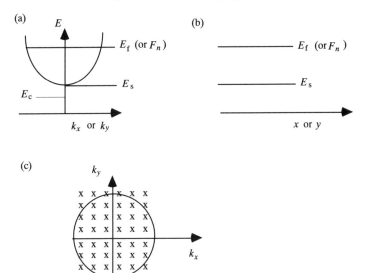

Fig. 1.2.1. (*a*) Dispersion relation for a free electron gas in two dimensions. Note that the bottom of the two-dimensional subband, E_s lies a little higher than the conduction band edge E_c in the bulk material. (*b*) Band diagram in real space showing only the bottom of the band and the Fermi energy, E_f. Under non-equilibrium conditions there is no common Fermi energy but one can often talk in terms of a quasi-Fermi level F_n. (*c*) Assuming periodic boundary conditions, k_x and k_y take on quantized values depending on the dimensions of the sample.

Density of states

Let us first calculate the total number of states $N_T(E)$ having an energy less than E. To do this we have to count the number of states contained within a circle of radius k (see Fig. 1.2.1c), where

$$E = E_s + \frac{\hbar^2 k^2}{2m}$$

We need to know the values of k_x–k_y that are allowed by the boundary conditions. For a large area conductor the real boundary conditions have minimal effect on the result, so that we could just as well use any boundary condition that is convenient. It is common to use the periodic boundary conditions which require k_x and k_y to take on quantized values depending on the dimensions L_x and L_y of the sample (n_x and n_y are integers):

$$k_x = n_x(2\pi/L_x) \quad \text{and} \quad k_y = n_y(2\pi/L_y) \tag{1.2.5}$$

Thus the area in the k_x–k_y plane 'occupied' by an individual state is given by (S: area of the conductor)

$$\frac{2\pi}{L_x} \times \frac{2\pi}{L_y} = \frac{4\pi^2}{S}$$

while the area enclosed by the circle is πk^2. Hence (for $E > E_s$)

$$N_T(E) = 2 \text{ (for spin)} \times \frac{\pi k^2}{4\pi^2/S} = S\frac{k^2}{2\pi} = \frac{mS}{\pi\hbar^2}(E - E_s)$$

This is the total number of states. The density of states per unit area per unit energy is given by

$$N(E) \equiv \frac{1}{S}\frac{d}{dE}N_T(E) = \frac{m}{\pi\hbar^2}\vartheta(E - E_s) \qquad (1.2.6)$$

where ϑ is the unit step function (see list of symbols). Thus the two-dimensional density of states is constant for all energies exceeding the subband energy E_s. It is approximately equal to $2.9 \times 10^{10}/\text{cm}^2$ meV if $m = 0.07$ times the free electron mass (a value typical of GaAs).

Degenerate and non-degenerate conductors

At equilibrium the available states in a conductor are filled up according to the Fermi function

$$f_0(E) = \frac{1}{1 + \exp\left[(E - E_f)/k_BT\right]} \qquad (1.2.7)$$

where E_f is the Fermi energy. Away from equilibrium the system has no common Fermi energy, but often we can talk in terms of a local quasi-Fermi level which can vary spatially and which can be different for different groups of states (such as electrons and holes) even at the same spatial location. We will generally use F_n to denote quasi-Fermi levels and reserve E_f for the equilibrium Fermi energy.

There are two limits in which the Fermi function inside the band ($E > E_s$) can be simplified somewhat making it easier to perform numerical calculations (see Fig. 1.2.2). One is the high temperature or the non-degenerate limit ($\exp[E_s - E_f]/k_BT \gg 1$) where

$$f_0(E) \approx \exp\left[-(E - E_f)/k_BT\right] \qquad (1.2.8)$$

(a) Non-degenerate limit

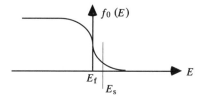

(b) Degenerate limit

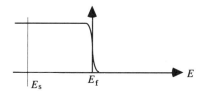

Fig. 1.2.2. The Fermi function inside the band $(E > E_s)$ can be approximated by (a) Eq.(1.2.8) in the non-degenerate limit and (b) by Eq.(1.2.9) in the degenerate limit.

The other is the low temperature or the degenerate limit $(\exp[E_s - E_f]/k_B T \ll 1)$ where

$$f_0(E) \approx \vartheta(E_f - E) \qquad (1.2.9)$$

In this book we will mainly be discussing degenerate conductors.

To relate the equilibrium electron density n_s (per unit area) to the Fermi energy we make use of the relation

$$n_s = \int N(E) f_0(E) dE$$

For degenerate conductors it is easy to perform the integral to obtain

$$n_s = N_s(E_f - E_s) \quad \text{where} \quad N_s \equiv m/\pi\hbar^2 \qquad (1.2.10)$$

where we have made use of Eqs.(1.2.6) and (1.2.9).

At low temperatures the conductance is determined entirely by electrons with energy close to the Fermi energy. The wavenumber of such electrons is referred to as the Fermi wavenumber (k_f):

$$E_f - E_s = \frac{\hbar^2 k_f^2}{2m} \quad \Rightarrow \quad \hbar k_f = \sqrt{2m(E_f - E_s)} \qquad (1.2.11)$$

Using Eq.(1.2.10) we can express the Fermi wavenumber in terms of the electron density:

$$k_f = \sqrt{2\pi n_s} \qquad (1.2.12)$$

The corresponding velocity is the Fermi velocity $v_f = \hbar k_f/m$.

1.3 Characteristic lengths

A conductor usually shows ohmic behavior if its dimensions are much larger than certain characteristic lengths, namely, (1) the de Broglie wavelength, (2) the mean free path, and (3) the phase-relaxation length. We will discuss these one by one. In addition to these characteristic lengths, the screening length can also play a significant role especially in low-dimensional conductors as we will see in Section 2.3 (see Fig. 2.3.3).

Wavelength (λ)

We have seen (Eq.(1.2.12)) that the Fermi wavenumber k_f goes up as the square root of the electron density. The corresponding wavelength goes down as the square root of the electron density:

$$\lambda_f = 2\pi/k_f = \sqrt{2\pi/n_s} \qquad (1.3.1)$$

For an electron density of $5 \times 10^{11}/cm^2$, the Fermi wavelength is about 35 nm. At low temperatures the current is carried mainly by electrons having an energy close to the Fermi energy so that the Fermi wavelength is the relevant length. Other electrons with less kinetic energy have longer wavelengths but they do not contribute to the conductance.

Mean free path (L_m)

An electron in a perfect crystal moves as if it were in vacuum but with a different mass. Any deviation from perfect crystallinity such as impurities, lattice vibrations (phonons) or other electrons leads to 'collisions' that scatter the electron from one state to another thereby changing its momentum. The momentum relaxation time τ_m is related to the collision time τ_c by a relation of the form

$$\frac{1}{\tau_m} \rightarrow \frac{1}{\tau_c}\alpha_m$$

where the factor α_m (lying between 0 and 1) denotes the 'effectiveness' of an individual collision in destroying momentum. For example if the collisions are such that the electrons are scattered only by a small angle then very little momentum is lost in an individual collision. The factor α_m is then very small so that the momentum relaxation time is much longer than the collision time. For a more detailed discussion of scattering times in semiconductors see, for example, Chapter 4 of S. Datta (1989), *Quantum Phenomena*, Modular Series on Solid-state Devices, vol. VIII, eds. R. F. Pierret and G. W. Neudeck, (New York, Addison–Wesley).

The mean free path, L_m, is the distance that an electron travels before its initial momentum is destroyed; that is,

$$L_m = v_f \, \tau_m \qquad (1.3.2)$$

where τ_m is the momentum relaxation time and v_f is the Fermi velocity. The Fermi velocity is given by

$$v_f = \frac{\hbar k_f}{m} = \frac{\hbar}{m}\sqrt{2\pi n_s} \rightarrow 3 \times 10^7 \, \text{cm}/\text{s} \quad \text{if} \quad n_s = 5 \times 10^{11}/\text{cm}^2$$

Assuming a momentum relaxation time of 100 ps we obtain a mean free path of $L_m = 30 \, \mu$m.

Phase-relaxation length (L_φ)

Let us first discuss what is meant by the phase-relaxation time (τ_φ). We will then relate it to the phase-relaxation length. In analogy with the momentum relaxation time we could write

$$\frac{1}{\tau_\varphi} \rightarrow \frac{1}{\tau_c}\alpha_\varphi$$

where the factor α_φ denotes the effectiveness of an individual collision in destroying phase. The destruction of phase is, however, a little more subtle than the destruction of momentum. A more careful discussion is required to define what the effectiveness factor α_φ is for different types of scattering processes.

One way to visualize the destruction of phase is in terms of a thought experiment involving interference. For example suppose we split a beam of electrons into two paths and then recombine them as shown in Fig. 1.3.1. In a perfect crystal the two paths would be identical resulting in constructive interference. By applying a magnetic field perpendicular to the plane containing the paths, one can change their relative phase (see for example, R. P. Feynman, (1965), *Lectures on Physics*, vol.II, Section 15–5, (New York, Addison–Wesley) thereby changing the interference

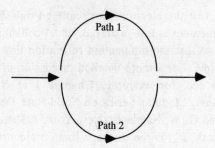

Fig. 1.3.1. A conceptual interference experiment involving the splitting of a beam of electrons and later recombining them.

alternately from constructive to destructive and back. Now suppose we are not in a perfect crystal but in a real one with collisions due to impurities, phonons etc. We would expect the interference amplitude to be reduced by a factor

$$\exp\left[-\tau_t/\tau_\varphi\right]$$

where τ_t is the transit time that the electron spends in each arm of the interferometer. Let us use this thought experiment to define the phase-relaxation time τ_φ and hence the effectiveness factor α_φ. Actually it is more than a thought experiment. The experiment with the mesoscopic ring referred to earlier (see Fig. 0.2) could be viewed as a laboratory implementation of this thought experiment.

Consider first what happens if we introduce impurities and defects randomly into each arm. The two arms are then no longer identical so that the interference may not be constructive at zero magnetic field. But the point is that as long as the impurities and defects are static there is a definite phase-relationship between the two paths and as we increase the magnetic field we would go through alternate cycles of constructive and destructive interference whose amplitude is unaffected by the length of each arm. We thus conclude that for static scatterers

$$\tau_\varphi \rightarrow \infty \quad (\text{that is } \alpha_\varphi \rightarrow 0)$$

This is confirmed by the fact that the resistance of a ring-shaped conductor (see Fig. 0.2) is observed to oscillate as a function of the magnetic field even though each arm is hundreds of mean free paths long.

The situation is different when we take into account the effect of a dynamic scatterer like lattice vibrations (phonons). The phase-relationship between the scattered waves in the two arms now varies randomly with time so that there is no stationary interference pattern. At a fixed value of the magnetic field the scattered waves show random variations from constructive to destructive interference which time-average to zero. Interference can only be observed between the unscattered components, whose amplitude decreases exponentially with the length of each arm.

Another important source of phase-randomizing collisions is electron–electron interactions. Electrons scatter off other electrons (which are not stationary) due to their mutual Coulomb repulsion. Interestingly the mean free path (L_m) is not affected by electron–electron scattering processes. This is because such processes do not lead to any loss in the net momentum. Any momentum lost by one electron is picked up by another. Consequently the effectiveness factor α_m is zero for such processes though α_φ is non-zero.

Impurity scattering too can be phase-randomizing if the impurity has an internal degree of freedom so that it can change its state. For example, magnetic impurities have an internal spin that fluctuates with time. Collisions with such impurities thus cause phase-relaxation. This has been observed in measurements on mesoscopic gold rings, where the oscillations in the resistance in a magnetic field were destroyed when manganese impurities were introduced by ion implantation. Interestingly the oscillations are quenched only at low magnetic fields but are still observed at high magnetic fields. The reason is that at high magnetic fields the energies of the different spin states separate due to the spin splitting. Once this splitting exceeds k_BT, spin fluctuations are suppressed and the magnetic impurity behaves like an ordinary rigid impurity with no internal degree of freedom (see the article by A. Benoit *et al.* in Anderson Localization, eds. T. Ando and H. Fukuyama, vol. 28 of Springer Proceedings in Physics, p.346).

The basic point is that rigid scatterers do not contribute to phase-relaxation; only fluctuating scatterers do. This is also in agreement with what one might have expected from a more philosophical argument. As Feynman describes in his lectures (see *Lectures on Physics*, vol. III, Section 1-1, (New York, Addison–Wesley 1965)), the interference between two paths is destroyed whenever we perform an experiment that allows us to tell which path the electron took. If on one of the paths an electron interacts with a scatterer causing it to change its state, then by measuring the state of the scatterer *we can tell which path the electron took*. Consequently such interactions destroy interference related effects. For a more detailed discussion of the equivalence of the two viewpoints see A. Stern, Y. Aharonov and Y. Imry (1990). *Phys. Rev. A*, **41**, 3436. See also the article by A. Leggett (1989) in *Nanostructure Physics and Fabrication*, eds. M. A. Reed and W. P. Kirk (New York, Academic Press).

The next question is: given a fluctuating scatterer, what is the associated phase-relaxation time? One might be tempted to guess that it should simply be equal to the collision time:

$$\tau_\varphi \to \tau_c$$

However, this is not necessarily true. For example, suppose we have a phonon that affects both arms of the ring equally (Fig. 1.3.1). In that case the phase associated with the two arms is randomized in a correlated way so that their phase difference is unaffected. Such a process should not affect the interference. Thus we would expect long wavelength phonons to be less effective in destroying phase. It has been argued (see B. L. Altshuler, A. G. Aronov and D. E. Khmelnitsky (1982). *J. Phys. C.*, **15**, 7367) that for a phonon with energy $\hbar\omega$, the mean squared energy spread of an electron after a time τ_φ is obtained by multiplying the square of the energy change per collision by the number of collisions

$$(\Delta\varepsilon)^2 = (\hbar\omega)^2 \, (\tau_\varphi/\tau_c)$$

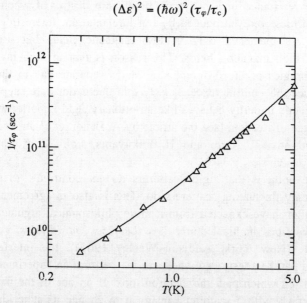

Fig. 1.3.2. Inverse phase-relaxation time (deduced from weak localization data) as a function of temperature for a GaAs sample having $n_s = 1.6 \times 10^{11}/\mathrm{cm}^2$, $\mu = 27\,000\ \mathrm{cm}^2/\mathrm{V}\ \mathrm{s}$. Reproduced with permission from Fig. 2 of K. K. Choi, D. C. Tsui and K. Alavi (1987), *Phys. Rev. B*, **36**, 7751.

The phase-relaxation time τ_φ is defined as the time after which the mean squared spread in the phase is of order one; that is,

$$\Delta\varphi \sim (\Delta\varepsilon)\tau_\varphi \sim 1 \quad \rightarrow \quad \tau_\varphi \sim (\tau_c/\omega^2)^{1/3}$$

showing that low frequency phonons are less effective in relaxing phase.

Usually at low temperatures the dominant source of phase-relaxation is electron–electron scattering. An electron is scattered by the fluctuating potential that it feels due to the other electrons. The frequency of electron–electron scattering depends on the excess energy of the electrons relative to the Fermi energy. An electron with a small excess energy Δ $(= E - E_f)$ has very few states to scatter down into since most states below it are already full. For this reason, the scattering is strongly suppressed by the exclusion principle as Δ tends to zero. The precise dependence of the phase-relaxation time on Δ depends on the dimensionality of the conductor. In a 2-DEG it has the form (see A. Yacoby *et al.* (1991), *Phys. Rev. Lett.* **66**, 1938 and references therein)

$$\frac{\hbar}{\tau_\varphi} \sim \frac{\Delta^2}{E_f}\left[\ln\left(\frac{E_f}{\Delta}\right) + \text{constant}\right]$$

Since the average excess energy of thermal electrons is $\sim k_BT$, the dependence of the phase-relaxation time on temperature is obtained simply by replacing Δ with k_BT.

An additional complication arises in low-mobility conductors having $\hbar/\tau_m > k_BT$. The phase-relaxation time due to electron–electron scattering then has an additional component which in two dimensions depends linearly on temperature (see K. K. Choi *et al.* (1987), *Phys. Rev. B*, **36**, 7751 and references therein). In low-mobility conductors the phase-relaxation time is usually obtained from weak localization experiments. This is an interference experiment that is conceptually more complicated than the simple two-arm interference in a ring-shaped conductor. We will discuss it in detail in Chapter 5. Here we simply show the measured phase-relaxation time as a function of the temperature to give the reader a feel for the order of magnitude of the times involved in low-mobility semiconductors (Fig. 1.3.2).

Now that we have discussed the phase-relaxation time (τ_φ), let us see how it is related to the phase-relaxation length (L_φ). The obvious approach is to multiply by the Fermi velocity:

$$L_\varphi = v_f\tau_\varphi \qquad (1.3.3)$$

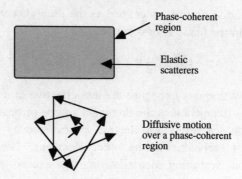

Fig. 1.3.3. For $\tau_\varphi \gg \tau_m$, transport within a phase-coherent region is diffusive since there are many elastic scatterers.

This is true if the phase-relaxation time is of the same order or shorter than the momentum relaxation time, that is, if $\tau_\varphi \sim \tau_m$, which is often the case with high-mobility semiconductors.

But with low-mobility semiconductors or polycrystalline metal films the momentum relaxation time can be considerably shorter than the phase-relaxation time: $\tau_\varphi \gg \tau_m$. We then have the situation depicted in Fig. 1.3.3. The motion of electrons over a phase-relaxation time is not ballistic. After an interval of time τ_m the velocity is completely randomized so that the electronic trajectory over a length of time τ_φ can be visualized as the sum of a number ($= \tau_\varphi/\tau_m$) of short trajectories each of length $\sim v_f \tau_m$. Since the individual trajectories are directed in random directions (θ), the root mean squared distance traveled by the electron in a particular direction is obtained by summing the squares of their lengths:

$$L_\varphi^2 = \frac{\tau_\varphi}{\tau_m} \left(v_f \tau_m \right)^2 \left\langle \cos^2 \theta \right\rangle \quad \text{where} \quad \left\langle \cos^2 \theta \right\rangle \equiv \int_{-\pi}^{+\pi} \frac{d\theta}{2\pi} \cos\theta = \frac{1}{2}$$

Hence
$$L_\varphi^2 = v_f^2 \tau_m \tau_\varphi / 2 \tag{1.3.4a}$$

As we will see later (see Eq.(1.7.8)) the diffusion coefficient is given by

$$D = v_f^2 \tau_m / 2$$

so that
$$L_\varphi^2 = D\tau_\varphi \tag{1.3.4b}$$

1.4 Low-field magnetoresistance

Drude model

Conductivity measurement in a weak magnetic field (generally referred to as a Hall measurement) is one of the basic tools used to characterize semiconducting films. This is because it allows us to deduce the carrier density n_s and the mobility μ individually, while the zero field conductivity only tells us the product of the two.

As we discussed for the zero field case (see Eq.(1.1.1)), at steady-state, the rate at which the electrons receive momentum from the external field is exactly equal to the rate at which they lose momentum due to scattering forces:

$$\left[\frac{d\mathbf{p}}{dt}\right]_{\text{scattering}} = \left[\frac{d\mathbf{p}}{dt}\right]_{\text{field}}$$

that is, (τ_m: momentum relaxation time)

$$\frac{m\mathbf{v}_d}{\tau_m} = e\left[\mathbf{E} + \mathbf{v}_d \times \mathbf{B}\right] \tag{1.4.1}$$

We can rewrite Eq.(1.4.1) in the form

$$\begin{bmatrix} m/e\tau_m & -B \\ +B & m/e\tau_m \end{bmatrix} \begin{pmatrix} v_x \\ v_y \end{pmatrix} = \begin{pmatrix} E_x \\ E_y \end{pmatrix}$$

where v_x, v_y are the x- and y- components of the drift velocity and E_x, E_y are the x- and y- components of the electric field. The current density $\mathbf{J}$ (per unit width) is related to the electron density n_s (per unit area) by the relation $\mathbf{J} = e\mathbf{v}_d n_s$, so that we can write

$$\begin{bmatrix} m/e\tau_m & -B \\ +B & m/e\tau_m \end{bmatrix} \begin{pmatrix} J_x/en_s \\ J_y/en_s \end{pmatrix} = \begin{pmatrix} E_x \\ E_y \end{pmatrix}$$

Rearranging we obtain

$$\begin{pmatrix} E_x \\ E_y \end{pmatrix} = \sigma^{-1} \begin{bmatrix} 1 & -\mu B \\ +\mu B & 1 \end{bmatrix} \begin{pmatrix} J_x \\ J_y \end{pmatrix} \tag{1.4.2}$$

where $\sigma \equiv |e| n_s \mu$ and $\mu \equiv |e| \tau_m / m$. Noting that the resistivity tensor is defined by the relation

$$\begin{pmatrix} E_x \\ E_y \end{pmatrix} = \begin{bmatrix} \rho_{xx} & \rho_{xy} \\ \rho_{yx} & \rho_{yy} \end{bmatrix} \begin{pmatrix} J_x \\ J_y \end{pmatrix}$$

we can write from Eq.(1.4.2)

$$\rho_{xx} = \sigma^{-1} \qquad (1.4.3a)$$

$$\rho_{yx} = -\rho_{xy} = (\mu B)/\sigma = B/|e|n_s \qquad (1.4.3b)$$

Thus this simple Drude model predicts that the longitudinal resistance is constant while the Hall resistance increases linearly with magnetic field.

Experiment

Experimentally the resistivity tensor is measured by preparing a rectangular sample, setting up a uniform current flow along the x-direction and measuring the longitudinal voltage drop $V_x = (V_1 - V_2)$ and the transverse (or Hall) voltage drop $V_H = (V_2 - V_3)$ (Fig. 1.4.1). Since $J_y = 0$, we can write

$$E_x = \rho_{xx}J_x \quad \text{and} \quad E_y = \rho_{yx}J_x$$

It is easy to see that $I = J_x W$, $V_x = E_x L$ and $V_H = E_y W$. Hence the resistivities ρ_{xx} and ρ_{yx} are related to the longitudinal and transverse voltages by

$$\rho_{xx} = \frac{V_x}{I}\frac{W}{L} \quad \text{and} \quad \rho_{yx} = \frac{V_H}{I}$$

Figure 1.4.2 shows the measured longitudinal voltage V_x and the transverse voltage V_H for a modulation-doped GaAs film using a rectangular Hall bridge with $W = 0.38$ mm and $L = 1$ mm and a current of $I = 25.5$ μA. At low magnetic fields the longitudinal voltage is nearly

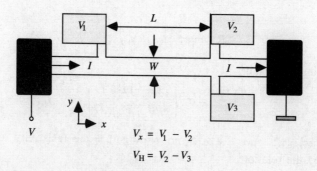

$$V_x = V_1 - V_2$$
$$V_H = V_2 - V_3$$

Fig. 1.4.1. Rectangular Hall bar for magnetoresistance measurements. The magnetic field is in the z-direction perpendicular to the plane of the conductor.

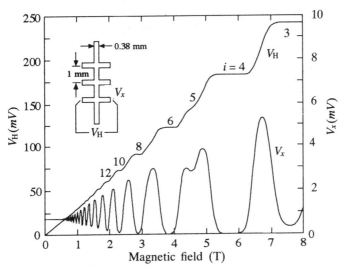

Fig. 1.4.2. Measured longitudinal and transverse voltages for a modulation-doped GaAs film at $T = 1.2$ K ($I = 25.5$ μA). Reproduced with permission from Fig. 1 of M. E. Cage, R. F. Dziuba and B. F. Field (1985), *IEEE Trans. Instrum. Meas.* **IM-34**, 301. © 1985 IEEE

constant while the Hall voltage increases linearly in agreement with the predictions of the semiclassical Drude model described above. At high fields, however, the longitudinal resistance shows pronounced oscillatory behavior while the Hall resistance exhibits plateaus corresponding to the minima in the longitudinal resistance. These features are usually absent at room temperature or even at 77 K but quite evident at cryogenic temperatures of 4 K and below. To understand these features we need to go beyond the Drude model and discuss the formation of Landau levels which is a quantum mechanical effect. That is what we will do in the next section.

Before we proceed it should be mentioned that we can obtain the carrier density n_s and the mobility μ from the measured low-field resistivities ρ_{xx} and ρ_{yx} using Eqs.(1.4.3a,b):

$$n_s = \left[|e| \frac{d\rho_{yx}}{dB} \right]^{-1} = \frac{I/|e|}{dV_H/dB} \qquad (1.4.4a)$$

$$\mu = \frac{1}{|e|n_s\rho_{xx}} = \frac{I/|e|}{n_s V_x W/L} \qquad (1.4.4b)$$

For this reason, Hall measurement is a basic characterization tool for semiconducting films (see Exercise E.1.2 at the end of this chapter).

1.5 High-field magnetoresistance

A comparison of the experimental data in Fig. 1.4.2 with the predictions of the Drude model (see Eqs.(1.4.3a,b)) shows clear disagreement at high magnetic fields. There are oscillations in the longitudinal resistivity ρ_{xx} which are referred to as Shubnikov–deHaas (or SdH) oscillations. Such oscillations are not unique to 2-D conductors and were first observed in bulk metals back in 1930. However, in 2-D semiconductors the effect is much larger. The minimum longitudinal resistivity ρ_{xx} is very nearly zero and there appear plateaus in the Hall resistivity ρ_{yx} whenever ρ_{xx} goes through a minimum. We will discuss these plateaus later in Chapter 4. For the moment let us concentrate on the SdH oscillations.

Origin of SdH oscillations

We will now try to explain the basic phenomenon underlying the SdH oscillations and show how we can deduce the carrier density from the period of the oscillations. Basically, the SdH oscillations arise because at high magnetic fields, the step-like density of states associated with a 2-DEG (see Eq.(1.2.6))

$$N_s(E) = \frac{m}{\pi \hbar^2} \vartheta(E - E_s)$$

breaks up into a sequence of peaks spaced by $\hbar \omega_c$ where $\omega_c = eB/m$ is the cyclotron frequency (B: magnetic field):

$$N_s(E,B) \approx \frac{2eB}{h} \sum_{n=0}^{\infty} \delta\left[E - E_s - \left(n + \tfrac{1}{2}\right)\hbar\omega_c\right] \qquad (1.5.1)$$

This is illustrated in Fig. 1.5.1. The spikes are ideally true delta functions. But in practice scattering processes spread them out in energy. As we change the magnetic field B, the energies of the Landau levels change. The resistivity ρ_{xx} goes through one cycle of oscillation as the Fermi energy moves from the center of one Landau level to the center of the next Landau level. This affords a simple method for calculating the electron density from the oscillations in ρ_{xx} which we will now describe.

It can be seen from Eq.(1.5.1) that for a given electron density n_s, we can calculate the number of occupied Landau levels simply by dividing by $2eB/h$. For example, if $B = 2$ T then $2eB/h = 9.6 \times 10^{10}/\text{cm}^2$. Hence if $n_s = 5 \times 10^{11}/\text{cm}^2$, then

$$\frac{n_s}{2eB/h} = 5.2$$

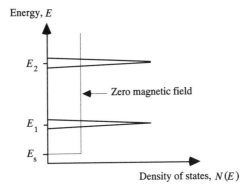

Fig. 1.5.1. Density of states $N_s(E,B)$ vs. energy E for a 2-DEG in a magnetic field. Note that $E_n = E_s + (n + 0.5)\hbar\omega_c$.

Hence five Landau levels are fully occupied while the sixth one is partially occupied. As we change the magnetic field B the number of occupied Landau levels changes. The resistivity ρ_{xx} goes through a maximum every time this number is a half-integer and the Fermi energy lies at the center of a Landau level. Hence the magnetic field values B_1 and B_2 corresponding to two successive peaks must be related by

$$\frac{n_s}{2eB_1/h} - \frac{n_s}{2eB_2/h} = 1$$

so that

$$n_s = \frac{2e}{h}\frac{1}{(1/B_1) - (1/B_2)}$$

We could choose many different values B_1 and B_2 corresponding to any pair of successive peaks. They should all yield approximately the same result for the carrier density. The usual procedure is to plot the positions of the maxima in ρ_{xx} as a function of $1/B$. They should lie in a straight line and the slope of this straight line gives the electron density (see Exercise E.1.2 at the end of this chapter).

Now that we understand the basic factor underlying SdH oscillations, let us try to explain why the density of states develops peaks at high magnetic fields as shown in Fig. 1.5.1.

Why do discrete states form at high magnetic fields?

A proper answer to this question requires us to start from the Schrödinger equation including a vector potential to represent the magnetic field and

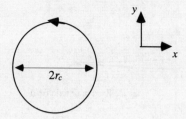

Fig. 1.5.2. Circular path described by a classical particle in a (z-directed) magnetic field.

calculate its eigenfunctions and eigenvalues. This is done in Section 1.7. However, the basic idea is easy to see using a simple classical argument.

Classically, an electron in a magnetic field goes round in a circular orbit as sketched in Fig. 1.5.2 (this is straightforward to show starting from Newton's law: $m\mathrm{d}\mathbf{v}/\mathrm{d}t = e\mathbf{v} \times \mathbf{B}$). The radius of the circle is proportional to the electron velocity, v:

$$r_\mathrm{c} = v/\omega_\mathrm{c} \tag{1.5.2}$$

where $\omega_\mathrm{c} = eB/m$. Classically the electron can have any velocity v and thus move on a circular path of any radius r_c. Quantum mechanically, however, the circumference must be an integer number (n) of de Broglie wavelengths:

$$2\pi r_\mathrm{c} = nh/mv \tag{1.5.3}$$

From Eqs.(1.5.2) and (1.5.3) we find that the kinetic energy can only have discrete values given by $mv^2/2 = n\hbar\omega_\mathrm{c}/2$. Hence we would expect that

$$E_n = E_\mathrm{s} + n(\hbar\omega_\mathrm{c}/2)$$

This is a little different from the correct answer obtained from a proper quantum mechanical treatment

$$E_n - E_\mathrm{s} = \left(n + \tfrac{1}{2}\right)\hbar\omega_\mathrm{c} \tag{1.5.4}$$

These energy levels E_n with different values of n are referred to as *Landau levels*. The density of states is peaked at energies corresponding to the Landau levels

$$N_\mathrm{s}(E,B) \approx N_0 \sum_{n=0}^{\infty} \delta\left[E - E_\mathrm{s} - \left(n + \tfrac{1}{2}\right)\hbar\omega_\mathrm{c}\right]$$

as stated earlier in Eq.(1.5.1). The correct prefactor N_0 can be obtained by noting that the magnetic field causes all the states in an energy range $\hbar\omega_c$ to be concentrated into a single Landau level. Since the 2-D density of states is $m/\pi\hbar^2$ it seems reasonable to expect that

$$N_0 = \hbar\omega_c \times (m/\pi\hbar^2) = 2eB/h$$

How high does the magnetic field need to be before the resistivity is affected by the formation of Landau levels? The answer is that an electron should be able to complete at least a few orbits before losing its momentum due to scattering; that is,

$$\omega_c^{-1} \ll \tau_m$$

We would arrive at the same criterion if we argue that the peaks in the density of states will be evident only if their spacing $\hbar\omega_c$ is much greater than the broadening caused by scattering:

$$\hbar\omega_c \gg \hbar/\tau_m$$

Noting that $\omega_c = |e|B/m$ and $\mu = |e|\tau_m/m$ we can rewrite this criterion in the form

$$B \gg \mu^{-1}$$

Thus the SdH oscillations arising from the formation of Landau levels are visible at lower magnetic fields for high mobility samples. If the mobility is 10^6 cm^2/V s (or in MKS units 10^2 m^2/V s) then the magnetic field has to be in excess of 10^{-2} T or 100 G for quantum effects to be manifest. The electron density deduced from the SdH data is usually somewhat smaller than that deduced from the low field data. This is because the low field data gives the total electron density including for example any leakage paths through the AlGaAs. But leakage paths are typically associated with low mobilities and do not contribute any observable SdH effect at reasonable magnetic fields.

1.6 Transverse modes (or magneto-electric subbands)

In this section we will discuss the concept of transverse modes or subbands which will appear repeatedly in this book. These are analogous to the transverse modes (TE$_{10}$, TM$_{11}$ etc.) of electromagnetic waveguides. In narrow conductors, the different transverse modes are well separated in energy and such conductors are often called *electron waveguides*.

We consider a rectangular conductor that is uniform in the x-direction and has some transverse confining potential $U(y)$ (see Fig. 1.6.1). The motion of electrons in such a conductor is described by the effective mass equation (see Eq.(1.2.2))

$$\left[E_s + \frac{(i\hbar\nabla + e\mathbf{A})^2}{2m} + U(y)\right]\Psi(x,y) = E\Psi(x,y)$$

We assume a constant magnetic field B in the z-direction perpendicular to the plane of the conductor. This can be represented by a vector potential of the form

$$\mathbf{A} = \hat{x}By \quad \Rightarrow \quad A_x = By \quad \text{and} \quad A_y = 0$$

so that Eq.(1.2.2) can be rewritten as

$$\left[E_s + \frac{(p_x + eBy)^2}{2m} + \frac{p_y^2}{2m} + U(y)\right]\Psi(x,y) = E\Psi(x,y) \qquad (1.6.1)$$

where
$$p_x \equiv -i\hbar\frac{\partial}{\partial x} \quad \text{and} \quad p_y \equiv -i\hbar\frac{\partial}{\partial y}$$

The solutions to Eq.(1.6.1) can be expressed in the form of plane waves (L: length of conductor over which the wavefunctions are normalized)

$$\Psi(x,y) = \frac{1}{\sqrt{L}}\exp[ikx]\chi(y) \qquad (1.6.2)$$

where the transverse function $\chi(y)$ satisfies the equation

$$\left[E_s + \frac{(\hbar k + eBy)^2}{2m} + \frac{p_y^2}{2m} + U(y)\right]\chi(y) = E\chi(y) \qquad (1.6.3)$$

Note that the choice of vector potential is not unique for the given magnetic field. For example we could choose $A_x = 0$ and $A_y = -Bx$. The solutions would then look very different though the physics of course must remain the same. It is only with our choice of gauge, that the solutions have the form of plane waves in the x-direction. We will use this gauge in all our discussions.

We are interested in the nature of the transverse eigenfunctions and the eigenenergies for different combinations of the confining potential U and the magnetic field B. In general for arbitrary confinement potentials $U(y)$

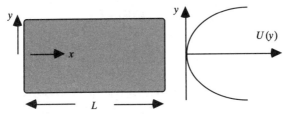

Fig. 1.6.1. A rectangular conductor assumed to be uniform in the x-direction and having some transverse confining potential $U(y)$.

there are no analytical solutions. However, for a parabolic potential (which is often a good description of the actual potential in many electron waveguides)

$$U(y) = \frac{1}{2} m \omega_0^2 y^2$$

analytical solutions can be written down and this is what we will discuss in this section. Later in Chapter 4 we will discuss an approximate solution that can be used at high magnetic fields for arbitrary confining potentials. An interesting discussion of the relation between the quantum mechanical solutions and the classical trajectories can be found in Section 12 of Ref.[1.1].

Confined electrons ($U \neq 0$) in zero magnetic field ($B = 0$)

Consider first the case of zero magnetic field, so that Eq.(1.6.3) reduces to

$$\left[E_s + \frac{\hbar^2 k^2}{2m} + \frac{p_y^2}{2m} + \frac{1}{2} m \omega_0^2 y^2 \right] \chi(y) = E \chi(y) \qquad (1.6.4)$$

The eigenfunctions of Eq.(1.6.4) are well-known (see any quantum mechanics text such as L. I. Schiff (1968), *Quantum Mechanics*, Third Edition, (New York, McGraw–Hill) Section 13). The eigenenergies and eigenfunctions are given by

$$\chi_{n,k}(y) = u_n(q) \quad \text{where} \quad q = \sqrt{m\omega_0/\hbar}\, y \qquad (1.6.5a)$$

$$E(n,k) = E_s + \frac{\hbar^2 k^2}{2m} + \left(n + \tfrac{1}{2}\right)\hbar\omega_0, \quad n = 0,1,2,\dots \qquad (1.6.5b)$$

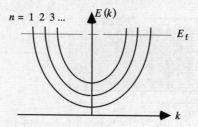

Fig. 1.6.2. Dispersion relation, $E(k)$ vs. k. for electric subbands arising from electrostatic confinement in zero magnetic field. Different subbands are indexed by n.

where
$$u_n(q) = \exp\left[-q^2/2\right]H_n(q)$$

$H_n(q)$ being the nth Hermite polynomial. The first three of these polynomials are

$$H_0(q) = \frac{1}{\pi^{1/4}}, \quad H_1(q) = \frac{\sqrt{2}q}{\pi^{1/4}} \quad \text{and} \quad H_2(q) = \frac{2q^2 - 1}{\sqrt{2}\pi^{1/4}}$$

The velocity is obtained from the slope of the dispersion curve:

$$v(n,k) = \frac{1}{\hbar}\frac{\partial E(n,k)}{\partial k} = \frac{\hbar k}{m} \qquad (1.6.5c)$$

The dispersion relation is sketched in Fig. 1.6.2. States with different index n are said to belong to different subbands just like the subbands that arise from the confinement in the z-direction (see Section 1.2). The spacing between two subbands is equal to $\hbar\omega_0$. The tighter the confinement, the larger ω_0 is, and the further apart the subbands are. Usually the confinement in the z-direction is very tight (~ 5–10 nm) so that the corresponding subband spacing is large (~ 100 meV) and only one or two subbands are customarily occupied. Indeed, in all our discussions we will assume that only one z-subband is occupied. But the y-confinement is relatively weak and the corresponding subband spacing is often quite small so that a number of these are occupied under normal operating conditions. The subbands are often referred to as *transverse modes* in analogy with the modes of an electromagnetic waveguide.

Unconfined electrons (U = 0) in non-zero magnetic field (B ≠ 0)
Next we consider unconfined electrons ($U = 0$) in a magnetic field. This

is the case that we discussed qualitatively in the last section. In this case Eq.(1.6.1) reduces to

$$\left[E_s + \frac{p_y^2}{2m} + \frac{(eBy + \hbar k)^2}{2m} \right] \chi(y) = E\chi(y)$$

which can be rewritten in the form

$$\left[E_s + \frac{p_y^2}{2m} + \frac{1}{2} m\omega_c^2 (y + y_k)^2 \right] \chi(y) = E\chi(y) \qquad (1.6.6)$$

where

$$y_k \equiv \frac{\hbar k}{eB} \quad \text{and} \quad \omega_c \equiv \frac{|e|B}{m} \qquad (1.6.7)$$

Eq.(1.6.6) is basically a one-dimensional Schrödinger equation with a parabolic potential just as we had before. The only difference is that the parabola is centered at $y = -y_k$ instead of $y = 0$. Thus the eigenenergies and eigenfunctions look very similar to the results for electric subbands (see Eqs.(1.6.5a,b,c)):

$$\chi_{n,k}(y) = u_n(q - q_k) \qquad (1.6.8a)$$

$$E(n,k) = E_s + (n + \tfrac{1}{2})\hbar\omega_c, \quad n = 0, 1, 2, \ldots \qquad (1.6.8b)$$

where

$$q = \sqrt{m\omega_c/\hbar}\, y \quad \text{and} \quad q_k = \sqrt{m\omega_c/\hbar}\, y_k$$

The mathematics describing Landau levels (or magnetic subbands) indexed by n is thus very similar to the mathematics describing the electric subbands for a parabolic confining potential. However, despite the formal similarity the physical content is completely different. The difference is

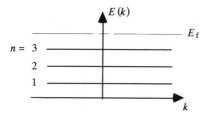

Fig. 1.6.3. Dispersion relation, $E(k)$ vs. k. for Landau levels (or magnetic subbands) in an unconfined system in non-zero magnetic field. Different Landau levels are indexed by n.

easily appreciated if we look at the velocity associated with these states which is obtained from the slope of the dispersion curve (sketched in Fig. 1.6.3):

$$v(n,k) = \frac{1}{\hbar}\frac{\partial E(n,k)}{\partial k} = 0!$$ (1.6.8c)

Although the eigenfunctions have the form of plane waves exp $[ikx]$, these waves have no group velocity because the energy E is independent of k. If we were to construct a wavepacket out of these states localized in x it would not move. This is in keeping with what we would expect from classical dynamics which predicts that an electron in a magnetic field will describe closed circular orbits in the x–y plane that do not move in any particular direction. The spatial extent of each wavefunction in the y-direction is approximately

$$\sqrt{\frac{\hbar}{m\omega_c}} = \frac{\sqrt{\hbar\omega_c/m}}{\omega_c} \rightarrow \frac{v}{\omega_c}$$

This is equal to the radius of the classical orbit that an electron would describe if it had an energy of $\hbar\omega_c/2$.

An important difference between the eigenfunctions corresponding to electric subbands (Eq.(1.6.5a)) and those corresponding to magnetic subbands (Eq.(1.6.8a)) is that in the latter case the wavefunctions shift along the transverse coordinate y as we change the wavevector k in the longitudinal direction. This is depicted in Fig. 1.6.4. One question that often comes up is the following: how many electrons can fit into one Landau level? In Section 1.5 we obtained the answer heuristically by arguing that this number, N, must equal the two-dimensional density of states multiplied by the energy spacing between two Landau levels:

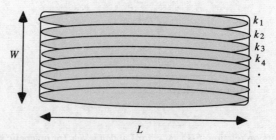

Fig. 1.6.4. Pictorial representation of the eigenfunctions (in a magnetic field) corresponding to a fixed index n but different values of k.

$$N = \frac{mS}{\pi\hbar^2} \times \hbar\omega_c = \frac{|e|BS}{\pi\hbar}$$

We can obtain this result more rigorously by noting that the allowed values of k are spaced by $2\pi/L$, which means that the corresponding wavefunctions are spaced by

$$\Delta y_k = \frac{\hbar\Delta k}{|e|B} = \frac{2\pi\hbar}{|e|BL}$$

along the y-coordinate. Hence the total number of states is given by

$$N = 2 \text{ (for spin)} \times \frac{W}{\Delta y_k} = \frac{|e|BS}{\pi\hbar}$$

in agreement with the heuristic result.

Confined electrons (U ≠ 0) in non-zero magnetic field (B ≠ 0)

Finally we consider the general case with a confining potential and a non-zero magnetic field. The eigenstates then form magneto-electric sub-bands which reduce to electric subbands when $B = 0$ and to magnetic subbands when $U = 0$. We start from Eq.(1.6.3) with a parabolic potential

$$\left[E_s + \frac{p_y^2}{2m} + \frac{(eBy + \hbar k)^2}{2m} + \frac{1}{2}m\omega_0^2 y^2 \right]\chi(y) = E\chi(y)$$

and rewrite it in the form

$$\left(E_s + \frac{p_y^2}{2m} + \frac{1}{2}m\frac{\omega_0^2\omega_c^2}{\omega_{c0}^2} y_k^2 + \frac{1}{2}m\omega_{c0}^2\left[y - \frac{\omega_c^2}{\omega_{c0}^2} y_k \right]^2 \right)\chi(y) = E\chi(y)$$

where

$$\omega_{c0}^2 \equiv \omega_c^2 + \omega_0^2 \qquad (1.6.9)$$

Once again, Eq.(1.6.9) is basically a one-dimensional Schrödinger equation with a parabolic potential and the eigenenergies and eigenfunctions look very similar to the results for electric subbands (Eqs.(1.6.5a,b,c)) and for magnetic subbands (Eqs.(1.6.8a,b,c)):

$$\chi_{n,k}(y) = u_n\left[q - \frac{\omega_c^2}{\omega_{c0}^2} q_k \right] \qquad (1.6.10a)$$

where $\qquad q = \sqrt{m\omega_{c0}/\hbar}\, y$ and $q_k = \sqrt{m\omega_{c0}/\hbar}\, y_k$

$$E(n,k) = E_s + \frac{1}{2} m \frac{\omega_0^2 \omega_c^2}{\omega_{c0}^2} y_k^2 + \left(n + \tfrac{1}{2}\right)\hbar\omega_{c0}$$

$$= E_s + \left(n + \tfrac{1}{2}\right)\hbar\omega_{c0} + \frac{\hbar^2 k^2}{2m} \frac{\omega_0^2}{\omega_{c0}^2} \qquad (1.6.10b)$$

The velocity is given by

$$v(n,k) = \frac{1}{\hbar} \frac{\partial E(n,k)}{\partial k} = \frac{\hbar k}{m} \frac{\omega_0^2}{\omega_{c0}^2} \qquad (1.6.10c)$$

The dispersion relation and the velocity are sketched in Fig. 1.6.5. It would seem that the effect of the magnetic field is simply to increase the mass by a factor that depends on the relative magnitudes of the confinement parameter ω_0 and the cyclotron frequency ω_c:

$$m \rightarrow m\left[1 + \frac{\omega_c^2}{\omega_0^2}\right]$$

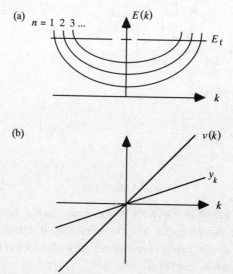

Fig. 1.6.5. Magneto-electric subbands in a parabolic potential: (a) Dispersion relation, $E(k)$ vs. k. for different subbands indexed by n. (b) Velocity, $v(k)$ vs. k and transverse location y_k vs. k for any subband n.

For zero magnetic field, the cyclotron frequency is zero and we recover the purely electric subbands discussed earlier. As the magnetic field is increased, the cyclotron frequency gets larger and the mass increases making the dispersion relations look nearly flat as expected from our discussion of magnetic subbands.

There is, however, a more profound change in the eigenstates due to the magnetic field, which is not apparent from this description. To see this, we have to look at the spatial location of the eigenstates as a function of k. We know that the wavefunction corresponding to a state (n,k) is centered around $y = y_k$ where

$$y_k = \hbar k/eB \quad \Rightarrow \quad y_k = v(n,k)\frac{\omega_0^2 + \omega_c^2}{\omega_c \omega_0^2}$$

as shown in Fig. 1.6.5b. The point is that the transverse location of the wavefunction is proportional to its velocity. As the magnetic field is increased, *states carrying current along +x shift to one side of the sample while states carrying current in the other direction shift to the other side of the sample.* This seems reasonable from a classical viewpoint since the Lorentz force $e\mathbf{v} \times \mathbf{B}$ is opposite for electrons moving in opposite directions. Increasing the magnetic field thus causes a reduction in the spatial overlap between the forward and backward propagating states, resulting in a suppression of the backscattering due to imperfections. The effect can be spectacular as we will see in Chapter 4.

1.7 Drift velocity or Fermi velocity?

The current density $\mathbf{J}$ in a homogeneous conductor is usually expressed as the product of the electron density and the drift velocity $\mathbf{v}_d$:

$$\mathbf{J} = en_s\mathbf{v}_d \tag{1.7.1}$$

This conveys the impression that all the conduction electrons drift along and contribute to the current. However, this picture is somewhat misleading for a degenerate gas at low temperatures ($k_BT \ll E_f - E_s$). If we were to measure the current carried by the electrons at different energies (such energy-resolved measurements can be done) we would find that the net current is non-zero only within a few k_BT of the quasi-Fermi energy F_n. This leads to a major conceptual simplification because it means that to understand the conduction properties at low temperatures we do not need to worry about the dynamics of the entire sea of electrons; it is sufficient

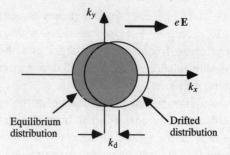

Fig. 1.7.1. At low temperatures all the states lying within a circle of radius k_f are occupied at equilibrium. In the presence of an electric field the circle is shifted in the direction of eE.

to understand the dynamics of electrons having energies close to the Fermi energy. This is the central point we wish to convey in this section.

It is easy to see why the current flows entirely within a few $k_B T$ of the quasi-Fermi energy. Let us define a distribution function $f(\mathbf{k})$ that tells us the probability that a state $\mathbf{k}$ is occupied. At equilibrium $f(\mathbf{k}) = 1$ for all the states lying inside a circle of radius k_f (that is, for $k < k_f$). An electric field makes the entire distribution shift by $\mathbf{k}_d$ as shown in Fig. 1.7.1:

$$\left[f(\mathbf{k}) \right]_{E \neq 0} = \left[f(\mathbf{k} - \mathbf{k}_d) \right]_{E=0} \tag{1.7.2}$$

where
$$\frac{\hbar \mathbf{k}_d}{m} = \mathbf{v}_d = \frac{e\mathbf{E}\tau_m}{m} \quad \Rightarrow \quad \mathbf{k}_d = \frac{e\mathbf{E}\tau_m}{\hbar} \tag{1.7.3}$$

Now, the point is that, deep inside the Fermi sea ($k \ll k_f$) nothing much happens, assuming that the field is small enough that the shift k_d is

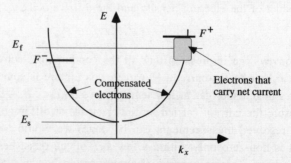

Fig. 1.7.2. States that carry current along $+x$ direction are filled up to a higher energy (F^+) than the states that carry current along $-x$ direction (F^-). A net current flows only in the range of energies lying between F^- and F^+.

small compared to k_f. The states were full without the field and they are still full when the field is applied. It is only near $+k_f$ that states that were empty become filled due to the field; while, near $-k_f$ the states that used to be full become empty. Thus although from a single-particle point of view the electric field gives all the electrons a drift velocity, from a collective point of view the electric field only moves a few electrons from $-k_f$ to $+k_f$. We could rewrite the current density in a slightly different form to reflect this point of view:

$$J = e\left[n_s \frac{v_d}{v_f}\right]v_f \qquad (1.7.4)$$

From this point of view, the current is carried by a small fraction of the total electrons ($n_s v_d/v_f$) which move with the Fermi velocity.

Quasi-Fermi level separation

An approximate way to visualize the shift in the distribution function $f(\mathbf{k})$ is to define a quasi-Fermi level F^+ for the electrons that move in the same direction as the force $e\mathbf{E}$ and another quasi-Fermi level F^- for the electrons that move in the opposite direction (see Fig. 1.7.2). All the states ($+k_x$ and $-k_x$) below F^- are completely full and carry no current. But in the energy range between F^+ and F^- the states with $-k_x$ are empty while those with $+k_x$ are full and it is the electrons in these states that carry the current. We can estimate F^+ and F^- by noting that

$$F^+ \sim \frac{\hbar^2(k_f + k_d)^2}{2m} \quad \text{and} \quad F^- \sim \frac{\hbar^2(k_f - k_d)^2}{2m}$$

where k_d is given by Eq.(1.7.3). Assuming that $k_f \gg k_d$, we can write

$$F^+ - F^- \sim \frac{2\hbar^2 k_f k_d}{m} = 2eEv_f\tau_m = 2eEL_m \qquad (1.7.5)$$

This is a very reasonable result: the separation of quasi-Fermi levels is proportional to the energy that an electron gains in the electric field in a mean free path.

Einstein relation

One way to calculate the current is to write the current density as $\mathbf{J} = en_s\mathbf{v}_d$ and then relate the drift velocity to the electric field as we did in Section 1.4. We will now describe a different way to calculate the

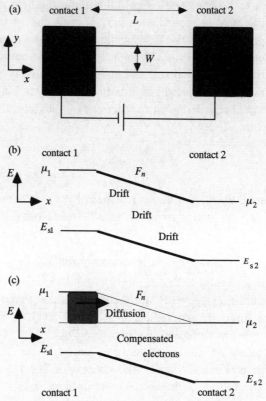

Fig. 1.7.3. (*a*) A conductor of length L is sandwiched between two contacts and an external bias is applied between the contacts. (*b*), (*c*) Band diagram under bias. Two ways to visualize the flow of current. Note that E_s is the bottom of the band while F_n is the quasi-Fermi level.

current that highlights the fact that the current at low temperatures is carried by a few electrons near the Fermi energy. For simplicity we will assume 'zero' temperature.

Consider a rectangular piece of conductor, of length L and width W, stretched between two large conducting pads (which we will refer to as the contacts) as shown in Fig. 1.7.3a. A bias V is applied across the contacts creating an electric field $\mathbf{E} = \hat{x}V/L$ in the conductor. Fig. 1.7.3b shows the band diagram under bias. The bottom of the subband E_s (which follows the electrostatic potential energy) acquires a constant slope proportional to the electric field:

$$\mathbf{E} = \nabla E_s / \left| e \right|$$

In a homogeneous conductor the electron density is constant everywhere. Since the electron density is a function of $(F_n - E_s)$ this means that the quasi-Fermi level F_n and the band-edge E_s run parallel to each other in a homogeneous conductor as shown. Note that we are using F_n to denote the average quasi-Fermi level of all the electrons in the states with $+k$ as well as $-k$; that is, $F_n = (F^+ + F^-)/2$. Usually for low fields the difference (see Eq.(1.7.5)) between F^+ and F^- is quite small anyway.

At zero temperature, the electrons in the states below μ_2 do not contribute to the current flow and can be ignored. Let us focus our attention in the energy range between μ_1 and μ_2. In this energy range, we have an electron density of ($N_s \equiv m / \pi \hbar^2$ is the 2-D density of states)

$$N_s \, (\mu_1 - \mu_2)$$

in contact 1 and no electrons in contact 2, *so that there is a concentration gradient from contact 1 to contact 2*. This sets up a diffusion current which can be calculated from the diffusion equation:

$$\mathbf{J} = -eD \, \nabla n = e^2 DN_s \frac{\mu_1 - \mu_2}{|e|L} \, \hat{x} \quad \rightarrow \quad e^2 DN_s \mathbf{E}$$

Comparing with the relation $\mathbf{J} = \sigma \mathbf{E}$, we obtain an expression for the conductivity:

$$\sigma = e^2 \, N_s \, D \qquad (1.7.6)$$

This is the Einstein relation for degenerate conductors.

Drift or diffusion?

Note that in this example there is no concentration gradient (and hence no diffusion current) if we consider the entire sea of electrons, since the total electron density is constant everywhere. The current is then purely due to drift ($\mathbf{J} = en_s \mathbf{v}_d$) as shown in Fig. 1.7.3b. This point of view yields an expression for the conductivity in terms of the mobility (see Eq.(1.4.2)):

$$\sigma = |e| n_s \mu \qquad (1.7.7)$$

But if we focus our attention on the energy range $\mu_1 > E > \mu_2$, then there is a concentration gradient and the current is purely due to diffusion as shown in Fig. 1.7.3c. This point of view yields Eq.(1.7.6) for the conductivity in terms of the diffusion coefficient. To ensure consistency between the two points of view the diffusion coefficient and the mobility must be related as follows:

$$\frac{|e|D}{\mu} = \frac{n_s}{N_s} = E_f - E_s \quad \Rightarrow \quad D = \frac{1}{2}v_f^2\tau_m \qquad (1.7.8)$$

It is important to note that Eqs.(1.7.6) and (1.7.7) express two very different perspectives on conductivity. If we double the electron concentration in a conductor, its conductivity will double. Eq.(1.7.7) 'explains' this increase in conductivity in terms of the increase in the electron density n_s; the mobility remains unchanged. But Eq.(1.7.6) 'explains' it in terms of an increase in the diffusion coefficient (since the Fermi velocity v_f is larger); the density of states N_s remains unchanged. From one point of view there are more electrons that move. From another point of view the same number of electrons move faster.

This curious difference in viewpoints does not arise for non-degenerate conductors. The Einstein relation in this case is obtained from Eq.(1.7.8) by replacing $(E_f - E_s)$ with k_BT (this correspondence between degenerate and non-degenerate systems seems to hold quite generally)

$$\frac{|e|D}{\mu} = k_BT$$

so that the diffusion coefficient is proportional to the mobility and not to the conductivity as in degenerate conductors (see Eq.(1.7.6)).

Zero temperature conductance is a Fermi surface property

The point of view depicted in Fig. 1.7.3c clearly shows that the current is carried by a few electrons near the Fermi energy that diffuse from 1 to 2. At zero temperature, electronic transport occurs in the energy range $\mu_1 > E > \mu_2$. At non-zero temperatures, the Fermi function (see Fig. 1.2.2) does not go from one to zero abruptly. Instead it changes over an energy range of the order of a few k_BT. Consequently transport occurs over an energy range

$$\mu_1 + \text{a few } k_BT > E > \mu_2 - \text{a few } k_BT$$

For linear response we conceptually let the bias tend to zero; that is, $\mu_1 \sim \mu_2 \sim E_f$. Transport then occurs over a few k_BT around the Fermi energy. If we let the temperature tend to zero then transport occurs right at the Fermi energy. This means that to understand the conduction properties at low temperatures it is sufficient to consider the diffusion of electrons right at the Fermi energy. We do not need to worry about the

dynamics of the entire sea of electrons. This is true even in the presence
of magnetic fields, but with one small caveat, which we will now discuss.

Is conductance a Fermi surface property when a magnetic field is present?

To appreciate the problem, we recall that in a magnetic field states carrying current in opposite directions are spatially separated as discussed in
Section 1.6 (see Fig. 1.6.5). The states carrying current from left to right
are localized along one edge while the states carrying current from right
to left are localized along the other edge. Consequently even at equilibrium, the local current density in a conductor is not zero (see Fig. 1.7.4).
There is no net current across any cross-section AA', but there are local
currents flowing at different points along the cross-section.

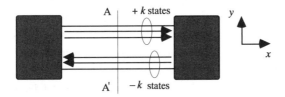

Fig. 1.7.4. Local circulating currents in a conductor at equilibrium do not contribute to
the net current flow across any plane AA' in the conductor. An applied bias could
change the circulating current pattern thus contributing to the conductivity.

Consider the conductivity tensor defined by the relation

$$\delta \mathbf{J} = \sigma \, \delta \mathbf{E} \qquad (1.7.9)$$

where $\delta \mathbf{J}$ is the change in the local current density in response to an electric field $\delta \mathbf{E}$. An applied bias gives rise to an electric field inside the
sample which changes the eigenstates of the sample and hence the local
current density within the sample. According to Eq.(1.7.9) this change in
the circulating current pattern in response to an applied electric field corresponds to a non-zero conductivity. All the electrons (even those in filled
bands far below the conduction band) can contribute to the circulating
current patterns and hence to the *conductivity*. We could miss these contributions if we only take the electrons at the Fermi energy into account.
But what we miss has no effect on the conductance as measured in a
transport experiment because the circulating currents do not contribute to

any net flow across any cross-section. For a more formal discussion of this point see H. U. Baranger and A. D. Stone (1989), *Phys. Rev. B*, **40**, 8169.

The statement that 'zero temperature conductance is a Fermi surface property' thus needs to be qualified when a magnetic field is present. It is true if we are talking about the conductance as measured between two contacts. But it is not true if we are talking about the conductivity as defined by Eq.(1.7.9).

Summary

This chapter summarizes a few basic concepts. A great majority of mesoscopic conductors are patterned out of the two-dimensional electron gas (2-DEG) formed at a GaAs–AlGaAs interface. What makes this 2-DEG special is the long mean free path (~ tens of microns) that have been demonstrated at temperatures below 10 K (Section 1.1). Most mesoscopic experiments involve conduction by electrons in the conduction band (rather than holes in the valence band) with energies low enough that an effective mass description is sufficiently accurate. Also most experiments are carried out at low temperatures where the Fermi function is strongly degenerate (Section 1.2). Mesoscopic behavior is usually observed when the sample size is smaller than one or more of several length scales: the de Broglie wavelength, mean free path, phase-relaxation length and screening length (Section 1.3). The variation of resistance with magnetic field at low fields is understood readily in terms of a classical model (Section 1.4) but to understand the variation at high fields we need to invoke the wave nature of electrons (Section 1.5). Electronic states in narrow conductors evolve from purely electric subbands (arising from electrostatic confinement) to purely magnetic subbands or Landau levels as the magnetic field is increased. These subbands are often referred to as transverse modes by analogy with the transverse modes (TE_{10}, TM_{11} etc.) of electromagnetic waveguides (Section 1.6). The conductance of degenerate conductors is determined by the dynamics of the electrons having energies close to the Fermi energy. It is not necessary to worry about the entire underlying sea of electrons (Section 1.7).

Exercises

E.1.1 Consider the sample used to obtain the data shown in Fig. 1.3.2. Calculate the mean free path L_m and the phase-relaxation length L_φ (at $T = 1$ K).

E.1.2 (a) Use the low-field data $(0 < B < 0.5$ T$)$ in Fig. 1.4.2 to calculate the electron density and mobility from Eqs.(1.4.4a,b).
(b) Use the high-field data $(1$ T $< B < 4$ T$)$ in Fig. 1.4.2 to construct a 'Landau plot' of the peak number i versus $1/B$ and deduce the electron density.

E.1.3 Consider a narrow conductor etched out of a wide conductor, as shown in Fig. E.1.3.

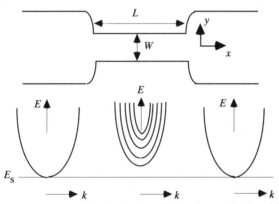

Fig. E.1.3. Narrow conductor etched out of a wide conductor. In the wide regions the transverse modes are essentially continuous, but in the narrow region the modes are well-separated in energy.

The wide conductor can be treated simply as a two-dimensional conductor so that it is easy to calculate the location of the Fermi energy E_f relative to the bottom of the band E_s using the 2-D density of states (see Eq.(1.2.10)). Assuming $m = 0.07m_0$

$$m/\pi\hbar^2 \approx 2.9 \times 10^{10}/\text{cm}^2 \text{ meV}$$

If the electron density in the wide conductor is $5 \times 10^{11}/\text{cm}^2$ then

$$E_{fs} \equiv E_f(2\text{D}) - E_s = \frac{n_s}{m/\pi\hbar^2} = 17.2 \text{ meV}$$

In the narrow region we cannot use the 2-D density of states if the width W is small enough; it is necessary to take into account the discreteness of the transverse modes. Plot the electron density versus Fermi energy for $W = 1000$ Å, assuming two different models for the confining potential:

(a) Hard wall potential:
$$U(y) = 0, \ -W/2 < y < W/2$$
$$= \infty, \text{otherwise}$$

(b) Parabolic potential:
$$U(y) = \frac{1}{2} m\omega_0^2 y^2$$

with ω_0 chosen such that

$$U(y = \pm W/2) = E_{fs} \implies \hbar\omega_0 \approx 3.9 \text{ meV}$$

Note that these are rather simplified models. For an accurate description of the confining potential it is necessary to solve the Schrödinger and Poisson equations numerically as described in, for example, S. E. Laux, D. J. Frank and F. Stern (1988), *Surface Science*, **196**, 101 and J. H. Davies and J. A. Nixon (1989), *Phys. Rev. B*, **39**, 3423.

E.1.4 Consider the structure in E.1.3 and assume a parabolic confining potential. Calculate the number of transverse modes as a function of the magnetic field, assuming (a) constant Fermi energy and (b) constant electron density.

Note that if the Fermi energy remains constant then the conductor can be completely depleted as the magnetic field is increased (see for example, B. J. van Wees *et al.* (1988), *Phys. Rev. B*, **38**, 3625). But if the electron density is assumed to remain constant then the number of modes cannot decrease to zero; at least one mode always remains occupied (see for example, K. F. Berggren, G. Roos and H. van Houten (1988), *Phys. Rev. B*, **37**, 10118).

Further reading

[1.1] Beenakker, C. W. J. and van Houten H. (1991). 'Quantum transport in semiconductor nanostructures' in *Solid State Physics*, **44**, eds. H. Ehrenreich and D. Turnbull, (New York, Academic Press) (see Part I).

For a review of modulation-doped heterostructures see

[1.2] Drummond, T. J., Masselink, W. T. and Morkoc, H. (1986). 'Modulation-doped GaAs/(Al,Ga)As heterojunction field-effect transistors: MODFET's', *Proc. IEEE*, **74**, 773.

[1.3] Melloch, M. R. (1993). 'Molecular beam epitaxy for high electron mobility modulation-doped two-dimensional electron gases', *Thin Solid Films*, **231**, 74.

Both [1.2] and [1.3] provide extensive bibliographies.

For a review of the work on 2-D semiconductors prior to 1981, see

[1.4] Ando, T., Fowler, A. B. and Stern, F. (1982). 'Electronic properties of two-dimensional systems', *Rev. Mod. Phys.*, **54**, 437.

2
Conductance from transmission

2.1 Resistance of a ballistic conductor
2.2 Landauer formula
2.3 Where is the resistance?
2.4 What does a voltage probe measure?
2.5 Non-zero temperature and bias
2.6 Exclusion principle?
2.7 When can we use the Landauer–Büttiker formalism?

Our purpose in this chapter is to describe an approach (often referred to as the Landauer approach) that has proved to be very useful in describing mesoscopic transport. In this approach, the current through a conductor is expressed in terms of the *probability that an electron can transmit through it*. The earliest application of current formulas of this type was in the calculation of the current–voltage characteristics of tunneling junctions where the transmission probability is usually much less than unity (see J. Frenkel (1930), *Phys. Rev.*, **36**, 1604 or W. Ehrenberg and H. Honl (1931), *Z. Phys.*, **68**, 289). Landauer [2.1] related the linear response conductance to the transmission probability and drew attention to the subtle questions that arise when we apply this relation to conductors having transmission probabilities close to unity. For example, if we impress a voltage across two contacts to a ballistic conductor (that is, one having a transmission probability of unity) the current is finite indicating that the resistance is not zero. But can a ballistic conductor have any resistance? If not, where does this resistance come from? These questions were clarified by Imry [2.2], enlarging upon earlier notions due to Engquist and Anderson [2.3]. Büttiker extended the approach to describe multi-terminal measurements in magnetic fields and this formulation (generally referred to as the

Landauer–Büttiker formalism) has been widely used in the interpretation of mesoscopic experiments. We will present these concepts in Sections 2.1–2.4, though not always in the historical order in which they were developed. For more details we refer the reader to Refs. [2.1]–[2.4] and references therein.

To simplify the discussion in Sections 2.1–2.4, we assume 'zero temperature' so that there is current flow only in the energy range $\mu_1 > E > \mu_2$. We further assume that the transmission characteristics are independent of energy in this range of energies so that the entire energy range can be viewed as a single energy channel. However, in general, current flow takes place over a large range of energies and the transmission characteristics could vary widely over this range. In Section 2.5 we generalize the discussion to include multiple energy channels and derive a current–voltage relation that can be used for high temperature and large applied bias. As we mentioned earlier, this current–voltage relation has been used ever since the 1930s to describe tunneling junctions (for a review see C. B. Duke (1969), 'Tunneling in solids', in *Solid State Physics*, eds. H. Ehrenreich and D. Turnbull, (New York, Academic Press)). More recently it has been used to calculate the I–V characteristics of resonant tunneling devices following the pioneering work of R. Tsu and L. Esaki (see *Appl. Phys. Lett.*, **22**, 562 (1973)).

The approach described in this chapter is intuitively very appealing because it seems obvious that the conductance of a sample ought to be proportional to the ease with which electrons can transmit through it. But is it accurate to describe the current flow in degenerate conductors at low temperatures in terms of a single-particle transmission coefficient? Doesn't the exclusion principle affect the transmission? This is the question we address in Sections 2.6 and 2.7. We show that if transport through the conductor is coherent then the exclusion principle has no effect on the transmission. If transport is not coherent then, in general, the exclusion principle can affect the transmission in a complicated way. This would seem to limit the usefulness of the Landauer–Büttiker formalism to coherent transport. However, even for non-coherent transport, the exclusion principle-related factors disappear if we neglect the *vertical flow* of electrons from one energy to another. Luckily it turns out that the vertical flow often has little effect on the total current through a conductor.

We end with a brief summary in Section 2.8.

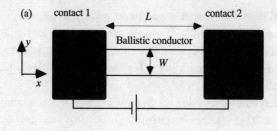

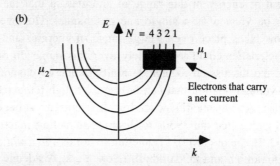

Fig. 2.1.1. (*a*) A conductor is sandwiched between two contacts across which an external bias is applied. The contacts are assumed to be 'reflectionless', that is, electrons can exit from the conductor into the contact with negligible probability of reflection. (*b*) Dispersion relations for the different transverse modes (or subbands) in the narrow conductor. For reflectionless contacts, the quasi-Fermi level for the $+k$ states is μ_1 while that for the $-k$ states is μ_2.

2.1 Resistance of a ballistic conductor

Consider a piece of conductor stretched between two large contact pads as shown in Fig. 2.1.1a. If the dimensions of the conductor were large then we know that its conductance would be given by $G = \sigma W/L$, where the conductivity σ is a material parameter independent of the sample dimensions. If this ohmic scaling relation were to hold as the length (L) is reduced, then we would expect the conductance to grow indefinitely. Experimentally, however, it is found that the measured conductance approaches a limiting value G_C, when the length of the conductor becomes much shorter than the mean free path ($L \ll L_m$). Note that this has nothing to do with quantum mechanics. Indeed in our discussion we will assume that the phase-relaxation length is short enough that interference-related effects can be neglected.

Where does this resistance come from? After all, a ballistic conductor (that is, a conductor with no scattering) should have zero resistance. This resistance arises from the interface between the conductor and the contact pads which are very dissimilar materials [2.2]. For this reason we will refer to this resistance (G_C^{-1}) as the *contact resistance*. The current is carried in the contacts by infinitely many transverse modes (this concept was introduced in Section 1.6), but inside the conductor by only a few modes. This requires a redistribution of the current among the current-carrying modes at the interface leading to the interface resistance. Could we get rid of this contact resistance simply by making the contacts identical to the conductors? Yes, but then the measurement we are talking about would not make sense. The contacts have to be 'infinitely' more conducting than the conductor in order to justify our assumption that the applied voltage drops entirely across the conductor.

'Reflectionless contacts'

To calculate the contact resistance G_C^{-1} we consider a ballistic conductor and calculate the current through it for a given applied bias $\mu_1 - \mu_2$. It is straightforward to calculate this current if we assume that the contacts are 'reflectionless', that is, the electrons can enter them from the conductor without suffering reflections. Numerical calculations indicate that as long as the energy is not too close to the bottom of the band, an electron can exit from a narrow conductor into a wide contact with negligible probability of reflection (see A. Szafer and A. D. Stone (1989), *Phys. Rev. Lett.*, **62**, 300). This is what we mean by a 'reflectionless' contact. We use the quotes as a reminder that the reflection is negligible only when transmitting from the narrow conductor to the wide contact. Going the other way from the contact to the conductor, the reflections can be quite large.

For 'reflectionless' contacts, we have a simple situation: $+k$ states in the conductor are occupied only by electrons originating in the left contact while $-k$ states are occupied only by electrons originating in the right contact. This is because electrons originating in the right contact populate the $-k$ states and empty without reflection into the left contact while electrons originating in the left contact populate the $+k$ states and empty without reflection into the right contact (note that k denotes the wavenumber in the x-direction).

We will now argue that the quasi-Fermi level F^+ for the $+k$ states is always equal to μ_1 even when a bias is applied (Fig. 2.1.1b). Suppose both contacts are at the same potential μ_1. There is no question then that the Fermi level for the $+k$ states (or any other state) is equal to the potential μ_1. Now if we change the potential at the right contact to μ_2, this can have no effect on the quasi-Fermi level F_1^+ for the $+k$ states since *there is no causal relationship between the right contact and the +k states*. No electron originating in the right contact ever makes its way to a $+k$ state. Similarly we can argue that the quasi-Fermi level F^- for the $-k$ states in lead 2 is always equal to μ_2. Hence at low temperatures the current is equal to that carried by all the $+k$ states lying between μ_1 and μ_2.

Calculating the current

To calculate the current we note that the states in the narrow conductor belong to different transverse modes or subbands as discussed in Section 1.6. Each mode has a dispersion relation $E(N,k)$ as sketched in Fig. 2.1.1b with a cut-off energy

$$\varepsilon_N = E(N, k = 0)$$

below which it cannot propagate. The number of transverse modes at an energy E is obtained by counting the number of modes having cut-off energies smaller than E:

$$M(E) \equiv \sum_N \vartheta(E - \varepsilon_N) \tag{2.1.1}$$

We can evaluate the current carried by each transverse mode (numbered by 'N' in Fig. 2.1.1b) separately and add them up.

Consider a single transverse mode whose $+k$ states are occupied according to some function $f^+(E)$. A uniform electron gas with n electrons per unit length moving with a velocity v carries a current equal to env. Since the electron density associated with a single k-state in a conductor of length L is $(1/L)$ we can write the current I^+ carried by the $+k$ states as

$$I^+ = \frac{e}{L} \sum_k v f^+(E) = \frac{e}{L} \sum_k \frac{1}{\hbar} \frac{\partial E}{\partial k} f^+(E)$$

Assuming periodic boundary conditions (see Fig. 1.2.1 and related discussion) and converting the sum over k into an integral according to the usual prescription

$$\sum_k \;\rightarrow\; 2 \text{ (for spin)} \times \frac{L}{2\pi} \int dk$$

we obtain
$$I^+ = \frac{2e}{h}\int_\varepsilon^\infty f^+(E)\,\mathrm{d}E$$

where ε is the cut-off energy of the waveguide mode. We could extend this result to multi-moded waveguides and write the current, I^+, carried by the $+k$ states in a conductor as

$$I^+ = \frac{2e}{h}\int_{-\infty}^{+\infty} f^+(E)M(E)\,\mathrm{d}E \tag{2.1.2}$$

where the function $M(E)$ (defined in Eq.(2.1.1)) tells us the number of modes that are above cut-off at energy E. Note that this is a general result independent of the actual dispersion relation $E(k)$ of the waveguide: the current carried per mode per unit energy by an occupied state is equal to $2|e|/h$ (which is about 80 nA/meV).

Contact resistance

Assuming that the number of modes M is constant over the energy range $\mu_1 > E > \mu_2$, we can write

$$I = \frac{2e^2}{h}M\frac{(\mu_1 - \mu_2)}{e} \quad \Rightarrow \quad G_C = \frac{2e^2}{h}M \tag{2.1.3}$$

so that the contact resistance (which is the resistance of a ballistic waveguide) is given by

$$G_C^{-1} \equiv \frac{(\mu_1 - \mu_2)/e}{I} = \frac{h}{2e^2 M} \approx \frac{12.9\,\mathrm{k\Omega}}{M}$$

Note that the contact resistance goes down inversely with the number of modes. The contact resistance of a single-moded conductor is ~ 12.9 kΩ, which is certainly not negligible! This is the resistance one would measure if a single-moded ballistic conductor were sandwiched between two conductive contacts.

Usually we are concerned with wide conductors having thousands of modes so that the contact resistance is very small and tends to go unnoticed. To calculate the number of modes $M(E)$ we need to know the cut-off energies for the different modes ε_N. As we have seen in Section 1.6, the details depend on the confining potential $U(y)$ and the magnetic field. However, for wide conductors in zero magnetic field the precise nature of the confining potential is not important. We can estimate the number of modes simply by assuming periodic boundary conditions. The allowed values of k_y are then spaced by $2\pi/W$ (see Fig. 1.2.1), with each value of k_y corresponding to a distinct transverse mode. At an energy E_f ($= \hbar^2 k_f^2/2m$), a mode can propagate only if $-k_f < k_y < k_f$. Hence the number of propagating modes can be written as

$$M = \text{Int}\left[\frac{k_f W}{\pi}\right] = \text{Int}\left[\frac{W}{\lambda_f/2}\right]$$

where $\text{Int}(x)$ represents the integer that is just smaller than x. Assuming a Fermi wavelength of 30 nm, the number of modes in a 15 μm wide field-effect transistor is approximately 1000, so that the contact resistance is about 12.5 Ω.

Experimental results

The contact resistance can be measured directly using point contacts to create a constriction (much shorter than a mean free path) in a conductor. This was done in the late 1960s in metals (see Yu. V. Sharvin and N. I. Bogatina (1969), *Sov. Phys. JETP*, **29**, 419). In metals the Fermi wavelength is extremely short, of the order of the distance between atoms. Consequently the number of modes M ($\sim k_f W/\pi$) is quite large and the contact resistance is relatively small. In semiconductors on the other hand, the Fermi wavelength is typically $\sim$ 30 nm so that the factor $k_f W$ can easily be made quite small.

The first experiment on semiconductors was reported independently by two groups in 1988 (see Fig. 2.1.2). As the width W of the constriction was reduced the conductance went down in discrete steps each of height $(2e^2/h)$. This is just what we would expect from Eq.(2.1.3) since M is an integer denoting the number of subbands or transverse modes in the constriction at the Fermi energy. Although the width of a conductor changes continuously the number of modes changes in discrete steps. This discreteness is not evident if the conductor is many thousands of wavelengths wide, since a very small fractional change in W changes M by many integers.

The results shown in Fig. 2.1.2 not only provide a striking demonstration of the existence of a contact or interface resistance but also serve to emphasize the reality of transverse modes when dealing with narrow conductors. A narrow conductor can be viewed as an 'electron waveguide' analogous to electromagnetic waveguides. The current is carried by a

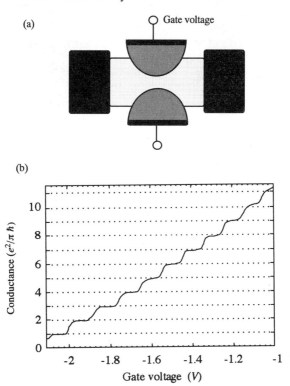

Fig. 2.1.2. Quantized conductance of a ballistic waveguide. (*a*) A negative voltage on a pair of metallic gates (called the split-gate configuration) is used to deplete and narrow down the constriction progressively. (*b*) Measured conductance vs. gate voltage. The measured resistance also includes a series resistance due to the wide regions connecting the constriction to the contacts. This series resistance is measured separately by removing the negative voltage on the gates and is subtracted off before plotting. Reproduced with permission from B. J. van Wees *et al.* (1988), *Phys. Rev. Lett.*, **60**, 848. Similar results were reported simultaneously by D. Wharam *et al.* (1988), *J. Phys. C*, **21**, L209.

discrete number of transverse modes (like the TE_{10} and the TE_{11} modes of electromagnetic waveguides) and the contact resistance is inversely proportional to this number, M.

Where is the voltage drop?

We can easily see that the resistance G_C^{-1} is associated with the interfaces and not with the conductor itself by sketching the variation of the

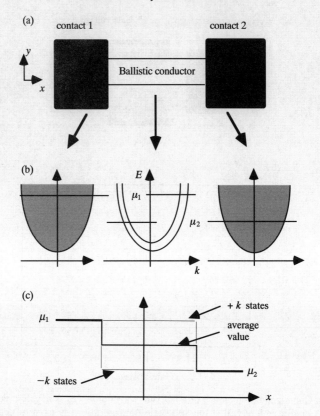

Fig. 2.1.3. (*a*) A ballistic conductor connected to two wide contacts. (*b*) Inside the wide contacts there is a very high density of transverse modes (indicated by shading) and both $+k$ and $-k$ states have nearly the same quasi-Fermi level. In the narrow ballistic conductor there are only a few modes and the $+k$ and $-k$ states have different quasi-Fermi levels. (*c*) Variation of the electrochemical potential from one contact to the other. Note that k stands for k_x.

electrochemical potential from one contact to the other (see Fig. 2.1.3). Inside the wide contacts there is a very high density of transverse modes (indicated by shading), so that both $+k$ and $-k$ states have nearly the same quasi-Fermi level even when there is a current flowing. But inside the narrow ballistic conductor there are only a few modes and the quasi-Fermi levels for the $+k$ and the $-k$ states are noticeably different. Indeed as we have discussed, the quasi-Fermi level for the $+k$ states follows μ_1 while that for the $-k$ states follows μ_2. Now if we sketch the *average* quasi-Fermi level we find that it drops equally at the two interfaces but is flat across the conductor as shown in Fig. 2.1.1c. Since a voltage drop is associated with resistance we conclude that there are two equal resistances at the two interfaces but none inside the conductor.

It might seem somewhat arbitrary to look at the average quasi-Fermi level for the $+k$ and the $-k$ states and call it the 'voltage'. We could for example look at the quasi-Fermi level for just the $+k$ states. We then find that the voltage drops entirely at the right interface but there is still no voltage drop across the conductor. No matter how we choose to define the 'voltage' there is no drop across the conductor. All the drop is at the interfaces. How it is divided between the two interfaces, however, depends on our definition of 'voltage'. Another possibility is to associate the voltage drop with the electrostatic potential. As we will see in Section 2.4, the electrostatic potential generally follows the average quasi-Fermi level. The only difference is that it cannot change abruptly at the interfaces. Instead the change is smeared out over a screening length.

It is important to note that the contact resistance arises because on one side the current is carried by infinitely many modes, while on the other side it is carried by a few modes. The details of the geometry are not important as long as the contacts are 'reflectionless' as explained earlier. However, the contact resistance can be smaller than $(h/2e^2M)$ if the number of modes in the contact is finite (see Exercise E.2.1 at the end of this chapter).

2.2 Landauer formula

To summarize our discussions in Section 2.1, the conductance of large samples obeys an ohmic scaling law: $G = \sigma W/L$. But as we go to smaller dimensions there are two corrections to this law. Firstly there is an interface resistance independent of the length L of the sample. Secondly the conductance does not decrease linearly with the width W. Instead it depends on the number of transverse modes in the conductor and goes down in discrete steps. The Landauer formula (which we will now derive) incorporates both of these features:

$$G = \frac{2e^2}{h} MT$$

The factor T represents the average probability that an electron injected at one end of the conductor will transmit to the other end. If the transmission probability is unity, we recover the correct expression for the resistance of a ballistic conductor *including the contact resistance* (see Eq.(2.1.3)).

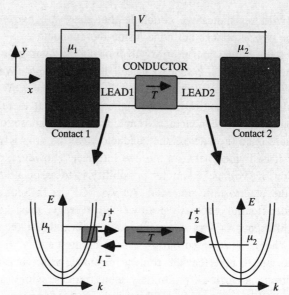

Fig. 2.2.1. A conductor having a transmission probability of T is connected to two large contacts through two leads. 'Zero' temperature is assumed such that the energy distributions of the incident electrons in the two leads can be assumed to be step functions. Note that k stands for k_x.

Consider a conductor connected to two large contacts by two leads as shown in Fig. 2.2.1. The leads are assumed to be ballistic conductors, each having M transverse modes. T is the average probability that an electron injected in lead 1 will transmit to lead 2. We assume that electrons can exit from the conductor into the contacts without any reflection (that is, the contacts are 'reflectionless', as explained in the last section). Then the $+k_x$ states in lead 1 are occupied only by electrons coming in from the left contact and hence these states must have an electrochemical potential of μ_1. Similarly we can argue that the $-k_x$ states in lead 2 are occupied only by electrons coming in from the right contact and hence must have an electrochemical potential of μ_2.

Assuming zero temperature, current flow takes place entirely in the energy range between μ_1 and μ_2. The influx of electrons from lead 1 is given by (see Eq.(2.1.3))

$$I_1^+ = (2e/h)M\big[\mu_1 - \mu_2\big]$$

The outflux from lead 2 is simply the influx at lead 1 times the transmission probability T:

$$I_2^+ = (2e/h)MT\big[\mu_1 - \mu_2\big]$$

The rest of the flux is reflected back to contact 1:

$$I_1^- = (2e/h)M(1-T)\left[\mu_1 - \mu_2\right]$$

The net current I flowing at any point in the device (and in the external circuit) is given by

$$I = I_1^+ - I_1^- = I_2^+ = (2e/h)MT\left[\mu_1 - \mu_2\right]$$

Hence the conductance is equal to

$$G = \frac{I}{(\mu_1 - \mu_2)/|e|} = \frac{2e^2}{h}MT \tag{2.2.1}$$

as stated earlier. We could view the Landauer formula as a mesoscopic version of the Einstein relation (see Eq.(1.7.6))

$$\sigma = e^2 N_s D \quad \Leftrightarrow \quad G = \frac{2e^2}{h}MT$$

with the conductivity replaced by the conductance, the density of states replaced by the number of transverse modes (or subbands) and the diffusion constant replaced by the transmission probability:

$$\sigma \to G \quad N_s \to M \quad D \to T$$

Should we include the contacts?

As we have discussed in Section 2.1, the conductance given by Eq.(2.2.1) includes the contact resistance; that is, it gives the total conductance as measured between two planes deep inside the *contacts* (marked '1' and '2' in Fig. 2.2.2) rather than between two planes in the *leads* (marked '1L' and '2L'). A common question that arises is whether we ought to calculate the transmission probabilities between '1' and '2' instead of between '1L' and '2L'. The answer is that we could do that but it is not necessary to do so as long as the contacts are 'reflectionless', that is, electrons can exit from the leads into the contacts with negligible probability of reflection. Before we explain the reason for this, let us point out an interesting implication of this observation in calculating the conductance of a ballistic conductor.

For a ballistic conductor, it is quite trivial to write down the transmission probability between points 1L and 2L in the leads (Fig. 2.2.2): $T = 1$ (for all modes m). Hence from Eq.(2.2.1), the conductance is equal to

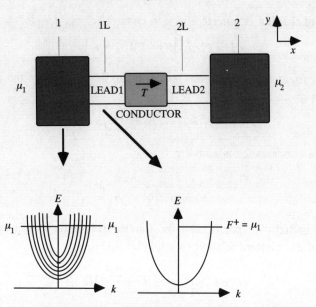

Fig. 2.2.2. In the wide contacts the current carried per mode is small so that $+k$ and $-k$ states have about the same quasi-Fermi level. If the contacts are 'reflectionless' then the $+k$ states in lead 1 are in equilibrium with contact 1 and have a quasi-Fermi level μ_1 while the $-k$ states in lead 2 (not shown) are in equilibrium with contact 2.

$(2e^2/h)M$, M being the number of modes in the narrow constriction. But suppose we decide to calculate the transmission between planes '1' and '2' inside the contacts. The number of modes in the wide contacts, M_W, is much larger but the average transmission probability per mode is much smaller: $T \sim M/M_W$ such that the conductance will still come out the same, namely, $(2e^2/h)M$. But this result is not at all obvious. To calculate the transmission probability between planes 1 and 2 in the contacts is a formidable task and it is not apparent what the answer should be. For this reason the quantized conductance of a ballistic conductor was not widely anticipated before its experimental discovery in 1988. It is really quite surprising that although the relation $G = (2e^2/h)MT$ gives us the conductance measured between planes '1' and '2' (see Fig. 2.2.2), we do not need to evaluate the quantity 'MT' between '1' and '2'. We can save ourselves a lot of work by evaluating it between '1L' and '2L' and still *obtain the same answer!* Let us explain why.

It is usually argued that a lead with only a few modes cannot be assumed to be in local equilibrium with a known electrochemical potential

μ. In any conductor the potentials F^+ and F^- for the incoming and outgoing electrons have to be slightly different in order for a net current to flow:

$$I = \frac{2e}{h} M \left(F^+ - F^- \right)$$

A contact has a very large number of modes M that ideally approaches infinity. Consequently the current per mode is infinitesimal and one can assume $F^+ = F^-$. But a lead has only a few modes and one cannot neglect the difference between F^+ and F^-. Indeed the energy distribution in the lead may not even be described by a Fermi function. For this reason it has been argued that in applying the Landauer formula one ought to calculate the transmission probability between two planes '1' and '2' located deep inside the contacts (Fig. 2.2.2) where the electron energy distribution is known for sure.

If we look back at our derivation of the Landauer formula it will be apparent that what we need is the energy distribution of the *incoming* electrons at the lead; the energy distribution of the outgoing electrons is irrelevant. We will now argue that, with 'reflectionless' contacts, the incoming states in each lead are in thermal equilibrium with the corresponding contact so that we can calculate the transmission between two planes '1L' and '2L' located in the leads (Fig. 2.2.2) and apply the Landauer formula without any ambiguity.

As we mentioned in Section 2.1, numerical calculations indicate that as long as the energy is not too close to the bottom of the band, an electron can exit from a narrow lead into a wide contact with negligible probability of reflection. Since the contacts are 'reflectionless', an electron originating in the right contact never makes its way to a $+k$ state in lead 1. This is because electrons originating in the right contact, that transmit through the conductor, populate the $-k$ states in lead 1 and empty without reflection into the left contact. Since the $+k$ states in the conductor are populated only by electrons originating from the left contact, these states remain in equilibrium with the left contact even when a bias is applied. Consequently the quasi-Fermi level F^+ for the $+k$ states in lead 1 is always equal to μ_1 (Fig. 2.2.2). Similarly an electron originating in the left contact never makes its way to a $-k$ state in lead 2, so that the quasi-Fermi level F^- for the $-k$ states in lead 2 is always equal to μ_2.

Note that we cannot use this argument for the outgoing states in the leads, that is, the $-k$ states in lead 1 or the $+k$ states in lead 2. These states are populated partially by electrons originating in contact 1 and partially by electrons originating in contact 2. Consequently the energy distribution of the electrons in the outgoing states is not known a priori. But we do not need the energy distribution in these states in deriving the Landauer formula. Only the energy distribution of the incoming stream is needed. If the contacts are not 'reflectionless' then we would have some ambiguity regarding the incoming stream too, since the electrons in the outgoing stream would get reflected into the incoming stream. But as we have mentioned above, reflections are usually quite negligible for all electrons except for those near the bottom of the band and under these conditions the incoming stream in each lead remains in thermal equilibrium with the corresponding contact. This allows us to neglect the details of the lead–contact interface and simply calculate the transmission probability from one lead to another.

Ohm's law

The Landauer formula incorporates the correct properties of the resistance of small conductors, namely (1) the length-independent interface resistance associated with the contacts and (2) the discrete steps related to the transverse modes in narrow conductors. We would now like to show that for large conductors we recover the familiar Ohm's law.

For a wide conductor with many modes the number of modes is proportional to the width: $M \sim k_f W/\pi$, so that Eq.(2.2.1) can be written as

$$G = e^2 W N_s \left(v_f T / \pi \right)$$

Next we make use of the following result which we will derive shortly treating the electrons as purely classical particles neglecting any quantum interference: the transmission probability through a conductor of length L is given by

$$T = \frac{L_0}{L + L_0}$$

$$(2.2.2)$$

where L_0 is a characteristic length of the order of a mean free path.

$$G = \frac{W}{L + L_0} e^2 N_s \left(v_f L_0 / \pi \right)$$

Identifying the diffusion coefficient as $v_f L_0 / \pi$ and making use of the Einstein relation $\sigma = e^2 N_s D$ we obtain

$$G = \frac{\sigma W}{L + L_0} \quad \Rightarrow \quad G^{-1} = \frac{L + L_0}{\sigma W}$$

We could write this as a series combination of an 'actual' resistance obeying Ohm's law and a 'contact' or 'interface' resistance ($G^{-1} = G_S^{-1} + G_C^{-1}$):

$$G_S^{-1} = \frac{L}{\sigma W} \quad \text{and} \quad G_C^{-1} = \frac{L_0}{\sigma W}$$

This completes the proof. We will now derive the expression for the transmission stated above (see Eq.(2.2.2)).

To prove that $T(L) = L_0/(L + L_0)$

Consider two conductors with transmission probabilities T_1 and T_2 connected in series as shown in Fig. 2.2.3a. The problem is to find the probability of transmission T_{12} through the series combination. It might seem that the answer is simply

$$T_{12} = T_1 T_2 \qquad \text{(WRONG)}$$

If this were true then the transmission probability through a chain of scatterers would go down exponentially with the length of the chain: $T(L) = \exp[L/L_0]$ and we would not get Ohm's law. The point is that the term $T_1 T_2$ only gives us the probability for direct transmission without any multiple reflections (see Fig. 2.2.3b). To obtain the total probability

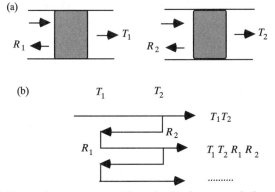

Fig. 2.2.3 (*a*) Two resistors connected in series having transmission probabilities T_1 and T_2 and reflection probabilities R_1 and R_2. (*b*) If we neglect all phase information then the net transmission through two scatterers can be calculated by summing the probabilities of transmitting without any reflection, with two reflections, with four reflections etc.

of transmission T_{12} we need to add the probabilities of all the multiply reflected paths as well. We will treat the electrons as classical particles and not worry about the phase relationships among the different paths. This is appropriate if the phase-relaxation length is much shorter than the distance between the scatterers.

The transmission probability T_{12} is obtained by summing the probabilities for transmission with zero reflections, with two reflections, with four reflections and so on (see Fig. 2.2.3b):

$$T_{12} = T_1 T_2 + T_1 T_2 R_1 R_2 + T_1 T_2 R_1^2 R_2^2 + \ldots$$

$$= \frac{T_1 T_2}{1 - R_1 R_2} \tag{2.2.3}$$

We can rewrite this result in the form (noting that $T_1 = 1 - R_1$ and $T_2 = 1 - R_2$)

$$\frac{1 - T_{12}}{T_{12}} = \frac{1 - T_1}{T_1} + \frac{1 - T_2}{T_2}$$

This simple calculation shows that when we place two scatterers in cascade the quantity [(1–T)/T] has an additive property. This means that the transmission probability $T(N)$ of N scatterers in series, each having a transmission probability of T, is given by

$$\frac{1 - T(N)}{T(N)} = N \frac{1 - T}{T} \quad \Rightarrow \quad T(N) = \frac{T}{N(1 - T) + T}$$

Now the number of scatterers N in a conductor of length L can be written as vL where v is the linear density of scatterers. Hence we can write

$$T(L) = \frac{L_0}{L + L_0} \quad \text{where} \quad L_0 \equiv \frac{T}{v(1 - T)}$$

as stated earlier in Eq.(2.2.2). It is easy to see that the length L_0 is of the order of a mean free path. A mean free path is the average distance an electron can travel before it is scattered. Since the probability of scattering by an individual scatterer is $(1 - T)$ we can write (assuming $T \sim 1$)

$$(1 - T)v L_m \sim 1 \quad \rightarrow \quad L_m \sim \frac{1}{v(1 - T)} \sim L_0$$

The fact that the quantity [(1 – T)/T] has an additive property suggests that the resistance of an individual scatterer is proportional to it. This agrees with what we obtain if we write the total resistance as a series combination of a contact resistance (G_C^{-1}) and the 'actual' resistance of the conductor:

$$G^{-1} = \frac{h}{2e^2 M} \frac{1}{T} = \frac{h}{2e^2 M} + \frac{h}{2e^2 M} \frac{1-T}{T}$$

$$\Rightarrow \quad G_C^{-1} + \text{'actual' resistance}$$

2.3 Where is the resistance?

The conductance formula $G = (2e^2/h)MT$ clearly shows that scatterers give rise to resistance by reducing the transmission probability. This is intuitively very satisfying because we all feel that the resistance of a sample ought to be related to the ease with which electrons can transmit through it. However, it raises interesting questions regarding the nature and meaning of resistance on a microscopic scale which we will now discuss.

The basic issue can be illustrated with a simple example. Consider a waveguide having M modes containing just one scatterer with a transmission probability of T (see Fig. 2.3.1a). We can view the total resistance given by Eq.(2.2.1) as an interface or contact resistance G_C^{-1} in series with a 'scatterer' resistance G_S^{-1}:

$$G^{-1} = \frac{h}{2e^2 M} \frac{1}{T} \quad \rightarrow \quad G_C^{-1} + G_S^{-1}$$

where $\quad G_C^{-1} \quad \rightarrow \quad \dfrac{h}{2e^2 M} \quad$ and $\quad G_S^{-1} \quad \rightarrow \quad \dfrac{h}{2e^2 M} \dfrac{1-T}{T}$

For clarity we will focus just on the resistance due to the scatterer; we will not worry about the contact resistance. The scatterer resistance G_S^{-1} is clearly determined entirely by the properties of the scatterer, namely, its transmissivity T. But can we associate this resistance just with the scatterer? What about the potential drop (I/G_S) associated with this resistance? Does it occur right across the scatterer? What happens to the Joule heat (I^2/G_S) associated with this resistance? Is it dissipated at the scatterer? The scatterer could be rigid and elastic having no internal degrees of freedom to dissipate energy. In that case the heat must be dissipated elsewhere. Is it dissipated before or after the scatterer?

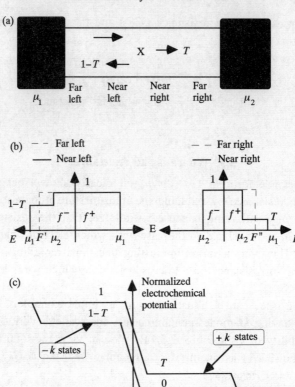

Fig. 2.3.1 (*a*) Waveguide with one scatterer having a transmission *T*. (*b*) Energy distribution of the electrons at different locations along the waveguide. (*c*) Normalized electrochemical potential.

Energy distribution of electrons

To answer these questions we need to consider the energy distribution of the carriers in the positive and negative k-states both to the left and to the right of the scatterer. These are sketched in Fig. 2.3.1b. In this discussion we will neglect interference effects and treat the electrons as semiclassical particles. The $+k$ states to the left of the scatterer are occupied only by electrons coming in from the left contact (again assuming that electrons can exit freely into the left contact without suffering any reflection). Hence these states have the same electrochemical potential as the left contact, that is, μ_1. At low temperatures we can write the distribution function $f^+(E)$ for the $+k$ states as (left of scatterer) $\qquad\qquad f^+(E) \cong \vartheta(\mu_1 - E)$ $\qquad\qquad$ (2.3.1a)

Similarly the $-k$ states to the right of the scatterer have the same electrochemical potential as the right contact, that is, μ_2. Hence the distribution function $f^-(E)$ for the $-k$ states is given by

(right of scatterer) $\qquad\qquad f^-(E) \cong \vartheta(\mu_2 - E)$ $\qquad\qquad$ (2.3.1b)

The situation is a little more complicated when we consider the occupation of the $-k$ states before the scatterer or the $+k$ states after the scatterer. All states are, of course, completely filled below μ_2. But in the energy range between μ_1 and μ_2, the $+k$ states to the right of the scatterer are filled only partially (with probability T) by the transmitted electrons so that

(near right) $\quad f^+(E) \cong \vartheta(\mu_2 - E) + T\{\vartheta(\mu_1 - E) - \vartheta(\mu_2 - E)\}$ $\quad$ (2.3.2a)

Similarly for the $-k$ states to the left of the scatterer

(near left) $\quad f^-(E) \cong \vartheta(\mu_2 - E) + (1 - T)\{\vartheta(\mu_1 - E) - \vartheta(\mu_2 - E)\}$ $\quad$ (2.3.2b)

These distributions are highly non-equilibrium distributions and are only valid very near the scatterer. If we move more than a few energy relaxation lengths away from the scatterer the electrons will settle down to lower energies and a Fermi distribution will be established:

(far left) $\qquad\qquad f^-(E) \cong \vartheta(F' - E)$ $\qquad\qquad$ (2.3.3a)

(far right) $\qquad\qquad f^+(E) \cong \vartheta(F'' - E)$ $\qquad\qquad$ (2.3.3b)

The appropriate electrochemical potentials F' and F'' are determined by noting that the total number of electrons integrated over all energy must remain the same although the distributions change from those in Eq.(2.3.2a,b) to those in Eq.(2.3.3a,b):

$$F' = \mu_2 + (1 - T)\left[\mu_1 - \mu_2\right] \qquad\qquad (2.3.4a)$$

$$F'' = \mu_2 + T\left[\mu_1 - \mu_2\right] \qquad\qquad (2.3.4b)$$

Here we are assuming that the energy relaxation processes help establish equilibrium among the $+k$ states and among the $-k$ states but do not cause any transfer of electrons between the two groups. This helps to keep the discussion simple. If the inelastic processes responsible for energy relaxation also cause backscattering from $+k$ states to $-k$ states then they would introduce additional resistance. In our simplified model they relax energy without relaxing momentum and thus do not contribute to the resistance.

Spatial variation of the electrochemical potential

Now we are ready to investigate the questions posed earlier regarding the spatial location of the resistance. First let us see how the electrochemical potential varies across the scatterer. For the $+k$ states it is equal to μ_1 everywhere to the left of the scatterer and equal to F'' to the far right of the scatterer. Immediately to the right of the scatterer (before energy relaxation has taken place) the distribution is strongly distorted from a Fermi function so that strictly speaking the electrochemical potential is not well-defined. However, we could define a potential such that when we integrate the corresponding Fermi function over energy we obtain the correct number of electrons. The potential to the immediate right of the scatterer will then be the same as that to the far right after energy relaxation. Thus we have for the $+k$ states

$$F^+ = \mu_1 \qquad \qquad \text{(left)}$$

$$= F'' = \mu_2 + T\left[\mu_1 - \mu_2\right] \qquad \text{(right)}$$

We should remember, however, that immediately to the right of the scatterer, before energy relaxation has taken place, the electrons are 'hot' and the electrochemical potential only gives the correct number of electrons. *It does not describe the energy distribution of the electrons.*

It is convenient to define normalized potentials μ^+, μ^- which are obtained from F^+, F^- simply by setting $\mu_2 = 0$ and $\mu_1 = 1$:

$$\mu^+ = 1 \text{ (left)} \quad \text{and} \quad \mu^+ = T \text{ (right)}$$

Similarly the normalized potential for the $-k$ states is given by

$$\mu^- = 1 - T \text{ (left)} \quad \text{and} \quad \mu^- = 0 \text{ (right)}$$

Figure 2.3.1c shows the variation of the normalized potentials across the scatterers. There is a sharp drop across the scatterer just as we would expect for a localized resistance. The normalized potential drop across the scatterer is equal to $(1 - T)$ for both the positive and the negative k-states. The actual potential drop eV_S is simply the normalized drop multiplied by the applied potential $(\mu_1 - \mu_2)$:

$$eV_S = (1-T)\big[\mu_1 - \mu_2\big]$$

What happened to the rest of the applied bias $T\,(\mu_1 - \mu_2)$? It is dropped at the interfaces with the contacts. These are precisely the potential drops we would expect for a current

$$I = (2e/h)MT\big[\mu_1 - \mu_2\big]$$

flowing through the scatterer resistance in series with the contact resistance:

$$G_S^{-1} = \frac{h}{2e^2 M}\frac{1-T}{T} \quad \text{and} \quad G_C^{-1} = \frac{h}{2e^2 M}$$

As we noted earlier (see Fig. 2.1.3) there is an ambiguity with respect to the spatial location of the contact resistance G_C^{-1}. If we were to measure the potential for the $+k$ states then the potential drop associated with G_C^{-1} would occur at the interface with the right contact and we might conclude that the contact resistance was located at the right interface. But if we were to measure the potential for the $-k$ states then we would conclude that it was located at the left interface. Another possibility is to look at the average of μ^+ and μ^-. The resistance is then equally divided between the two interfaces.

Where is the heat dissipated?

It is clear from the above discussion that the potential drop I/G_S indeed occurs right across the scatterer and from this consideration we might view the resistance G_S^{-1} as being spatially located at the scatterer. What about the Joule heat I^2/G_S? Assuming that the scatterer is rigid with no internal degrees of freedom it cannot dissipate the heat. The dissipation has to occur through inelastic processes such as phonon emission that can remove energy from the electrons. Earlier when discussing the energy distribution of the electrons we mentioned that the energy distribution for the $+k$ states to the right of the scatterer (and that for the $-k$ states to the left) would evolve from a highly non-equilibrium distribution to a Fermi function as we go away from the scatterer (see Fig. 2.3.1b). This evolution of the electron energy distribution requires the dissipation of heat and it takes place over a distance of the order of an energy relaxation length from the scatterer.

To describe this in more quantitative terms we note that the heat dissipation P_D is given by the spatial gradient of the energy current I_U:

$$P_D = \frac{d}{dz} I_U \qquad (2.3.5)$$

The energy current can be calculated if we know the energy distribution $i(E)$ of the current that flows at any spatial location.

$$I_U = \frac{1}{e} \int E i(E) dE$$

Noting that the net current I is given by

$$I = \int i(E) dE$$

we could define an average energy U of the current as

$$U \equiv \frac{\int E i(E) dE}{\int i(E) dE} = \frac{e I_U}{I} \qquad (2.3.6)$$

From Eqs.(2.3.5) and (2.3.6) we can write (note that the current I is spatially constant)

$$P_D = \frac{I}{e} \frac{dU}{dz} \qquad (2.3.7)$$

Thus if we are interested in the power dissipation then we should look at the average energy U of the current distribution which can be very different from the electrochemical potential because of the highly non-equilibrium energy distributions that arise around a scatterer over a distance of the order of an energy relaxation length.

To obtain U we first calculate the current per unit energy given by

$$i(E) = \frac{2eM}{h} \left(f^+(E) - f^-(E) \right)$$

where $f^+(E)$ and $f^-(E)$ are the distribution functions for the $+k$ and the $-k$ states respectively. Using the distribution functions from Fig. 2.3.1b we obtain the current distributions shown in Fig. 2.3.2a. The average energy of the current (U) is straightforward to calculate:

$$\begin{aligned} U &= (F' + \mu_1)/2 & \text{(far left)} \\ &= (\mu_1 + \mu_2)/2 & \text{(near left and near right)} \\ &= (F'' + \mu_2)/2 & \text{(far right)} \end{aligned}$$

The normalized average energy of the current (obtained from U by setting

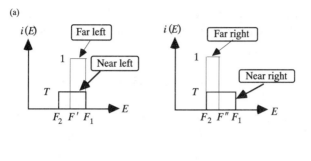

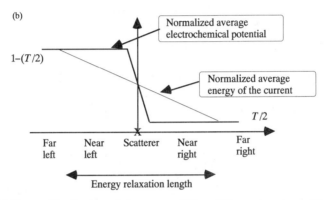

Fig. 2.3.2. (*a*) Energy distribution of the current *i(E)* at different locations. (*b*) The normalized average energy of the current. Also shown for comparison is the normalized average electrochemical potential of the +*k* and −*k* states shown in Fig. 2.3.1c.

$\mu_1 = 1$ and $\mu_2 = 0$) is shown in Fig. 2.3.2b. Also shown for comparison is the average electrochemical potential μ (obtained by averaging μ^+ and μ^- in Fig. 2.3.1c). The point to note is that while the electrochemical potential drops sharply across the scatterer, the average energy of the current changes slowly over an energy relaxation length which is the distance required to dissipate the Joule heat associated with G_S^{-1}.

Thus when we consider how the Joule heat associated with G_S^{-1} is dissipated it would seem that the resistance G_S^{-1} is not localized at the scatterer. One viewpoint we could adopt is that the resistance G_S^{-1} is localized at the scatterer and that the Joule heat associated with G_S^{-1} goes locally into the electron system. It heats up the electron flux away from the scatterer (to the left and to the right). Eventually of course this excess energy is dissipated to the lattice. Where and how this occurs is determined by the inelastic processes that are inevitably present. But this energy relaxation has little effect on the resistance G_S^{-1}, which is determined entirely by the momentum relaxation caused by the scatterer.

Resistivity dipoles

We have seen that the quasi-Fermi energy F_n drops sharply at the scatterer (see Fig. 2.3.2b). The reason is easy to see. Immediately to the left of the scatterer there are lots of electrons arriving from contact 1 but this number drops sharply as we cross the scatterer because only a fraction (T) can cross the scatterer. Since the quasi-Fermi energy is a measure of the number of electrons, it drops sharply (almost discontinuously) at the scatterer.

How does the electrostatic potential V vary across a scatterer? The bottom of the band E_s follows the electrostatic potential energy eV and thus gives an inverted (because the electronic charge is negative) picture of the electrostatic potential. The electron density can be written as (see Eq.(1.2.10))

$$n_s = N_s(F_n - E_s) \quad \Rightarrow \quad \delta n_s = N_s(\delta F_n - \delta E_s) \tag{2.3.8}$$

In general E_s will follow any changes in the Fermi energy F_n keeping the electron density constant. *But E_s cannot follow a sharp change in F_n.* A change in E_s implies an electric field which requires a charge imbalance around the scatterer. This build-up of charge takes place over a screening length which is only a few angstroms in metals but can be hundreds of angstroms in typical semiconductors. As a result, E_s and F_n separate over a region whose dimensions are of the order of the screening length (see Fig. 2.3.3a). In this region, where $\delta E_s \neq \delta F_n$, the electron density will change. Electrons will pile up to the left of the scatterer while to the right there will be a deficit (see Fig.2.3.3b), thus forming a mesoscopic dipole around each obstacle known as a 'resistivity dipole'. This produces an extra electric field at the location of the scatterer as shown in Fig. 2.3.3c. A little reflection shows that this is just a mesoscopic version of what happens on a macroscopic scale if we sandwich a low conductivity material between two high conductivity materials. The point to note is that even a 'homogeneous conductor with a uniform conductivity' is really extremely inhomogeneous on a mesoscopic scale. The electric field and the current flow on this scale are very far from uniform (see R. B. S. Oakeshott and A. MacKinnon (1994), *J. Phys. Cond. Matt.*, **6**, 1513 for some interesting numerical results).

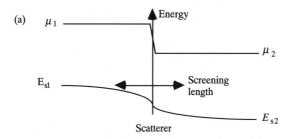

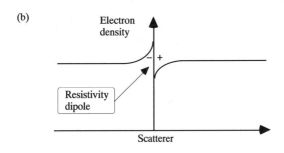

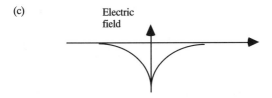

Fig. 2.3.3. Spatial variation of (*a*) conduction band-edge and electrochemical potential, (*b*) electron density and (*c*) electric field across a scatterer.

Screening length

The screening length can be an important length scale especially in low-conductance materials. To estimate the screening length we start from the Poisson equation

$$\nabla^2(\delta E_s) = -\frac{e^2(\delta n_s)}{\varepsilon d}$$

where ε is the dielectric constant, d is the thickness of the 2-D electron system. Making use of Eq.(2.3.8) we can write

$$\nabla^2(\delta E_s) = -\frac{e^2 N_s(\delta F_n - \delta E_s)}{\varepsilon d}$$

that is
$$\left(\nabla^2 - \beta^2\right)\delta E_s = -\beta^2 \delta F_n \qquad (2.3.9)$$

where $\beta^{-1} = \sqrt{\varepsilon d / e^2 N_s}$ is the screening length. We can write

$$\delta E_s \sim \delta F_n \otimes S$$

where the screening function S is the 'impulse response' of Eq.(2.3.9) and
the symbol $\otimes$ denotes convolution. The spatial variation of E_s looks much
like that of F_n, except that the sharp changes are smoothed out due to the
convolution with the screening function (see Fig. 2.3.3a). The details of
the solution are not important but the point is that the response decays
exponentially over a distance β^{-1} which can be identified as the screening length. We can express the screening length in the form

$$\beta^{-1} = \sqrt{a_B d}/2$$

where $a_B \equiv 4\pi\varepsilon\hbar^2/me^2$ is the Bohr radius. With $m = 0.07$ times the free
electron mass and $\varepsilon = 12.6$ times the permittivity of free space (values
typical of GaAs) the Bohr radius is about 10 nm. Since the thickness d is
also about 10 nm, a typical screening length is about 5 nm. Note that this
length is much shorter (typically a fraction of a nm) in metals due to the
high density of states.

2.4. What does a voltage probe measure?

The routine procedure for making resistance measurements in large
macroscopic conductors is to use a multi-terminal Hall bridge of the type
shown in Fig. 1.4.1. The current enters through one of the terminals and
leaves through another as in two-terminal structures. But there are additional terminals along the current path which are left floating (zero external current). These terminals serve as voltage probes that sense the local
electrochemical potential (or quasi-Fermi energy). Suppose we fabricate
a similar Hall bridge with two voltage probes located right across
a scatterer (see Fig. 2.4.1a). We have seen in the last section that the
electrochemical potentials for the $+k$ and the $-k$ states vary as shown in
Fig. 2.4.1b. If we assume that the probes will measure the local
electrochemical potential of either the $+k$ or the $-k$ states (or some
specified combination of the two) then we would expect that

$$\mu_{P1} - \mu_{P2} = (1 - T)\Delta\mu \quad \text{where} \quad \Delta\mu \equiv (\mu_1 - \mu_2)$$

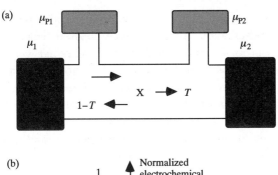

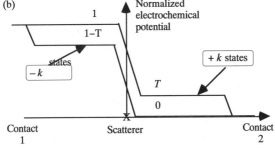

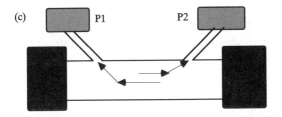

Fig. 2.4.1. (*a*) A four-probe arrangement designed to measure the potential drop across a scatterer. (*b*) Spatial variation of the electrochemical potential of $+k$ and $-k$ states. (*c*) If the probes are bent as shown they could show an apparent negative resistance if $T > 0.5$.

Since the current is given by

$$I = (2e/h)\, MT\, \Delta\mu$$

the resistance measured in a four-probe configuration should be

$$R_{4t} = \frac{(\mu_{P1} - \mu_{P2})/e}{I} = \frac{h}{2e^2 M} \frac{1-T}{T} \tag{2.4.1}$$

If we try to apply this result we run into three separate problems. Firstly, *mesoscopic probes are often invasive*, that is, they change what we are trying to measure. With macroscopic conductors, the probes represent a minor perturbation. Their presence does not change the current significantly. But for a small conductor, the probes can very well be the dominant source of scattering (and hence resistance). This, however, is a practical and not a fundamental problem. There is no fundamental reason why a voltage probe has to be strongly coupled to the conductor. There has been some work using weakly coupled scanning tunneling probes to observe resistivity dipoles around individual scatterers and it is likely that there will be more of such non-invasive microscopic measurements as nanotechnology progresses.

Secondly, *mesoscopic probes are seldom identical* so that the two voltage probes could very well couple differently to the $+k$ and $-k$ states. For example, suppose a probe were bent over to the right like P2 in Fig. 2.4.1c then it could couple much better to $+k$ states than to $-k$ states. This is because a small deflection would make an electron in a $+k$ state enter the probe but a large angle scattering is needed to make a $-k$ state enter. Hence probe P2 would register a potential close to that of the $+k$ state:

$$\mu_{P2} \approx T \Delta \mu$$

Similarly if a probe were bent over to the left like P1 in Fig. 2.4.1c it would couple more strongly to a $-k$ state and register a potential close to that of the $-k$ state:

$$\mu_{P1} \approx (1 - T)\Delta \mu$$

The measured resistance would then be

$$\frac{(\mu_{P1} - \mu_{P2})/e}{I} = \frac{h}{2e^2 M} \frac{1 - 2T}{T}$$

which can even be negative for conductors having $T > 0.5$!

In practice one has little control over the microscopic potential profiles that determine the coupling of the probes to the $+k$ and $-k$ states. We expect to measure the resistance given by Eq.(2.4.1) only if the two voltage probes couple identically to the $+k$ and $-k$ states. Otherwise the measured resistance will lie somewhere between the extremes

$$\frac{h}{2e^2 M} \frac{1}{T} \quad \text{and} \quad \frac{h}{2e^2 M} \frac{1 - 2T}{T}$$

This makes little difference if the conductor is many mean free paths long. For such conductors, the transmission probability is much less than one (see Eq.(2.2.2)) and there is hardly any difference between 1 and $(1 - 2T)$.

Finally *mesoscopic measurements are strongly affected by quantum interference effects* unless the distance of the probes from the scatterer is much greater than the phase-relaxation length. Consider for example a strongly reflecting scatterer with $T \ll 1$. In that case the electrochemical potentials for the $+k$ and $-k$ states are both nearly equal to one to the left of the scatterer and zero to the right of the scatterer (see Fig. 2.4.1b). We would expect that a probe to the left of the scatterer should measure a potential of approximately one (equal to that of the left reservoir). However, due to quantum interference it could measure any potential between zero and one depending on its distance from the scatterer. The reason is that the probe may not be able 'see' the electrons from the left reservoir due to destructive interference between the incident wave and the reflected wave (for a quantitative description of this effect see Exercise E.3.1 at the end of Chapter 3). As a result it could float to a potential closer to that of the *right* reservoir even though it is located to the *left* of a strongly reflecting scatterer! We can use Eq.(2.4.1) to describe the four-terminal resistance only if such interference effects are either absent (because of a short phase-relaxation length) or carefully eliminated (by averaging measurements over a wavelength or using 'directional couplers' to couple the probes so that they see only the $+k$ or the $-k$ states).

The issue of quantum interference serves as a strong reminder of the limitations of the semiclassical concepts that we have used for our discussion in the last section. We calculated the distribution functions for the $+k$ and the $-k$ states, f^+ and f^-, and used these distribution functions to derive electrochemical potentials (or quasi-Fermi energies) for the two groups of states. This is analogous to what device engineers routinely do when analyzing structures involving the transport of both electrons and holes. In a p–n junction, for example, electrons and holes often have very different quasi-Fermi energies. However, it is important to remember that in phase-coherent conductors the $+k$ and $-k$ states can be strongly correlated so that distribution functions only tell part of the story.

The problem can be appreciated by considering a simple analogy with optics. A beam of unpolarized light is a 50-50 mixture of photons that are polarized in the x-direction and photons that are polarized in the y-direction. But so is a beam of light that is polarized at 45 degrees to the x-axis. Yet the two are physically very different and there are many experiments that can distinguish between them. One way to represent this difference is by using a density matrix:

$$\begin{bmatrix} 0.5 & 0 \\ 0 & 0.5 \end{bmatrix} \qquad \begin{bmatrix} 0.5 & 0.5 \\ 0.5 & 0.5 \end{bmatrix}$$

(unpolarized)   (45-degree polarized)

The diagonal elements of the density matrix represent the usual distribution function while the off-diagonal elements represent phase-correlations. For unpolarized light, the x- and y-polarizations are uncorrelated so that the off-diagonal elements are zero. For 45-degree polarized light the off-diagonal elements are as large as the diagonal ones due to the perfect phase correlation between the x- and y-polarizations.

We have an analogous situation in phase-coherent conductors with $+k$ and $-k$ states playing the roles of x- and y-polarizations. The distribution function only gives us the diagonal elements. The rest of the story is contained in the off-diagonal elements which cannot be neglected unless the phase-relaxation length is much shorter than the other length scales. To describe the internal state of phase-coherent conductors we need concepts like density matrices or correlation functions which we will postpone till the last chapter of this book. We will now describe an approach that was developed by Büttiker following the work of Engquist and Anderson (see Refs. [2.3], [2.4]), which allows us to describe multi-terminal phase-coherent conductors directly in terms of measured currents and voltages, *completely bypassing any questions regarding the internal state of the conductor*. This makes it possible to handle a complex topic like quantum magnetotransport without the use of advanced concepts (as long as 'vertical flow' can be neglected, as explained in Sections 2.6 and 2.7).

Büttiker formula

Since 1985 many mesoscopic experiments have been conducted using miniature Hall bridges fabricated on both metallic and semiconducting samples. However, because of the reasons mentioned above, for a while there was serious confusion about how such four-terminal measurements should be interpreted. Indeed there was no consensus regarding the two-terminal resistance either, primarily because the importance of the contact resistance in this context was not recognized (although for pedagogical reasons we have chosen to start this chapter with a discussion of the contact resistance).

Büttiker found a simple and elegant solution to this problem. He noted that since there is really no qualitative difference between the current and voltage probes, one could treat all the probes on an equal footing and simply extend the two-terminal linear response formula

$$I = \frac{2e}{h}\overline{T}[\mu_1 - \mu_2]$$

by summing over all the terminals (indexed by p and q) as follows

$$I_p = \frac{2e}{h}\sum_q[\overline{T}_{q \leftarrow p}\mu_p - \overline{T}_{p \leftarrow q}\mu_q]$$

We can rewrite this in the form (with $V = \mu/e$)

$$I_p = \sum_q[G_{qp}V_p - G_{pq}V_q] \qquad (2.4.2a)$$

where
$$G_{pq} \equiv \frac{2e^2}{h}\overline{T}_{p \leftarrow q} \qquad (2.4.2b)$$

The arrows in the subscripts have been inserted just as a reminder that the electron transfer is backwards from the second subscript to the first one. We will generally write the subscripts without the arrows.

The coefficients G in Eq.(2.4.2) must satisfy the following 'sum rule', regardless of the detailed physics, in order to ensure that the current is zero when all the potentials are equal:

$$\sum_q G_{qp} = \sum_q G_{pq} \qquad (2.4.3)$$

This allows us to rewrite Eq.(2.4.2) in an equivalent form

$$I_p = \sum_q G_{pq}[V_p - V_q] \qquad (2.4.4)$$

The conductance coefficients (G) also obey the relation $(B$: magnetic field)

$$[G_{qp}]_{+B} = [G_{pq}]_{-B} \qquad (2.4.5)$$

Unlike the sum rule in Eq.(2.4.3), there is no simple reason why this relation has to be true regardless of the detailed physics. To prove this, one needs to assume a particular model for the transport. Experimentally, however, there is no evidence that this relation is ever violated (in the linear response regime) regardless of the nature of the transport. In

Chapter 3 we will prove this relation for coherent transport using the properties of the S-matrix.

A couple of comments before we proceed. Firstly we note that the potential V_P at a voltage probe can be written from Eq.(2.4.4) as (setting $I_P = 0$)

$$V_P = \frac{\sum\limits_{q \neq P} G_{Pq} V_q}{\sum\limits_{q \neq P} G_{Pq}}$$

This means that the potential measured at a floating terminal P is simply a weighted average of all other terminal potentials q and the weighting is determined by the conductance coefficient G_{Pq} which is proportional to the transmission function from the terminal q to the floating terminal P. The shape and construction of the probes affects the measured potential through the transmission functions. The probes in Fig. 2.4.1a and Fig. 2.4.1c will register different potentials and we can calculate these in a straightforward manner.

The second point is that if the magnetic field is zero then the coefficients are symmetric and Eq.(2.4.4) is precisely what we would get if we applied Kirchhoff's law to a network of conductors $G_{qp} (= G_{pq})$ connecting every terminal q to every other terminal p. This simple resistor model, however, cannot be used in a non-zero magnetic field since the conductance coefficients (G) are usually not symmetric, that is, $G_{qp} \neq G_{pq}$.

Let us now look at a three-terminal structure and a four-terminal structure to see how the Büttiker formula is actually applied to conductors having voltage probes.

Three-terminal device

For simplicity we first consider a three-terminal device where one of the terminals serves both as a voltage and as a current terminal. Suppose it is connected up as shown in Fig. 2.4.2a (or 2.4.2b). To calculate the resistance $R_{3t} = V/I$ (or the resistance $R'_{3t} = V'/I'$) we start from Eq.(2.4.4):

$$\begin{Bmatrix} I_1 \\ I_2 \\ I_3 \end{Bmatrix} = \begin{bmatrix} G_{12} + G_{13} & -G_{12} & -G_{13} \\ -G_{21} & G_{21} + G_{23} & -G_{23} \\ -G_{31} & -G_{32} & G_{31} + G_{32} \end{bmatrix} \begin{Bmatrix} V_1 \\ V_2 \\ V_3 \end{Bmatrix}$$

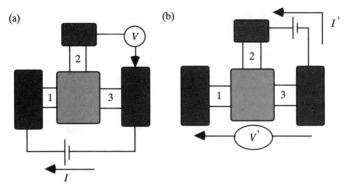

Fig. 2.4.2 (*a*) A three-terminal conductor with an external current I flowing from terminal 3 to 1. A voltage drop V is measured between the terminals 2 and 3. (*b*) The same conductor with the current and voltage terminals reversed.

Actually we do not need to solve the set of three equations since they are not independent. The sum rules ensure that Kirchoff's law is satisfied, namely, $I_1 + I_2 + I_3 = 0$. Also since the currents only depend on voltage differences between different terminals we can without loss of generality set one of the voltages to zero. We will set $V_3 = 0$. This allows us to truncate the third row and the third column of the matrix and write

$$\begin{Bmatrix} I_1 \\ I_2 \end{Bmatrix} = \begin{bmatrix} G_{12} + G_{13} & -G_{12} \\ -G_{21} & G_{21} + G_{23} \end{bmatrix} \begin{Bmatrix} V_1 \\ V_2 \end{Bmatrix}$$

Inverting we obtain

$$\begin{Bmatrix} V_1 \\ V_2 \end{Bmatrix} = \begin{bmatrix} R_{11} & R_{12} \\ R_{21} & R_{22} \end{bmatrix} \begin{Bmatrix} I_1 \\ I_2 \end{Bmatrix} \qquad (2.4.6)$$

where the matrix $[R]$ is defined as

$$[R] = \begin{bmatrix} G_{12} + G_{13} & -G_{12} \\ -G_{21} & G_{21} + G_{23} \end{bmatrix}^{-1} \qquad (2.4.7)$$

The resistance R_{3t} measured in the configuration shown in Fig. 2.4.2a is given by

$$R_{3t} = \frac{V}{I} = \left[\frac{V_2}{I_1} \right]_{I_2 = 0} = R_{21} \qquad (2.4.8a)$$

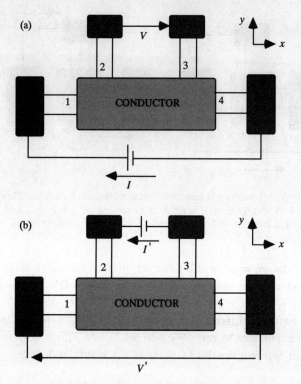

Fig. 2.4.3 (*a*) A four-terminal conductor with an external current *I* flowing from terminal 4 to 1. A voltage drop *V* is measured between the floating terminals 2 and 3. (*b*) The same conductor with the current and voltage terminals reversed.

while the resistance measured in the configuration shown in Fig.2.4.2b is given by

$$R'_{3t} = \frac{V'}{I'} = \left[\frac{V_1}{I_2}\right]_{I_1=0} = R_{12} \qquad (2.4.8b)$$

Four-terminal device

Next we consider a four-terminal device as shown in Fig. 2.4.3a or 2.4.3b. The basic procedure is the same as in a three-terminal device. We set the voltage at one of the terminals (say V_4) equal to zero and write out Eq.(2.4.4) for the currents at the other terminals in the form of a matrix equation:

$$\begin{Bmatrix} I_1 \\ I_2 \\ I_3 \end{Bmatrix} = \begin{bmatrix} G_{12} + G_{13} + G_{14} & -G_{12} & -G_{13} \\ -G_{21} & G_{21} + G_{23} + G_{24} & -G_{23} \\ -G_{31} & -G_{32} & G_{31} + G_{32} + G_{34} \end{bmatrix} \begin{Bmatrix} V_1 \\ V_2 \\ V_3 \end{Bmatrix}$$

Inverting we obtain

$$\begin{Bmatrix} V_1 \\ V_2 \\ V_3 \end{Bmatrix} = \begin{bmatrix} R_{11} & R_{12} & R_{13} \\ R_{21} & R_{22} & R_{23} \\ R_{31} & R_{32} & R_{33} \end{bmatrix} \begin{Bmatrix} I_1 \\ I_2 \\ I_3 \end{Bmatrix} \qquad (2.4.9)$$

where the matrix $[R]$ is given by

$$[R] = \begin{bmatrix} G_{12} + G_{13} + G_{14} & -G_{12} & -G_{13} \\ -G_{21} & G_{21} + G_{23} + G_{24} & -G_{23} \\ -G_{31} & -G_{32} & G_{31} + G_{32} + G_{34} \end{bmatrix}^{-1} \qquad (2.4.10)$$

The resistance R_{4t} measured in the configuration shown in Fig. 2.4.3a is given by

$$R_{4t} = \frac{V}{I} = \left[\frac{V_2 - V_3}{I_1} \right]_{I_2 = I_3 = 0} = R_{21} - R_{31} \qquad (2.4.11a)$$

while the resistance R'_{4t} measured in the configuration shown in Fig.2.4.3b is given by

$$R'_{4t} = \frac{V'}{I'} = \left[\frac{V_1}{I_2} \right]_{I_1 = 0, \, I_2 = -I_3} = R_{12} - R_{13} \qquad (2.4.11b)$$

Reciprocity

In a macroscopic rectangular Hall bridge the resistance R_{4t} would be related to the ρ_{xx} component of the resistivity tensor which was shown by Onsager (using thermodynamic arguments) to be symmetric in the magnetic field.

$$\rho_{xx}(B) = \rho_{xx}(-B) \quad \Rightarrow \quad R_{4t}(B) = R_{4t}(-B)$$

In mesoscopic Hall bridges it is found that the resistance R_{4t} fluctuates randomly as a function of the magnetic field (perpendicular to the plane of the conductor) due to random interference among multiple scatterers (we will discuss this in Chapter 5). The fluctuations have no particular symmetry in the magnetic field. This caused some consternation initially since based on our experience with macroscopic Hall bridges it was expected that $R_{4t}(B)$ should be equal to $R_{4t}(-B)$. But it was soon realized that the current flow on a mesoscopic scale is very irregular (due to random scatterers) and not directed along x or y even in a rectangular conductor. Consequently the measured resistance R_{4t} is some average of both ρ_{xx} and ρ_{xy} components of the resistivity tensor. Since the component ρ_{xy} is antisymmetric in the magnetic field (while the component ρ_{xx} is symmetric in the magnetic field) the measured resistance R_{4t} is neither symmetric nor antisymmetric. This is what we would see even with large conductors if we were to make measurements on an irregularly shaped Van der Pauw sample. The point is that the current flow pattern on a mesoscopic scale is quite irregular even though the shape of the conductor may be rectangular.

For arbitrarily shaped macroscopic conductors, it has long been known that if we reverse the magnetic field and reverse the current and voltage terminals (as we have done in going from the setup in Fig. 2.4.2a to the setup in Fig. 2.4.2b or from Fig. 2.4.3a to Fig. 2.4.3b), then the measured resistance will be the same as before; that is,

$$R_{3t}(+B) = R'_{3t}(-B) \quad \text{and} \quad R_{4t}(+B) = R'_{4t}(-B) \qquad (2.4.12)$$

This relation, known as the reciprocity relation, was originally derived for macroscopic conductors using thermodynamic arguments. But is it true for mesoscopic conductors? Experimentally the answer is yes (see Fig. 6 in R. A. Webb and S. Washburn (1988), *Physics Today*, **41**, 52).

We will now make use of the reciprocity of the conductance coefficients stated earlier

$$\left[G_{qp} \right]_{+B} = \left[G_{pq} \right]_{-B} \qquad \text{(same as Eq.(2.4.5))}$$

to show that the three-terminal and four-terminal resistances that we have calculated above from the Büttiker formula indeed obey the reciprocity condition set forth in Eq.(2.4.12). Looking at the definition of [R] (Eq.(2.4.7) for the three-terminal geometry and Eq.(2.4.10) for the four-terminal case) it is obvious that the reciprocity of the conductance coefficients implies that *reversing* the magnetic field *transposes* the matrix $[R^{-1}]$, that is,

$$\left[R^{-1} \right]_{+B} = \left[R^{-1} \right]^{\mathrm{T}}_{-B} \qquad (2.4.13)$$

except for one point. It may not be apparent that the diagonal elements of $[R^{-1}]$ are unchanged when the magnetic field is reversed. To see this we need to make use of the sum rule for the conductance matrix (see Eq.(2.4.3)):

$$\sum_q [G_{pq}]_{+B} = \sum_q [G_{qp}]_{+B} = \sum_q [G_{pq}]_{-B}$$

Since $G_{pp}(+B) = G_{pp}(-B)$ this implies that

$$\sum_{q \neq p} [G_{pq}]_{+B} = \sum_{q \neq p} [G_{pq}]_{-B}$$

showing that the diagonal elements of $[R^{-1}]$ are unchanged when the magnetic field is reversed.

Since $[R^{-1}]^T = [R^T]^{-1}$, it follows from Eq.(2.4.13) that *reversing* the magnetic field also *transposes* the matrix $[R]$:

$$[R]_{+B} = [R]^T_{-B} \tag{2.4.14}$$

Hence

$$[R_{31}]_{+B} = [R_{13}]_{-B} \quad \text{and} \quad [R_{21}]_{+B} = [R_{12}]_{-B}$$

Making use of Eqs.(2.4.11a,b) for the four-terminal resistance (or Eqs.(2.4.8a,b) for the three-terminal resistance) it is easy to see that the reciprocity condition in Eq.(2.4.12) is satisfied. The correct prediction of the reciprocity properties observed experimentally in four-terminal mesoscopic structures was the first important application of the Büttiker formula [2.4].

The Büttiker formula (Eq.(2.4.4)) provides a terminal description in terms of measured currents and voltages, completely bypassing any questions regarding the spatial variation of the potential inside a sample. It has been used widely by both theorists and experimentalists to understand and interpret mesoscopic four-probe resistance measurements. A number of interesting phenomena have been observed in narrow Hall bridges which are all related to the non-intuitive behavior of mesoscopic voltage probes discussed earlier. The common feature of all these phenomena is that they require us to view the probes as extensions of the waveguide itself and calculate the resistance of the composite probe–device configuration using Eq.(2.4.4) (see Exercises E.2.3 and E.2.4 at the end of this chapter). A fairly detailed review of these experiments can be found in Ref.[2.5] and will not be repeated here.

2.5 Non-zero temperature and bias

Our derivation of the Landauer formula in Section 2.2 was simplified by the assumption of zero temperature so that transport occurred only from contact 1 to 2 and not from 2 to 1 (see Fig. 2.5.1b). We also assumed that

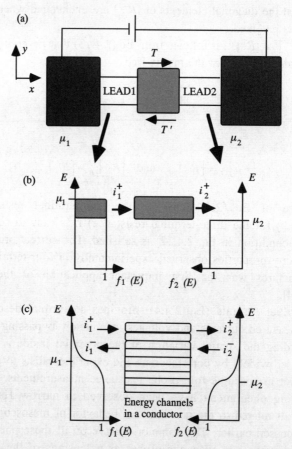

Fig. 2.5.1 (a) A conductor is connected to two large contacts through two leads. (b) Energy distributions of the incident electrons in the two leads at zero temperature. (c) Energy distributions at non-zero temperatures.

the current was carried by a single energy channel around the Fermi energy. This allowed us to write the current simply as

$$I = \frac{2e}{h} \overline{T}[\mu_1 - \mu_2]$$

where $\overline{T}$ denotes the product of the number of modes M and the transmission probability per mode T at the Fermi energy (assumed constant over the range $\mu_1 > E > \mu_2$).

In general, however, we have the situation depicted in Fig. 2.5.1c where transport takes place through multiple energy channels in the energy range

$$\mu_1 + (\text{a few } k_BT) > E > \mu_2 - (\text{a few } k_BT)$$

and each channel can have a different transmission $\overline{T}$. Under these conditions an expression for the current can be derived in much the same way as in Section 2.2 except that we now need to include injection from both contacts. The influx of electrons per unit energy from lead 1 is given by (see Fig. 2.5.1c)

$$i_1^+(E) = (2e/h)Mf_1(E)$$

while the influx from lead 2 is given by (M' is the number of modes in lead 2)

$$i_2^-(E) = (2e/h)M'f_2(E)$$

The outflux from lead 2 is written as

$$i_2^+(E) = T i_1^+(E) + (1 - T')i_2^-(E)$$

while the outflux from lead 1 is written as

$$i_1^-(E) = (1 - T) i_1^+(E) + T' i_2^-(E)$$

The net current $i(E)$ flowing at any point in the device is given by

$$i(E) = i_1^+ - i_1^- = i_2^+ - i_2^-$$
$$= T i_1^+ - T' i_2^-$$
$$= \frac{2e}{h}\left[M(E)T(E) f_1(E) - M'(E)T'(E)f_2(E)\right]$$

Defining the transmission function as $\overline{T}(E) = M(E) T(E)$ we can write

$$i(E) = \frac{2e}{h}\left[\overline{T}(E) f_1(E) - \overline{T}'(E) f_2(E)\right]$$

The total current can be written as

$$I = \int i(E)dE \quad \text{where} \quad i(E) = \frac{2e}{h} \bar{T}(E)\big[f_1(E) - f_2(E) \big] \qquad (2.5.1)$$

if we assume that $\bar{T}(E) = \bar{T}'(E)$. But why should the transmission function from 1 to 2 be equal to that from 2 to 1? One could argue that they ought to be equal in order to ensure that there is no current at equilibrium (that is, $i(E) = 0$ when $f_1(E) = f_2(E)$). However, this argument only proves that $\bar{T}(E)$ should equal $\bar{T}'(E)$ at equilibrium. As we move away from equilibrium, the applied bias could change the two transmission functions and make them unequal. Thus $\bar{T}(E) \neq \bar{T}'(E)$ in general. However, if we assume that there is no inelastic scattering (from one energy to another) inside the device, then it can be shown that $\bar{T}(E)$ is always equal to $\bar{T}'(E)$ for a two-terminal device even in the presence of magnetic fields. We will provide a more rigorous derivation of Eq.(2.5.1) in Section 2.6. For the moment let us accept it and work out some of its consequences.

Linear response

If both contacts are held at the same potential then $\mu_1 = \mu_2$ and Eq.(2.5.1) predicts zero current, as we would expect: $f_1(E) = f_2(E) \rightarrow I = 0$. For small deviations from this equilibrium state, the current is proportional to the applied bias. We can write from Eq.(2.5.1)

$$\delta I = \frac{2e}{h} \int \Big(\big[\bar{T}(E)\big]_{eq} \delta[f_1 - f_2] + [f_1 - f_2]_{eq} \delta\big[\bar{T}(E)\big] \Big) dE$$

The second term is clearly zero. We can simplify the first term by using a Taylor's series expansion to write

$$\delta[f_1 - f_2] \approx [\mu_1 - \mu_2]\left(\frac{\partial f}{\partial \mu}\right)_{eq} = \left(-\frac{\partial f_0}{\partial E}\right)$$

where $f_0(E)$ is the equilibrium Fermi function (see Eq.(1.2.7))

$$f_0(E) = \left[\frac{1}{\exp((E - \mu)/k_B T)) + 1} \right]_{\mu = E_f}$$

We then obtain the (non-zero temperature) linear response formula

$$G = \frac{\delta I}{(\mu_1 - \mu_2)/e} = \frac{2e^2}{h} \int \bar{T}(E)\left(-\frac{\partial f_0}{\partial E}\right) dE \qquad (2.5.2)$$

At low temperatures we can write

$$f_0(E) \approx \vartheta(E_f - E) \quad \rightarrow \quad -\frac{\partial f_0}{\partial E} \approx \delta(E_f - E)$$

so that Eq.(2.5.2) reduces to the zero temperature linear response result:

$$G = \frac{2e^2}{h} \overline{T}(E_f) \tag{2.5.3}$$

Note that in Eqs.(2.5.2) and (2.5.3) we have dropped the subscript 'eq' for clarity. It is implied that the quantities on the right are all evaluated at equilibrium: linear response is an equilibrium property.

When is the response linear?

From the above derivation it would seem that the response should be linear if the bias ($\mu_1 - \mu_2$) is much less than $k_B T$, so that the Taylor's series expansion is accurate. While this is certainly a sufficient condition it is not a necessary condition. If this were a necessary condition then there would be no linear response at zero temperature! Actually, the response is linear regardless of the temperature if the transmission function $\overline{T}(E)$ is approximately constant over the energy range where transport occurs, and can be assumed to be unaffected by the bias. We can then write from Eq.(2.5.1)

$$I = \frac{2e}{h} \overline{T}(E_f) \int \left[f_1(E) - f_2(E) \right] dE$$

At low temperatures it is easy to see that

$$\int \left[f_1(E) - f_2(E) \right] dE = \mu_1 - \mu_2$$

since $f_1(E) \approx \vartheta(\mu_1 - E)$ and $f_2(E) \approx \vartheta(\mu_2 - E)$ (see Eq.(1.2.9)). This result is actually valid at high temperatures as well, though it is not as easy to see. We thus obtain a linear relationship between the current and the applied bias

$$I = \frac{2e}{h} \overline{T}(E_f) \left[\mu_1 - \mu_2 \right]$$

for arbitrary temperatures, as long as the transmission function is independent of energy and unaffected by the bias.

We can arrive at a general criterion for linear response by rewriting Eq.(2.5.1) in the form (see Exercise E.2.7 at the end of this chapter)

$$I = \frac{1}{e}\int_{\mu_1}^{\mu_2}\hat{G}(E')dE' \qquad (2.5.4)$$

where the conductance function is defined as

$$\hat{G}(E') = \frac{2e^2}{h}\int\overline{T}(E)F_T(E-E')dE \qquad (2.5.5)$$

$F_T(E)$ being the thermal broadening function

$$F_T(E) \equiv -\frac{d}{dE}\left(\frac{1}{\exp(E/k_BT)+1}\right) = \frac{1}{4k_BT}\text{sech}^2\left(\frac{E}{2k_BT}\right) \qquad (2.5.6)$$

This new expression for the current (Eq.(2.5.4)) makes it easy to see when the current will respond linearly to the applied bias. It will respond linearly if the conductance function $\hat{G}(E)$ is *independent of energy* in the energy range $\mu_1 > E > \mu_2$. We can then write the current as

$$I = \hat{G}(E_f)\frac{\mu_1 - \mu_2}{e}$$

where the conductance is given by

$$G = \hat{G}(E_f) = \frac{2e^2}{h}\int\overline{T}(E)F_T(E-E_f)dE$$

$$= \frac{2e^2}{h}\int\overline{T}(E)\left[-\frac{d}{dE}\frac{1}{\exp[(E-E_f)/k_BT]+1}\right]dE$$

in agreement with the result obtained earlier (Eq.(2.5.2)).

At low temperatures when the phase-relaxation time is long, quantum interference can give rise to sharp resonant structures in the transmission characteristics of conductors having multiple sources of scattering. For this reason the transmission $\overline{T}(E)$ often changes rapidly with energy (or, in other words, the *correlation energy* is small) at low temperatures in mesoscopic conductors as sketched in Fig. 2.5.2. A good estimate for the correlation energy of a diffusive conductor of the type sketched in Fig. 1.3.3 is

$$\varepsilon_c \approx \frac{\hbar}{\tau_\varphi} \quad \rightarrow \quad 0.006 \text{ meV if } \tau_\varphi = 100 \text{ ps}$$

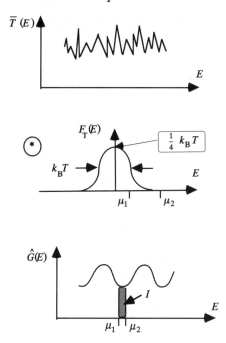

Fig. 2.5.2. The conductance function $\widehat{G}(E)$ is obtained by convolving the transmission $\overline{T}(E)$ with the thermal broadening function $F_T(E)$. The current is obtained by integrating the conductance function $\widehat{G}(E)$ over the range of the applied bias from μ_1 to μ_2.

If we were truly at zero temperature such that the thermal smearing function is a delta function then the response would be linear only if the applied bias were much less than this correlation energy. But the effect of temperature is to smear out the sharp features to produce a relatively smooth conductance function $\widehat{G}(E)$ which varies only on the scale of the thermal energy k_BT (~ 0.01 meV at $T = 0.1$ K). So the response is linear as long as the bias is much less than k_BT. In mesoscopic experiments the bias is always kept smaller than k_BT in order to ensure that the measurements are in the linear response regime.

Thus the criterion for linear response in systems with a small correlation energy is given by $(\mu_1 - \mu_2) \ll k_BT$. But if the correlation energy is large then the transmission function $\overline{T}$ is fairly smooth to start with and the smoothing action of the thermal function $F_T(E)$ is unnecessary. The response could then be linear for much larger values of bias. The general criterion for linear response is simply that the function obtained by volving the transmission function with the thermal smearing function be constant in the energy range where transport occurs, that is,

$$(\mu_1 - \mu_2) << k_BT + \varepsilon_c$$

It should be mentioned that as the bias is increased an electric field develops within the conductor which can change the transmission function $\bar{T}$ and thereby make the response non-linear. Thus the transmission is a function of the bias and we should really write it as $\bar{T}(E, \mu_1, \mu_2)$ and not just $\bar{T}(E)$. This aspect of the problem is absent when we consider linear response ($\mu_1, \mu_2 \to E_f$) but has to be taken into account when applying Eq.(2.5.1) to high bias situations. Indeed the earliest application of current formulas of this type was not in the calculation of linear response conductance but in the calculation of the $I-V$ characteristics of tunneling junctions where the lowering of the tunneling barrier by an applied bias causes an exponential increase in the transmission. More recently Eq.(2.5.1) has been used to calculate the $I-V$ characteristics of resonant tunneling devices (to be discussed in Chapter 6). In all these calculations it is important to include the effect of the electric field on the transmission. In some cases it may be adequate to assume a constant electric field ($= V/L$) inside the conductor. But in general one has to take into account the electron density inside the conductor and obtain the electric field from the Poisson equation, somewhat in the spirit of what we did when discussing screening lengths at the end of the last section (see Fig. 2.3.3).

Multi-terminal conductors

For multi-terminal devices we could argue as we did for two-terminal devices and write the current as (different terminals are indexed by p and q)

$$I_p = \int i_p(E)dE$$

$$\text{where} \quad i_p(E) = \frac{2e}{h} \sum_q \left[\bar{T}_{qp}(E)f_p(E) - \bar{T}_{pq}(E)f_q(E) \right] \quad (2.5.7a)$$

Here $\bar{T}_{pq}(E)$ represents the total transmission from terminal q to terminal p at the energy E and $f_p(E)$ is the Fermi function for terminal p. As in the two-terminal case (see discussion following Eq.(2.5.1)) we could argue that since there can be no current flowing at equilibrium, the transmission functions must satisfy the sum rule

$$\sum_q \bar{T}_{qp}(E) = \sum_q \bar{T}_{pq}(E)$$

Once again, far away from equilibrium, this relation is true only if there is no inelastic scattering inside the device. If we assume the sum rule to be true then the current can just as well be written in the form

$$i_p(E) = \frac{2e}{h} \sum_q \overline{T}_{pq}(E)\left[f_p(E) - f_q(E) \right]$$ (2.5.7b)

We have seen how the linear response formula Eq.(2.4.4) is applied to multi-terminal conductors with floating voltage probes. We set the current I_P at a voltage probe to zero and solve Eq.(2.4.4) to obtain the potential V_P at the probe. We could do the same with the general current expression (Eq.(2.5.7)) and set the current $i_P(E)$ *at every energy* equal to zero. But this is not necessarily justified. All that we know for sure at a floating probe is that the net current is zero:

$$I_P = \int i_P(E)dE = 0$$

A better approach is to adjust the electrochemical potential μ_P at the floating probe so that the net current I_P is zero.

We can linearize Eq.(2.5.7b) as we did earlier for two-terminal conductors to obtain ($f_0(E)$: equilibrium Fermi function)

$$I_p = \sum_q G_{pq}\left[V_p - V_q \right]$$ (2.5.8)

where

$$G_{pq} = \frac{2e^2}{h} \int \overline{T}_{pq}(E)\left(-\frac{\partial f_0}{\partial E} \right)dE$$ (2.5.9a)

At low temperatures

$$G_{pq} = \frac{2e^2}{h} \overline{T}_{pq}(E_f)$$ (2.5.9b)

All the work on multiterminal conductors so far has been based on this linear response formula (Eq.(2.5.8)). The author is not aware of any calculations based on Eq.(2.5.7) although its two-terminal counterpart (Eq.(2.5.1)) is widely used to analyze quantum devices like tunnel diodes and resonant tunneling diodes.

2.6 Exclusion principle?

A common question that arises is whether we ought to modify the current expression (see Eq.(2.5.7a))

$$i_p(E) = \frac{2e}{h} \sum_q \left[\overline{T}_{qp}(E) f_p(E) - \overline{T}_{pq}(E) f_q(E) \right] \qquad (2.6.1)$$

to account for the exclusion principle:

$$i_p(E) = \frac{2e}{h} \sum_q \left[\overline{T}_{qp} f_p (1 - f_q) - \overline{T}_{pq} f_q (1 - f_p) \right] \quad \text{(Wrong)} \qquad (2.6.2)$$

We can rewrite Eq.(2.6.2) in a form that clearly shows the difference with Eq.(2.6.1):

$$i_p(E) = \frac{2e}{h} \sum_q \left[\overline{T}_{qp} f_p - \overline{T}_{pq} f_q \right] - \left[\overline{T}_{qp} - \overline{T}_{pq} \right] f_p f_q$$

The second term is identically zero if $\overline{T}_{qp} = \overline{T}_{pq}$ and we recover our earlier result. As we will see, for two-terminal devices without inelastic scattering, $\overline{T}_{12}$ is always equal to $\overline{T}_{21}$. It thus makes no difference whether we include the $(1 - f)$ factors shown in Eq.(2.6.2). That is why the distinction between Eqs.(2.6.1) and (2.6.2) is not appreciated widely, even though these equations have been applied to tunneling (and later to resonant tunneling) devices ever since the 1930s. It was only in the late 1980s following the work of Büttiker [2.4] that these equations came to be applied to multiterminal devices in the presence of magnetic fields where $\overline{T}_{qp} \neq \overline{T}_{pq}$ in general. Eqs.(2.6.1) and (2.6.2) can then lead to very different predictions for the terminal currents.

One might wonder why we should care about Eq.(2.6.1) anyway since we argued in the last Section that Eq.(2.6.1) is difficult to apply to conductors with floating probes and what we normally use is the linear response equation (Eq.(2.5.8)). This equation of course has to be valid as long as the currents depend linearly on the voltage, since it is hard to imagine any other linear relationship. However, the expression for the conductance in terms of the transmission (see Eq.(2.5.9)) is obtained by linearizing Eq.(2.6.1). If we were to include the $(1 - f)$ factors shown in Eq.(2.6.2), the process of linearization would yield much more complicated expressions for the conductance.

We will show that if transport is coherent across the conductor, so that we can define a single wavefunction extending from one lead to another, then Eq.(2.6.1) is the correct expression for the current. On the other hand if transport is not coherent then in general neither Eq.(2.6.1) nor Eq.(2.6.2) is correct. The effect of the exclusion principle is more complicated. First let us see why for coherent transport the current is given by Eq.(2.6.1) without the exclusion principle factors.

Scattering states

The reason most people tend to choose Eq.(2.6.2) instinctively is that we tend to think of the current as arising from electronic transitions between an eigenstate localized in lead p and an eigenstate localized in lead q (note how the coordinates x_p are defined in Fig. 2.6.1):

$$|p,k\rangle \equiv \sin(kx_p) \quad \rightarrow \quad |q,k'\rangle \equiv \sin(k'x_q)$$

Clearly such transitions are blocked if the final state is occupied and it seems reasonable to include the $(1-f)$ factors as shown in Eq.(2.6.2). However, this view is accurate only if the leads are weakly coupled to the conductor. As the coupling between the leads and the conductor is made stronger, the lead eigenstates evolve into scattering states of the form shown in Fig. 2.6.1. A scattering state (q,k) consists of an incident wave in lead q, together with scattered waves in every lead p. It carries a current from lead q onto lead p as long as it is occupied. No transition from one state to another is needed, so that there is no reason to include the $(1-f)$ factors. This viewpoint leads naturally to Eq.(2.6.1) rather than Eq.(2.6.2), as we will now show.

To keep things simple we consider only one transverse mode in each lead. There is no loss of generality since we can always treat each individual mode as a separate terminal. The wavefunction in lead p due to a scattering state (q,k) is given by ($\delta_{pq} = 1$ if p = q, zero otherwise)

$$\Psi_p(q) = \delta_{pq}\chi_p^+(y_p)\exp\left[ik^+x_p\right] + s'_{pq}\,\chi_p^-(y_p)\exp\left[ik^-x_p\right] \qquad (2.6.3)$$

where χ_p^+ and χ_p^- are the transverse mode wavefunctions for the incident and scattered waves obtained by solving Eq.(1.6.3). If the vector potential is zero in the lead then the two functions are identical and $k^- = -k^+$.

A scattering state (q,k), if occupied, gives rise to a current $i_p(q)$ per unit energy in lead p which is given by

$$i_p(q) = \frac{2e}{h}\left(\delta_{pq} - T_{pq}\right) \qquad (2.6.4)$$

where the first term is the ingoing incident current, arising from the first term in Eq.(2.6.3), and the second term is the outgoing scattered current, arising from the second term in Eq.(2.6.3). A scattering state (q,k) is occupied only by electrons that come from the contact connected to lead q (assuming as we have always done that there is no reflection at the interface between the lead and the contact). Hence a scattering state

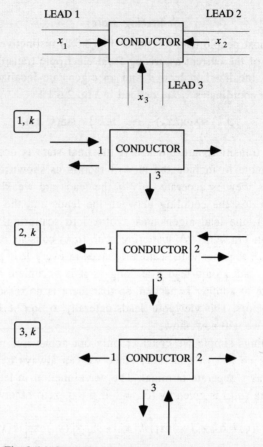

Fig. 2.6.1. Scattering states in a three-terminal structure.

(q,k) is occupied according to the Fermi function in contact q and the current in terminal p can be written as

$$I_p = \int \sum_q f_q(E) i_p(q) \mathrm{d}E \qquad (2.6.5)$$

There is no reason to include a $(1 - f_p)$ factor since no electronic trans-ition is involved in the flow of current. The Fermi function in the reservoir where a scattering state terminates is clearly irrelevant. From Eqs.(2.6.4) and (2.6.5) we obtain

$$I_p = \frac{2e}{h} \int \left[f_p - \sum_q T_{pq} f_q \right] \mathrm{d}E$$

We will show in Chapter 3 that for coherent transport the transmission coefficients obey the sum rule (see Eq.(3.1.5))

$$\sum_q T_{pq} = \sum_q T_{qp} = 1$$

so that we can write the current as

$$I_p = \frac{2e}{h} \int \left[\sum_q T_{pq}(f_p - f_q) \, dE \right] = \frac{2e}{h} \int \left[\sum_q T_{qp} f_p - T_{pq} f_q \right] dE$$

For simplicity we considered only one mode per lead. If we consider multiple modes in each lead we obtain the same expression but with the transmission coefficient T_{pq} replaced by the transmission function $\bar{T}_{pq}$, which is obtained by summing the transmission probabilities T_{mn} between every mode m in lead p and every mode n in lead q:

$$\bar{T}_{pq} = \sum_{m \in p} \sum_{n \in q} T_{mn}$$

as shown in Eq.(2.6.1).

One question that we have glossed over is whether the scattering states are orthogonal and can be filled up independently from different reservoirs as we have assumed in our discussion. It can be shown that the scattering states together with any bound states do form a complete orthonormal set (see A. M. Kriman, N. C. Kluksdahl and D. K. Ferry (1987), *Phys. Rev. B*, **36**, 5953). Recently the set of scattering states has been used to develop a formal theory not only for the current but also the current *fluctuations* in coherent conductors (see, for example, M. Büttiker (1992), *Phys. Rev. B*, **46**, 12485). The theory for such fluctuations is more subtle because we need to consider the correlations between the wavepackets incident from different leads and the exclusion principle cannot simply be ignored.

Can scattering states be defined in non-zero magnetic fields?

Let us briefly discuss a couple of subtle points that arise when there is a non-zero magnetic field in the leads. The first question is whether in the presence of a magnetic field the eigenstates in the leads can be expressed in the form of plane waves as assumed in Eq.(2.6.3). We have seen earlier that if we represent the magnetic field by a vector potential of the form

$$\mathbf{A} = \hat{x} B y$$

then the eigenstates in the leads do have the form of plane waves along the x-direction (see Section 1.7). But this would not be true if we were to use a different gauge (such as $\mathbf{A} = \hat{y}Bx$). The difficulty is that in general the x-direction is different in each of the leads. For example, in Fig. 2.6.1, x_3 is orthogonal to x_1 and x_2. So if we were to choose a gauge such that

$$\mathbf{A} \sim \hat{x}_1 By_1 \sim \hat{y}_3 Bx_3$$

then the eigenstates would have the form of plane waves along x_1 in lead '1' but not along x_3 in lead '3'. Thus we could not write scattering states of the form shown in Eq.(2.6.3). We need to choose our gauge such that the vector potential has the form

$$\mathbf{A} \sim \hat{x}_q By_q$$

in *every* lead q. It has been shown that this can be done even for conductors with multiple leads arranged arbitrarily (see Appendix E of H. U. Baranger and A. D. Stone (1989), *Phys. Rev. B*, **40**, 8169).

Are different transverse modes independent?

The second question has to do with whether we can obtain the net current by superposing the currents carried by different transverse modes, especially in the presence of a magnetic field. In other words, are we justified in going from Eq.(2.6.3) to (2.6.4)? To appreciate the problem, note that the general expression for calculating the current density $\mathbf{J}$ from the wavefunction Ψ is given by (see R. P. Feynman (1965), *Lectures on Physics*, Vol. III, p. 21–4, (New York, Addison–Wesley))

$$\mathbf{J} = \frac{e}{2m}\left(\Psi\big[(\mathbf{p} - e\mathbf{A})\Psi\big]^* - \Psi^*\big[(\mathbf{p} - e\mathbf{A})\Psi\big]\right) \qquad (2.6.6)$$

where $\mathbf{p} \equiv -i\hbar\nabla$ is the momentum operator. To obtain the total current carried by a lead we integrate the longitudinal (x-directed) current over the transverse coordinate (y):

$$I = \frac{e}{2m}\int\left[\Psi(p_x - eA_x)\Psi^* - \Psi^*(p_x - eA_x)\Psi\right]dy \qquad (2.6.7)$$

If the wavefunction consists of an incident wave

$$\Psi_i = \frac{1}{\sqrt{L}}\chi^+(y)\exp\big[ik^+x\big]$$

then the current is given by (we are assuming that $\chi(y)$ is real)

$$I_i = \frac{e}{mL} \int \left[\chi^+ \left(\hbar k^+ - eA_x \right) \chi^+ \right] dy$$

On the other hand if the wavefunction consists of only a scattered wave of amplitude s'

$$\Psi_s = s' \frac{1}{\sqrt{L}} \chi^-(y) \exp\left[ik^- x \right]$$

then the current is given by

$$I_s = |s'|^2 \frac{e}{mL} \int \left[\chi^- \left(\hbar k^- - eA_x \right) \chi^- \right] dy$$

But if the wavefunction consists of both an incident and a scattered wave

$$\Psi = \chi^+(y) \exp\left[ik^+ x \right] + s' \chi^-(y) \exp\left[ik^- x \right]$$

can we automatically assume that the current is given by $I = I_i - I_s$ as we did in writing Eq.(2.6.4)? A direct evaluation of the current from Eq.(2.6.7) would also give rise to cross-terms of the form

$$\frac{e}{mL} \int \left[\chi^+ \left(\frac{\hbar(k^+ + k^-)}{2} - eA_x \right) \chi^- \right] dy$$

However, we do not need to worry about these cross-terms. They are always zero, by virtue of the following orthogonality relation which is satisfied by any two transverse mode wavefunctions satisfying Eq.(1.6.3) (see Exercise E.2.6 at the end of this chapter):

$$\int \left[\chi_{m,k} \left(\frac{\hbar(k + k')}{2} - eA_x \right) \chi_{n,k'} \right] dy = \delta_{k,k'}$$

Non-coherent transport

Next we address the question about the effect of the exclusion principle on the transmission if transport is not coherent. A proper treatment of non-coherent transport requires advanced concepts that we will discuss in Chapter 8. However, the basic issues can be appreciated using a relatively simple phenomenological approach to visualize the role of phase-breaking processes in transport. The key insight is an observation due to Büttiker (see *IBM J. Res. Dev.*, **32** (1988), 63) that a voltage probe acts as

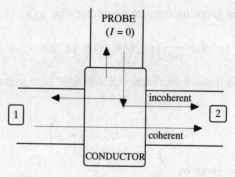

Fig. 2.6.2. A device with net current from 1 to 2. The voltage probe in between floats to an appropriate potential such that there is zero current at the probe.

a phase-breaking scatterer. This can be understood by considering the current flow from terminal 1 to 2 in a device having a voltage probe in between (see Fig. 2.6.2). We can calculate the S-matrix for such a device using the Schrödinger equation if we assume that transport is fully coherent between any pair of terminals. The point to note is that in this structure the net current flows from terminal 1 to 2 since there is no external current at a floating probe. This current has both a coherent and an incoherent component. A fraction of the electrons goes directly from 1 to 2 bypassing the probe entirely; this is the coherent component. The remaining electrons from 1 enter the probe and have their phases randomized before they are reinjected into the device. Some of these electrons then reach 2 (the rest are returned to 1); this is the incoherent component. The presence of the voltage probe thus introduces an incoherent component to the overall current flow from 1 to 2, much like a phase-breaking scatterer. We can reverse this argument to conclude that phase-breaking scatterers can be simulated by introducing conceptual voltage probes where none exist in the real structure.

A conceptual probe with the appropriate properties can thus be used to simulate the effect of phase-breaking processes. A partially coherent conductor or an incoherent conductor can be visualized as a fully coherent conductor with a conceptual probe stuck to it. The current in such a structure can be described by an equation of the form

$$i_p(E) = \frac{2e}{h} \sum_q \overline{T}_{pq}(E)\big[f_p(E) - f_q(E)\big] + \frac{2e}{h} \overline{T}_{p\varphi}(E)\big[f_p(E) - f_\varphi(E)\big] \quad (2.6.8a)$$

where the index q runs over all the real terminals while the index φ

denotes a fictitious phase-breaking probe attached to the conductor. The current at the fictitious probe is given by

$$i_\varphi(E) = \frac{2e}{h} \sum_q \overline{T}_{\varphi q}(E)\left[f_\varphi(E) - f_q(E)\right] \tag{2.6.8b}$$

We can use Eq.(2.6.8b) to express the distribution function at the fictitious probe, $f_\varphi(E)$, in terms of those at the real terminals ($f_q(E)$).

$$f_\varphi = \frac{1}{\overline{R}_\varphi}\left(i_\varphi + \sum_q \overline{T}_{\varphi q} f_q\right) \quad \text{where} \quad \overline{R}_\varphi \equiv \sum_q \overline{T}_{\varphi q} \tag{2.6.9}$$

Substituting this expression back into Eq.(2.6.8a) we obtain an expression for the current where the phase-breaking probes have been eliminated but their effect is incorporated into a set of effective transmission coefficients among the real terminals:

$$i_p = \frac{2e}{h} \sum_q \overline{T}_{pq}^{\text{eff}}\left[f_p - f_q\right] - \frac{2e}{h}\left[\overline{T}_{p\varphi}/\overline{R}_\varphi\right]i_\varphi \tag{2.6.10}$$

where

$$\overline{T}_{pq}^{\text{eff}} = \overline{T}_{pq} + \frac{\overline{T}_{p\varphi}\overline{T}_{\varphi q}}{\overline{R}_\varphi} \tag{2.6.11}$$

If we neglect the current at the fictitious probe, $i_\varphi(E)$, then Eq.(2.6.10) has exactly the same form as the current expression for coherent transport without any exclusion principle factors (Eq.(2.6.1)). However, there is no reason to assume that $i_\varphi(E) = 0$. Scattering processes take electrons out from one energy channel and reinject them at another channel. Consequently $i_\varphi(E)$ is negative at some energies (representing outscattering) and positive at some others (representing inscattering). All that we know for sure is that the net current integrated over all energies must be zero

$$\int i_\varphi(E)\mathrm{d}E = 0 \tag{2.6.12}$$

since the net outscattering must balance the net inscattering.

The function $i_\varphi(E)$ describes what we will refer to as 'vertical flow', namely, the flow of carriers from one energy channel to another. In general these vertical currents depend on the Fermi functions in the contacts $f_p(E)$ in a complicated manner due to the exclusion principle. This is, of course, not the case for non-degenerate systems, but the experiments of interest in mesoscopic physics are largely low temperature experiments involving degenerate electron systems. In such conductors the exclusion principle does affect the terminal currents through the vertical current $i_\varphi(E)$. Note, however, that the effect of the exclusion principle cannot be accounted for with simple ad hoc measures (such as inserting $(1 - f)$ factors as shown in Eq.(2.6.2)). A detailed microscopic theory is needed to calculate the vertical currents.

But could we not calculate the correct vertical currents $i_\varphi(E)$ from a microscopic theory and include it in the probe model? Yes, we could, and roughly speaking, that is what the Green's function formalism described in Chapter 8 does. We say 'roughly speaking' because the full story is somewhat more complicated. Our discussion has been based on a phenomenological model for dephasing which illustrates the basic issues quite well and often gives fairly accurate results. But it oversimplifies one detail. We assumed that the distribution function at the fictitious probe can be described by a simple function $f_\varphi(E)$ just like real probes. But it is easy to see that different points (r) inside the conductor will have different distribution functions so that the 'probe' covering the conductor should be described by a function of the form $f_\varphi(r,E)$. Actually the general situation is even more complicated because one needs to keep track of phase correlations between different points and the 'probe' has to be described by a correlation function (or density matrix) of the form $f_\varphi(r,r';E)$. This is explained in Section 8.7 after we discuss the detailed microscopic theory. The main point we wish to make here is that the vertical currents are difficult to calculate without a detailed microscopic theory. The simplest way to get around this problem is to neglect vertical currents altogether and set $i_\varphi(E) = 0$. This may sound like a rather drastic assumption but it actually works remarkably well in many cases as we will discuss in the next section.

2.7 When can we use the Landauer–Büttiker formalism?

We have seen that if transport is coherent across the conductor, then the current per unit energy is given by

$$i_p(E) = \frac{2e}{h} \sum_q \bar{T}_{pq}(E)\big[f_p(E) - f_q(E)\big] \qquad (2.7.1)$$

without any exclusion principle-related factors for the receiving contacts. This is a rigorous result. It is only for *non-coherent transport* that the transmission function is affected by the exclusion principle and the applicability of the Landauer–Büttiker formalism can be questioned. For non-coherent transport we can use a phenomenological model to express the current in the form (see Eq.(2.6.10))

$$i_p(E) = \frac{2e}{h} \sum_q \bar{T}_{pq}^{\text{eff}}(E) \big[f_p(E) - f_q(E) \big]$$

$$-\frac{2e}{h} \big[\bar{T}_{p\varphi}(E) / \bar{R}_\varphi(E) \big] i_\varphi(E)$$

(2.7.2)

As we have discussed, the vertical current $i_\varphi(E)$ is affected by the exclusion principle in a complicated way and it is difficult to do justice to this term without a detailed microscopic theory. The simplest way to get around this problem is simply to set $i_\varphi(E) = 0$. In this section we will explain the meaning and consequences of this assumption. As we will see, it actually works much better than one might expect.

Non-coherent elastic transport

Consider the transport of electrons across a device that is much longer than the phase-relaxation length. Typically the motion of an electron will involve both 'vertical' and 'lateral' flow as sketched in Fig. 2.7.1a. An electron will propagate coherently in one energy channel for a while, suffer an inelastic scattering which will transfer it to another energy channel, propagate coherently in the new channel, suffer another inelastic scattering and so on. In general, under these conditions, exclusion principle-related factors will enter the expression for the current in a complicated way, thus severely limiting its usefulness.

However, the exclusion principle-related factors disappear even for non-coherent transport, if we make a simplifying assumption. The assumption is that there is no net 'vertical' flow. Every electron that scatters out of a channel at E_1 into another channel at E_2, is balanced by another electron that scatters out of E_2 into E_1. Of course this means that there is no net exchange of energy between the electrons and the surrounding lattice and hence no dissipation or energy relaxation. However, our discussion in Section 2.3 shows that energy relaxation is only incidental to the problem of calculating the resistance. Momentum relaxation is what gives rise to resistance. Energy relaxation only helps to keep the energy distribution of the electrons close to equilibrium and often has no significant impact on the resistance. Thus this transport model (which we will refer to as non-

(a) Actual transport with inelastic scattering

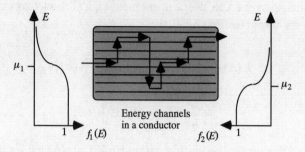

(b) Idealized transport without 'vertical flow'

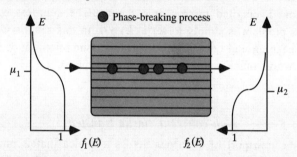

Fig. 2.7.1. (*a*) Transport with inelastic processes involves 'vertical' as well as 'lateral' flow. (*b*) If we assume that there is no net vertical flow then we can replace the real problem with an idealized one where transport is elastic but not necessarily coherent.

coherent elastic transport) can often be used to calculate the current flow through a conductor with reasonable accuracy.

The assumption of no net vertical flow allows us to visualize transport as proceeding via independent energy channels as depicted in Fig. 2.7.1b. An electron in a particular energy channel propagates coherently for a while, suffers a phase-breaking process, continues on in the same channel, suffers another phase-breaking process and so on. In this case too exclusion principle related factors can simply be ignored and we can express the current in terms of the transmission function just as we did for coherent transport. But the transmission of electrons from one contact to the other is not a single coherent process. Instead it is characterized by repeated phase-breaking along the way.

Does vertical flow have any effect on the resistance?

The next question we ask is what physics are we missing by neglecting vertical flow and adopting an elastic model for transport. Consider the following 'thought experiment'. We have a real sample with elastic and inelastic processes and we wish to calculate the current I through it for a particular bias. Let us replace it conceptually with another sample which has only elastic scatterers that provide the same momentum relaxation and phase relaxation as the real sample but do not provide any energy relaxation; in short, there is no vertical flow. The current, I_{EL}, through this sample can be calculated using the Landauer–Büttiker formalism. Is I_{EL} equal to I? There is no unique answer to this question. Let us consider a few possibilities.

Uniform transmission

There is one relatively simple situation where I_{EL} is clearly equal to I. Suppose the transmission function $\bar{T}(E)$ is independent of energy in the energy range where transport takes place, that is over the range

$$\mu_1 + (\text{a few } k_BT) > E > \mu_2 - (\text{a few } k_BT)$$

Then vertical flow has absolutely no effect on the resistance since all energy channels conduct equally well. We might as well lump them conceptually into a single channel as we were doing earlier in this chapter. Any vertical flow causes 'internal' transfers within this channel which has no effect on the current flow. This can be seen from Eq.(2.7.2). If we integrate over all energy then the term involving the vertical currents vanishes (making use of Eq.(2.6.12))

$$\int\left[\bar{T}_{p\varphi}(E)/\bar{R}_\varphi(E)\right]i_\varphi(E)\mathrm{d}E = \left[\bar{T}_{p\varphi}/\bar{R}_\varphi\right]\int i_\varphi(E)\mathrm{d}E = 0$$

so that we can obtain accurate results without including vertical flow ($I_{EL} = I$). Note, however, that this is only true for the net current and not the energy distribution of the current. We need the detailed energy distribution of the current in order to calculate things like the heat dissipated inside the conductor:

$$\text{Heat dissipated} \sim \int Ei_\varphi(E)\mathrm{d}E$$

Obviously we cannot calculate such quantities without doing justice to the problem of vertical flow. But usually we are most interested in the resistance of the conductor which is determined by the total current and this can be calculated neglecting vertical flow as long as the transmission characteristics are uniform over the range of energies where transport occurs. This transport regime can be described by the criterion

$$(\mu_1 - \mu_2) + (\text{a few } k_B T) \ll \varepsilon_c \qquad (2.7.3)$$

where ε_c (often called the correlation energy) is the energy range over which the transmission characteristics can be assumed uniform. *Zero temperature linear response* clearly belongs to this transport regime.

Non-uniform transmission

Next consider a transport regime such that the criterion in Eq.(2.7.3) is not satisfied. In many cases I_{EL} may still be quite close to the actual current I, but not always, as we will illustrate with a few examples.

Consider a conductor whose left hand side conducts well at energy E_A and whose right hand side conducts well at energy E_B (Fig. 2.7.2a). The transmission function at either energy is nearly zero since an electron cannot transmit across the sample. Hence the current I_{EL} calculated assuming purely elastic transport is approximately zero. But in the real sample a significant current can flow if the temperature is high enough, so that there are phonons present to scatter electrons inelastically from E_A to E_B. We can use a simple analogy to visualize this (see article by R. Landauer (1984), in *Localization, Interaction and Transport Phenomena*, eds. B. Kramer, Y. Bruynseraede and G. Bergmann, (Springer, Heidelberg)). Suppose we replace the energy coordinate E in Fig. 2.7.2a

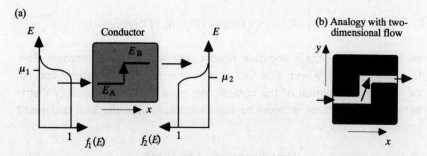

Fig. 2.7.2 (*a*) A conductor whose left hand side conducts well at energy E_A and whose right hand side conducts well at energy E_B. In this case $I_{EL} \sim 0$, but a non-zero current can flow in the real sample if phonons are present to take electrons from E_A to E_B.
(*b*) Analogy with two-dimensional flow.

by the transverse coordinate y as shown in Fig. 2.7.2b. Neglecting inelastic scattering is *analogous to isolating the upper channel from the lower channel* so that there is no transverse flow (analogous to the vertical flow in Fig. 2.7.2a). There would then be no current flow from left to right.

The above example involves the flow of electrons through energy levels that are localized in space. In this regime of 'strong localization' vertical flow can lead to a significant increase in the current over what we would expect if we were to neglect inelastic transitions. However, even when such strong localization is not involved, inelastic transitions can lead to a small increase in the current which may or may not be experimentally observable. Consider the current flow across a simple potential barrier as shown in Fig. 2.7.3. Below the barrier $T(E) = 0$ while above the

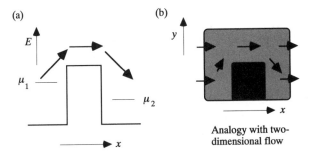

Fig. 2.7.3 (*a*) Transport of electrons over a potential barrier. (*b*) Analogy with two-dimensional flow: the actual conductance is a little smaller if we prevent transverse flow from occurring on either side of the obstacle.

barrier $T(E) \sim 1$. If we assume that electrons remain in the same energy channel as they cross the barrier, then the current is entirely due to the electrons in the tail of the Fermi function that start out with enough energy to cross the barrier. This is the standard thermionic emission theory. But the real current will be a little larger because electrons to the left of the barrier can absorb energy from the lattice (through inelastic processes) and rise to higher energies. The increase in current due to this effect has been observed in numerical simulations based on semiclassical transport (F. Venturi *et al.* (1991), *IEEE Trans.* **ED-38**, 611) as well as quantum transport (R. Lake and S. Datta (1992), *Phys. Rev. B*, **46**, 4757). It can be understood using an analogy with two-dimensional flow similar to that used in Fig. 2.7.2. The actual conductance of the structure in Fig. 2.7.3b will be a little smaller if we prevent transverse flow from occurring on either side of the obstacle.

Interestingly, the region to the left of the barrier would actually be cooled by the flow of current. This effect is analogous to the Peltier effect in macroscopic junctions made of dissimilar materials which is used to make thermoelectric coolers. Mesoscopic systems can be viewed as numerous junctions between microscopically dissimilar materials leading to Peltier-like effects. The heat exchanged with the reservoirs is thus partially reversible.

In summary, the basic point we are trying to make is that the Landauer–Büttiker formalism should be used with caution when the transmission functions vary widely over the energy range where transport occurs. Inelastic processes can then cause a significant amount of vertical flow among different energy channels and it can be quite inaccurate to replace the real problem with the idealized one as suggested in Fig. 2.7.1.

'Equilibrium current'

We end this section with an interesting aside not directly related to the rest of our discussion. We have seen that the conductance is given by ($f_0(E)$: equilibrium Fermi function)

$$
\begin{aligned}
G &= \frac{2e^2}{h} \int \overline{T}(E) \left[-\frac{\partial f_0(E)}{\partial E} \right] dE \\
&= \frac{2e^2}{hk_B T} \int \overline{T}(E) f_0(E) \left[1 - f_0(E) \right] dE
\end{aligned}
\tag{2.7.4}
$$

We could rewrite this in the form

$$
G = \frac{|e| I_{eq}}{k_B T}
\tag{2.7.5}
$$

where I_{eq} is defined as

$$
I_{eq} = \frac{2|e|}{h} \int \overline{T}(E) f_0(E) \left[1 - f_0(E) \right] dE
\tag{2.7.6}
$$

We can interpret I_{eq} as the current that 'flows' at equilibrium from contact 1 to contact 2 when the two are shorted together as shown in Fig. 2.7.4. It is of course balanced by an equal and opposite flow from 2 to 1. The conductance expression in Eq.(2.7.4) is only valid if we neglect vertical flow. But Eq.(2.7.5) seems to be valid quite generally even when vertical flow plays a significant role. Consider for example the conductor shown in Fig. 2.7.2 whose left hand side conducts well at E_A and whose

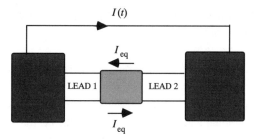

Fig. 2.7.4. A shorted conductor has 'equilibrium currents' I_{eq} flowing in either direction that give rise to the equilibrium noise.

right hand side conducts well at E_B. As we mentioned earlier the conductance of this sample can be significantly enhanced by the presence of inelastic processes. This is just what we might expect from Eq.(2.7.5), since inelastic processes would cause electrons to cross over from one side of the conductor to the other leading to equilibrium current fluctuations. The point is that in any conductor there are current fluctuations I_{eq} in both directions which balance out on the average at equilibrium. A small bias $\Delta\mu$ increases the flow in one direction to $I_{eq} + \Delta I$ such that

$$\frac{\Delta I}{I_{eq}} = \frac{\Delta\mu}{k_B T} \quad \Rightarrow \quad G = \frac{\Delta I}{\Delta\mu/|e|} = \frac{|e|I_{eq}}{k_B T}$$

This result seems to be quite general. Another example of this is a p–n junction diode where the current–voltage characteristics are given by

$$I = I_{eq}\left[\exp\left(\frac{eV}{k_B T}\right) - 1\right] \quad \Rightarrow \quad G = \left[\frac{dI}{dV}\right]_{V=0} = \frac{|e|I_{eq}}{k_B T}$$

From a practical point of view this relation between G and I_{eq} could be useful if there is a simple way to calculate I_{eq} in the presence of vertical flow.

Eq.(2.7.5) suggests an interesting relationship between the Landauer formula and the Nyquist–Johnson formula which relates the conductance to the correlation function for the equilibrium noise current:

$$G = \frac{1}{2k_B T}\int_{-\infty}^{+\infty}\langle I(t_0 + t)I(t_0)\rangle_{eq}\, dt \qquad (2.7.7)$$

The angle-brackets denote averaging over many values of t_0, or equivalently, averaging over many identical conductors (see, for example, A. van der Ziel (1978), *Noise in Solid State Devices*, Advances in Electronics and Electron Physics, vol. 46, (New York, Academic Press)). Eqs.(2.7.5) and (2.7.7) are consistent *only if*

$$\int_{-\infty}^{+\infty} \langle I(t_0 + t)I(t_0) \rangle_{eq} \, dt = 2|e|I_{eq}$$

which seems reasonable if we view the equilibrium noise as the shot noise due to independent uncorrelated equilibrium currents I_{eq} flowing in either direction. A proper theory of equilibrium and non-equilibrium noise, however, involves more subtle considerations (see, for example, R. Landauer and T. Martin (1992), *Phys. Rev. B*, **45**, 1742 and M. Büttiker (1992), *Phys. Rev. B*, **46**, 12845).

Summary

The basic results of the Landauer–Büttiker formalism are summarized here for convenience:

$$I_p = \int i_p(E)dE \quad \text{where} \quad i_p(E) = \frac{2e}{h} \sum_q \overline{T}_{pq}(E)\big[f_p(E) - f_q(E)\big] \quad (2.5.7)$$

Here $f_p(E)$ is the Fermi function for terminal p

$$f_p(E) = \left[\exp\left(\frac{E - \mu_p}{k_B T}\right) + 1\right]^{-1}$$

and $\overline{T}_{pq}(E)$ represents the total transmission (average transmission probability per mode times the number of modes in the leads) from lead q to lead p at the energy E. The transmission function obeys the sum rule

$$\sum_q \overline{T}_{qp}(E) = \sum_q \overline{T}_{pq}(E)$$

If the bias is small such that (ε_c: energy range over which the transmission function is nearly constant)

$$\Delta\mu \ll \varepsilon_c + (\text{a few } k_B T)$$

then we can linearize Eq.(2.5.7) to obtain ($V_p = \mu_p/e$)

$$I_p = \sum_q G_{pq}\big[V_p - V_q\big] \quad (2.5.8)$$

with
$$G_{pq} = \frac{2e^2}{h} \int \bar{T}_{pq}(E)\left(-\frac{\partial f_0}{\partial E}\right) dE \qquad (2.5.9a)$$

where $f_0(E)$ is the equilibrium Fermi function. At low temperatures $(k_B T \ll \varepsilon_c)$,

$$G_{pq} = \frac{2e^2}{h} \bar{T}_{pq}(E_f) \qquad (2.5.9b)$$

The Landauer–Büttiker formalism provides a rigorous framework for the description of mesoscopic transport as long as transport across the conductor is coherent. As we will see in Chapter 3, it is then straightforward to calculate the transmission functions from the Schrödinger equation. For non-coherent transport too the formulation is valid as long as transport through the conductor does not involve the vertical flow of electrons from one energy to another. The transmission functions can then be calculated using phenomenological approaches. If coherent effects can be neglected altogether then the transmission functions can be evaluated using a semi-classical approach.

But if vertical flow is present then the transmission functions are affected in a complicated way by the exclusion principle (in degenerate systems), thus severely limiting the utility of this formalism. Luckily, even if vertical flow is present it can be neglected (because it has no effect on the total current) if the transmission functions are approximately constant over the energy range where transport occurs:

$$(\mu_1 - \mu_2) + (\text{a few } k_B T) \ll \varepsilon_c$$

If this criterion is not satisfied then vertical flow may or may not have a significant effect on the current and the formalism should be used with caution.

Despite this limitation with respect to vertical flow, the Landauer–Büttiker formalism has a wide range of applicability. In the quantum Hall regime (see Chapter 4) the transmission functions are uniform over large ranges of energy so that vertical flow can be ignored. Localization phenomena (Chapter 5) are significantly affected by phase-breaking processes, but these are accounted for fairly well using simple phenomenological approaches that neglect vertical flow. Transport in the regime of strong localization, however, cannot be described without doing justice to the problem of vertical flow (see Fig. 2.7.2). Double-barrier tunneling (Chapter 6) is generally unaffected by phase-breaking processes though as we will see, the valley current cannot be understood without a proper description of vertical flow. Indeed in Chapter 8 we will use this example to illustrate the application of the non-equilibrium Green's function formalism which provides a general framework for the description of quantum transport with or without vertical flow. The price one pays for this generality is the increased conceptual complexity. What is particularly attractive about the Landauer–Büttiker formalism is that it allows one to handle a complex topic like quantum magnetotransport in degenerate conductors, armed with little more than elementary quantum mechanics. The only limitation is that there seems to be no way to include vertical flow without compromising its endearing simplicity.

Exercises

E.2.1 In Section 2.1 we calculated the contact resistance when a narrow conductor with N modes is connected to two very wide contacts. If the number of modes in the contacts is not infinite, but some finite number, M, then the left-moving and right-moving carriers inside the contacts

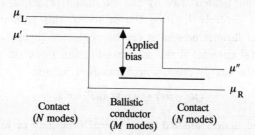

Fig. E.2.1. Spatial variation of the electrochemical potential for a ballistic conductor with M modes connected between two contacts having a finite number of modes (N).

have different electrochemical potentials as shown in Fig. E.2.1. Show that the contact resistance taking this into account is given by

$$R_c = \frac{h}{2e^2} \left[\frac{1}{M} - \frac{1}{N} \right]$$

Assume reflectionless contacts as in the text. For further discussions on the nature of the contact resistance at different types of interfaces see R. Landauer (1989), *J. Phys. Cond. Matter*, **1**, 8099 and M. C. Payne (1989), *J. Phys. Cond. Matter*, **1**, 4931.

E.2.2 Fig. 2.1.2 in the text shows the variation in the conductance as the width of a ballistic conductor is reduced. How will the conductance vary if the width is held constant and the magnetic field is increased? See Exercise E.1.4 regarding the variation in the number of modes as a function of the magnetic field. For experimental results, see van Wees *et al.* (1988), *Phys. Rev. B*, **38**, 3625.

E.2.3

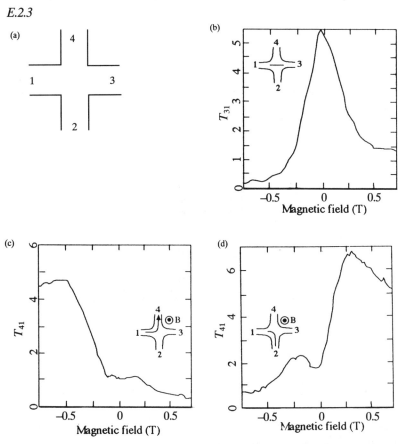

Fig. E.2.3. Cross junction: (*a*) Structure, (*b*) T_F vs. *B*, (*c*) T_L vs. *B* and (*d*) T_R vs. *B*. Adapted with permission from K. L. Shepard, M. L. Roukes and B. P. van der Gaag (1992), *Phys. Rev. B*, **46**, 9648.

Consider a cross junction as shown in Fig. E.2.3. Assuming that the four ports are completely symmetric, we can define a coefficient for forward transmission, one for right turning and one for left turning as follows:

$$\overline{T}_{13} = \overline{T}_{31} = \overline{T}_{42} = \overline{T}_{24} \equiv T_F$$

$$\overline{T}_{21} = \overline{T}_{32} = \overline{T}_{43} = \overline{T}_{14} \equiv T_R$$

$$\overline{T}_{41} = \overline{T}_{12} = \overline{T}_{23} = \overline{T}_{34} \equiv T_L$$

We have reproduced the measured values of T_F, T_L and T_R as a function of the magnetic field from *Phys. Rev. B*, **46**, 9653–4 (1992). See this reference for a description of how the transmission functions are measured.

(a) Suppose we measure the Hall resistance R_H by running a current from 1 to 3 and measuring the voltage between 2 and 4. Show that

$$R_H = \frac{h}{2e^2} \frac{T_R^2 - T_L^2}{(T_R + T_L)\left[T_R^2 + T_L^2 + 2T_F(T_F + T_R + T_L)\right]}$$

(b) Use the data provided above to calculate R_H vs. B numerically from the relation derived in part (a).

E.2.4

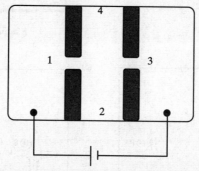

Fig. E.2.4. Terminals 1 and 3 are connected through two narrow apertures in series.

We would expect the conductance of two apertures in series to be half that of a single aperture: $G = M(2e^2/h) \times 0.5$. But if the two apertures are sufficiently close together then the electrons emerging from one aperture do not have the chance to spread out before they reach the second aperture. Consequently the conductance is the same as that of a single aperture: $G = M(2e^2/h)$. But if we turn on a magnetic field then the electrons get deflected and the conductance is reduced (see A. A. M. Staring *et al.* (1990), *Phys. Rev. B*, **41**, 8461). Use the Büttiker formula to show that the conductance is given by (see C. W. J. Beenakker and H. van Houten (1989), *Phys. Rev. B*, **39**, 10445)

$$G = (e^2/h)\left[M + T_F + \frac{(T_R - T_L)^2}{2T'_F + T_R + T_L} \right]$$

where

$$\overline{T}_{13} = \overline{T}_{31} \equiv T_F$$
$$\overline{T}_{42} = \overline{T}_{24} \equiv T'_F$$
$$\overline{T}_{21} = \overline{T}_{32} = \overline{T}_{43} = \overline{T}_{14} \equiv T_R$$
$$\overline{T}_{41} = \overline{T}_{12} = \overline{T}_{23} = \overline{T}_{34} \equiv T_L$$

E.2.5 Coherent inelastic transport In this book we will generally restrict ourselves to steady-state transport in the presence of a d.c. bias. However, it is interesting to note that if an alternating field is present within the conductor (but not inside the contacts) then we can define scattering states just like those in Eq.(2.6.3) but with different energies all coupled together. A scattering state (q,E) now consists of an incident wave with energy E in lead q, together with scattered waves with energy E_n ($= E \pm n\hbar\omega$, n being an integer) in every lead p (cf. Eq.(2.6.3)):

$$\Psi_p(q) = \delta_{pq} \exp\left[+ikx_q\right]\exp\left[-iEt/\hbar\right]$$
$$+ \sum_{p,n} s'_{pq}(E_n,E)\exp\left[-ik_n x_p\right]\exp\left[-iE_n t/\hbar\right]$$

Proceeding as before derive the following expression for the *d.c. current:*

$$I_p = \frac{2e}{h} \int \left[f_p(E) - \sum_{q,n} T_{pq}(E_n,E)f_q(E) \right] dE$$

The point is that there is no reason to include any exclusion principle-related factors in the expression for the current even if transport is inelastic, as long as it is coherent. But if there are phase-breaking processes within the conductor then we cannot define coherent scattering states that extend from one lead to another and the scattering state argument cannot be used.

E.2.6 Show that the following orthogonality relation

$$\int \left[\chi_{m,k}\left(\frac{\hbar(k + k')}{2} + eBy \right) \chi_{n,k'} \right] dy = \delta_{k,k'}$$

is satisfied by two functions $\chi_{m,k}(y)$ and $\chi_{n,k}(y)$ satisfying Eq.(1.6.3)

$$\left[E_s + \frac{(\hbar k + eBy)^2}{2m} + \frac{p_y^2}{2m} + U(y)\right]\chi_{m,k}(y) = E\chi_{m,k}(y)$$

$$\left[E_s + \frac{(\hbar k' + eBy)^2}{2m} + \frac{p_y^2}{2m} + U(y)\right]\chi_{n,k'}(y) = E\chi_{n,k'}(y)$$

which are appropriately normalized.

E.2.7 Starting from the current expression

$$I = \frac{2e}{h}\int \bar{T}(E)\big[f_1(E) - f_2(E)\big]dE \qquad \text{(see Eq.(2.5.1))}$$

derive the result stated in Eqs.(2.5.4)–(2.5.6) (see P. F. Bagwell and T. P. Orlando (1989), *Phys. Rev. B*, **40**, 1456).

Further reading

[2.1] Landauer, R. (1988). 'Spatial variation of currents and fields due to localized scatterers in metallic conduction', *IBM J. Res. Dev.*, **32**, 306. Landauer, R. (1992), 'Conductance from transmission: common sense points' *Physica Scripta*, **T42**, 110.

[2.2] Imry, Y. (1986). 'Physics of mesoscopic systems' in *Directions in Condensed Matter Physics*, eds. G. Grinstein and G. Mazenko (World Scientific Press, Singapore).

[2.3] Engquist, H. L. and Anderson, P. W. (1981). 'Definition and measurement of the electrical and thermal resistances', *Phys. Rev. B*, **24**, 1151.

[2.4] Büttiker, M. (1988). 'Symmetry of electrical conduction', *IBM J. Res. Dev.*, **32**, 317.

[2.5] Beenakker, C. W. J. and van Houten, H. (1991). 'Quantum transport in semiconductor nanostructures' in *Solid State Physics*, **44**, eds. H. Ehrenreich and D. Turnbull (New York, Academic Press) (see Part III).

3

Transmission function, *S*-matrix
and Green's functions

In Chapter 2 we have tried to establish that there exists a useful quantity called the transmission function in terms of which one can describe the current flow through a conductor. In this chapter we address the question of how the transmission function can be calculated for actual mesoscopic conductors. As we might expect, this chapter is somewhat mathematical and familiarity with matrix algebra is required. It could be skipped on first reading since it is not essential to know how to calculate the transmission function in order to appreciate mesoscopic phenomena, just as it is not necessary to understand the microscopic theory of diffusion or mobility in order to appreciate bulk transport phenomena. However, we will occasionally (especially in Chapter 5) use some of the concepts introduced here.

If the size of the conductor is much smaller than the phase-relaxation length then transport is said to be coherent and one can calculate the transmission function starting from the Schrödinger equation. A large majority of the theoretical work in this field is centered around this coherent transport regime where we can relate the transmission function to the *S*-matrix as discussed in Section 3.1.

When dealing with a large conductor it is often convenient to divide it conceptually into several sections whose S-matrices are determined individually. We discuss in Section 3.2 how the S-matrices of successive sections can be combined assuming complete coherence, complete incoherence or partial coherence among the sections. This is important because it affords a simple way to calculate the transmission function for a partially coherent conductor. For example, we could divide up the sample conceptually into little segments whose dimensions are of the order of a phase-relaxation length. The S-matrix for an individual segment is calculated assuming transport to be coherent within the segment. Different segments are then combined incoherently. Phenomenological approaches like this can often provide a satisfactory description of non-coherent transport without the use of advanced concepts that are needed for a proper microscopic theory of non-coherent transport (Chapter 8).

The Green's function $G^R(r,r')$ can be viewed as a generalized S-matrix that allows us to describe the reponse at any point r due to an excitation at point r'. This concept is really not essential for describing coherent transport but it provides a convenient method for calculating the S-matrix of arbitrarily shaped conductors. Besides it is helpful in relating the scattering theory to other formalisms. For this reason we introduce the concept of Green's functions in Section 3.3 and relate it to the S-matrix in Section 3.4. In Section 3.5 we show that a conductor connected to infinite leads can be replaced by a finite conductor with the effect of the leads incorporated through a 'self-energy' function. This provides a convenient method for evaluating the Green's function (and hence the transmission function) numerically. At the same time it serves to introduce the important concepts of self-energy and spectral function in a relatively simple context (Section 3.6).

This expression for the transmission function in terms of the Green's function and the self-energy also serves to relate the results of scattering theory to the results obtained from the Kubo formalism and from the transfer Hamiltonian formalism, both of which are widely used in the literature (Section 3.7). We will also use it later in Chapter 8 to relate to the nonequilibrium Green's function formalism, which is gaining increasing popularity in the field of mesoscopic transport.

We end this chapter with a discussion of the concept of Feynman paths (Section 3.8) which allows us to visualize the Green's function as a sum over an infinite number of paths connecting the initial and final states. This viewpoint is often useful in understanding the underlying physics which would otherwise be buried in the algebra. The Aharonov–Bohm effect is used as an illustration.

3.1 Transmission function and the *S*-matrix

In Chapter 2 we have seen how the current can be expressed in terms of the transmission function (see Eq.(2.5.7)). In this section we will discuss the relationship between the transmission function and the scattering matrix (or in short, the *S*-matrix) for coherent conductors. A coherent conductor can be characterized at each energy by an *S*-matrix that relates the outgoing wave amplitudes to the incoming wave amplitudes at the different leads. To be specific, if there is a total of three modes in the leads as shown in Fig. 3.1.1, we can write

$$
\begin{pmatrix} b_1 \\ b_2 \\ b_3 \end{pmatrix} = \begin{bmatrix} s_{11} & s_{12} & s_{13} \\ s_{21} & s_{22} & s_{23} \\ s_{31} & s_{32} & s_{33} \end{bmatrix} \begin{pmatrix} a_1 \\ a_2 \\ a_3 \end{pmatrix}
$$

At any given energy E, we will denote the number of propagating modes at lead p by $M_p(E)$. The total number of modes is obtained by summing the number of modes in each lead

$$
M_T(E) = \sum_p M_p(E)
$$

The scattering matrix is of dimensions $M_T \times M_T$.

In principle we can calculate the *S*-matrix starting from the (effective mass) Schrödinger equation (see Eq.(1.2.2))

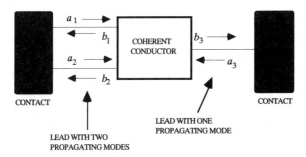

Fig. 3.1.1. A coherent device can be characterized by a scattering matrix at each energy. The scattering matrix relates the outgoing mode amplitudes b to the incoming mode amplitudes a.

$$\left[E_s + \frac{(i\hbar\nabla + e\mathbf{A})^2}{2m} + U(x,y)\right]\Psi(x,y) = E\Psi(x,y)$$

if we know the vector potential $\mathbf{A}$ and the potential energy $U(x,y)$ inside the conductor. In the following sections we will discuss how the S-matrix can be calculated for specific structures. However, these details are not important for the moment. The main point is that for a coherent conductor one can define and (if necessary) compute an S-matrix.

The transmission probability T_{nm} is obtained by taking the squared magnitude of the corresponding element of the S-matrix:

$$T_{m\leftarrow n} = |s_{m\leftarrow n}|^2 \qquad (3.1.1)$$

We are interested in the transmission function $\bar{T}_{pq}(E)$. As we explained in Section 2.6, this quantity is obtained by summing the transmission probability T_{nm} over all modes m in lead q and all modes n in lead p:

$$\bar{T}_{p\leftarrow q} = \sum_{m\in q} \sum_{n\in p} T_{n\leftarrow m} \qquad (3.1.2)$$

We have inserted the arrows in the subscripts just as a reminder that the direction of propagation is backwards from the second subscript to the first one. We will generally write the subscripts without the arrows.

Unitarity

We will now show that in order to ensure current conservation the S-matrix must be unitary. We can write in matrix notation

$$\{b\} = [S]\{a\}$$

where the matrix $[S]$ has dimensions $M_T \times M_T$, M_T being the total number of modes in all the leads, while $\{a\}$ and $\{b\}$ are column vectors representing the incoming and outgoing wave amplitudes in the different modes in the leads. We assume that the incoming and outgoing currents in a particular mode m are proportional to the squared magnitudes of the corresponding mode amplitudes a_m and b_m respectively. Current conservation then requires that (the superscript '+' denotes conjugate transpose)

$$\sum_m |a_m|^2 = \sum_m |b_m|^2$$

that is,
$$\{a\}^+\{a\} = \{b\}^+\{b\}$$

Since $\qquad\qquad \{b\} = [S]\{a\}$

we can write $\qquad \{a\}^+\{a\} = \{Sa\}^+\{Sa\} = \{a\}^+[S]^+[S]\{a\}$

Hence $\qquad\qquad [S]^+[S] = I = [S][S]^+ \qquad\qquad$ (3.1.3a)

so that in terms of the elements of the S-matrix we have

$$\sum_{m=1}^{M_T} |S_{mn}|^2 = 1 = \sum_{m=1}^{M_T} |S_{nm}|^2 \qquad\qquad (3.1.3b)$$

The first of these relations is obvious since the left-hand side represents the sum of the transmission probabilities for a given input mode n over all possible output modes m. This sum must equal one since the electron must go somewhere, or in other words, current must be conserved. But the second relation is not as obvious. Here we are summing the transmission probabilities over all possible inputs (for a fixed output). There seems to be no simple reason why this should equal one. However, as we can see, both the results follow from the unitarity of the S-matrix which is essential for current conservation.

An important point

An important point to note is that the current associated with a scattered wave is proportional to the square of the wavefunction multiplied by the velocity. For this reason, it is customary to define the S-matrix in terms of the 'current amplitude' which is equal to the wave amplitude times the square root of the velocity. Instead we could define a matrix [s'] in terms of the wave amplitudes so that

$$s'_{nm} = \sqrt{v_m / v_n}\, S_{nm} \qquad\qquad (3.1.4)$$

But the matrix [s'] is not unitary, unlike the matrix [s]. We can no longer invoke current conservation to write $\{a\}^+\{a\} = \{b\}^+\{b\}$ (a: incoming wave amplitude, b: outgoing wave amplitude) as we did in proving that the S-matrix must be unitary.

Sum rules

Using Eqs.(3.1.3) it is easy to see that

$$\sum_q \bar{T}_{qp} = \sum_{n \in p} \sum_{m=1}^{M_T} T_{mn} = \sum_{n \in p} 1 = M_p$$

$$\sum_q \bar{T}_{pq} = \sum_{n \in p} \sum_{m=1}^{M_T} T_{nm} = \sum_{n \in p} 1 = M_p$$

Hence the following sum rule is always satisfied by the transmission function:

$$\sum_q \bar{T}_{pq}(E) = \sum_q \bar{T}_{qp}(E) = M_p(E) \qquad (3.1.5)$$

where $M_p(E)$ is the number of modes in lead p.

For a device with N leads we can write down the transmission function $\bar{T}_{pq}(E)$ in the form of an $N \times N$ matrix. Assuming $N = 3$ (to be specific) we have a 3×3 matrix:

$\bar{T}_{pq}(E)$:	$q = 1$	$q = 2$	$q = 3$	
$p = 1$	xx	xx	xx	SUM = M_1
$p = 2$	xx	xx	xx	SUM = M_2
$p = 3$	xx	xx	xx	SUM = M_3
SUM =	M_1	M_2	M_2	

The sum rules require that the elements in each row and each column add up to equal the number of modes for that terminal.

It is interesting to note that the sum rules imply that for two-terminal devices the transmission function is always reciprocal, *even if a magnetic field is present*; that is, $\bar{T}_{12} = \bar{T}_{21}$. To see this, consider a two-terminal device

$\bar{T}_{pq}(E)$:	$q = 1$	$q = 2$	
$p = 1$	xx	xx	SUM = M_1
$p = 2$	xx	xx	SUM = M_2
SUM =	M_1	M_2	

Since $\bar{T}_{11} + \bar{T}_{12} = M_1 = \bar{T}_{11} + \bar{T}_{21}$ we must have $\bar{T}_{12} = \bar{T}_{21}$.

Sum rule for the conductance matrix

Using the definition for the conductance in terms of the transmission (see Eq.(2.5.9a)) we can obtain the following sum rule for the conductance matrix:

$$\sum_q G_{pq} = \sum_q G_{qp} = \frac{2e^2}{h} \int M_p(E)\left(-\frac{\partial f_0}{\partial E}\right) dE$$

$$= \frac{2e^2}{h} M_p(E_f) \text{ at low temperatures}$$

(3.1.6)

Thus for coherent transport, the unitarity of the S-matrix ensures that the sum rule for the conductance matrix (see Eq.(2.4.3)) is satisfied.

Reciprocity

We will now show that for coherent transport, the symmetry properties of the S-matrix ensure that the reciprocity relation (Eq.(2.4.5)) is satisfied. The basic property of the S-matrix that we will use is that reversing the magnetic field transposes (denoted by the superscript 't') the S-matrix

$$[S]_{+B} = \left[S^t\right]_{-B} \quad \text{that is,} \quad [s_{mn}]_{+B} = [s_{nm}]_{-B}$$

(3.1.7)

To prove this, suppose we have solved the Schrödinger equation (see Eq.(1.2.2))

$$\left[E_s + \frac{(i\hbar\nabla + e\mathbf{A})^2}{2m} + U(x,y)\right]\Psi(x,y) = E\Psi(x,y)$$

(3.1.8a)

to obtain the S-matrix connecting the outgoing amplitudes $\{b\}$ to the incoming amplitudes $\{a\}$ (see Fig. 3.1.1). If we take the complex conjugate of Eq.(3.1.8a)

$$\left[E_s + \frac{(-i\hbar\nabla + e\mathbf{A})^2}{2m} + U(x,y)\right]\Psi^*(x,y) = E\Psi^*(x,y)$$

and at the same time, reverse the vector potential $\mathbf{A}$ (and hence the magnetic field $\mathbf{B}$), we obtain

$$\left[E_s + \frac{(i\hbar\nabla + e\mathbf{A})^2}{2m} + U(x,y)\right]\Psi^*(x,y) = E\Psi^*(x,y)$$

(3.1.8b)

The differential operators appearing in Eqs.(3.1.8a) and (3.1.8b) are identical. This means that

$$\left[\Psi^*(x,y)\right]_{-B} = \left[\Psi(x,y)\right]_{+B}$$

Or in other words, if we know the solutions to the Schrödinger equation in a magnetic field $+B$, we can obtain a solution that is valid for $-B$ simply by taking its complex conjugate. Taking the complex conjugate, however, turns an incoming wave into an outgoing wave and vice versa. So if

$$\{b\} = [S]_{+B}\{a\} \quad \text{that is} \quad \left\{b^*\right\} = \left[S^*\right]_{+B}\left\{a^*\right\}$$

then we must have

$$\left\{a^*\right\} = [S]_{-B}\left\{b^*\right\} \quad \text{that is} \quad \left\{b^*\right\} = \left[S^{-1}\right]_{-B}\left\{a^*\right\}$$

Hence

$$\left[S^*\right]_{+B} = \left[S^{-1}\right]_{-B}$$

But

$$\left[S^{-1}\right]_{-B} = \left[S^+\right]_{-B} \quad \text{(Unitarity, see Eq.(3.1.3a))}$$

Hence

$$\left[S^*\right]_{+B} = \left[S^+\right]_{-B} \quad \Rightarrow \quad [S]_{+B} = \left[S^t\right]_{-B}$$

as stated above (see Eq.(3.1.7)).

Next we take the squared magnitude of both sides of Eq.(3.1.7) and sum over all modes m in lead p and over all modes n in lead q to obtain

$$\sum_{m \in p, n \in q} \left| S_{mn} \right|^2_{+B} = \sum_{m \in p, n \in q} \left| S_{nm} \right|^2_{-B}$$

Using Eq.(3.1.1),

$$\left[\overline{T}_{pq}\right]_{+B} = \left[\overline{T}_{qp}\right]_{-B}$$

Hence, using Eq.(2.5.9a),

$$\left[G_{pq}\right]_{+B} = \left[G_{qp}\right]_{-B}$$

which is the desired result. It should be mentioned that this property of reciprocity holds only for small bias, unlike the sum rules discussed in the last section which hold irrespective of the bias. The reason is that in order to prove Eq.(3.1.7) we need to assume that the *electrostatic potential U(x,y) inside the conductor remains unchanged* when we reverse the magnetic field. This is clearly not true for large bias since the Hall voltage reverses when the magnetic field reverses. But for an infinitesimally small bias the electrostatic potential inside the conductor is essentially the same as that at equilibrium and it can be shown that the potential distribution $U(x,y)$ is unchanged when the B-field is reversed.

A reminder

We have argued earlier (see Fig. 2.2.2) that although the relation $G = (2e^2/h)\overline{T}$ gives the conductance measured between two planes '1' and '2' inside the contacts, we do not need to evaluate the transmission function $\overline{T}$ (or the S-matrix) between the contacts. We can get the same answer (and save ourselves a lot of work) by evaluating the S-matrix between two planes '1L' and '2L' located in the leads, as long as the lead–contact interface is reflectionless. In this chapter we will often not draw the contacts explicitly since our focus is on the problem of calculating the S-matrix relating the wave amplitudes in different leads. It is important to note that we are calculating the transmission function between two *leads*, though the relation $G = (2e^2/h)\overline{T}$ gives the conductance between two *contacts* (see Fig. 3.1.1).

3.2 Combining *S*-matrices

Consider a four-terminal Hall bridge as shown in Fig. 3.2.1. We could in principle evaluate the overall S-matrix by directly solving the Schrödinger equation using the formulation to be described in the following sections. In practice, however, it is difficult to handle conductors whose dimensions exceed a few tens of wavelengths because of the size of the matrix that needs to be inverted. One solution to this problem is to divide the conductor into two (or more) sections as shown in Fig. 3.2.1, compute the individual S-matrices $[s^{(1)}]$ and $[s^{(2)}]$ and then combine them to obtain the composite S-matrix.

$$s = s^{(1)} \otimes s^{(2)}$$

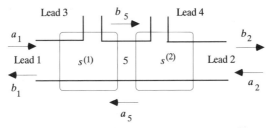

Fig. 3.2.1. A long structure can be divided into two parts whose S-matrices $s^{(1)}$ and $s^{(2)}$ can be combined to yield the overall S-matrix.

We will now describe the rules for combining S-matrices; that is, the meaning of the symbol $\otimes$.

Let us write the individual S-matrices in the form

$$\begin{Bmatrix} b_{13} \\ b_5 \end{Bmatrix} = \begin{bmatrix} r^{(1)} & t'^{(1)} \\ t^{(1)} & r'^{(1)} \end{bmatrix} \begin{Bmatrix} a_{13} \\ a_5 \end{Bmatrix} \quad \text{and} \quad \begin{Bmatrix} a_5 \\ b_{24} \end{Bmatrix} = \begin{bmatrix} r^{(2)} & t'^{(2)} \\ t^{(2)} & r'^{(2)} \end{bmatrix} \begin{Bmatrix} b_5 \\ a_{24} \end{Bmatrix}$$

where $\{a_{13}\}$ is a column vector representing the incoming wave amplitudes in all the various modes in terminals '1' and '3', $\{b_{13}\}$ is a column vector representing the outgoing wave amplitudes in all the various modes in terminals '1' and '3', etc. The matrices $[r]$ and $[r']$ describe the reflection amplitudes while the matrices $[t]$ and $[t']$ describe the transmission amplitudes; the subscripts 1 and 2 refer to the two sections respectively. Note that at terminal '5', the incoming amplitude $\{a_5\}$ for the first section is the same as the outgoing amplitude for the second section.

It is straightforward to eliminate a_5 and b_5 from the above equations to obtain the S-matrix for the composite structure:

$$\begin{Bmatrix} b_{13} \\ b_{24} \end{Bmatrix} = \begin{bmatrix} r & t' \\ t & r' \end{bmatrix} \begin{Bmatrix} a_{13} \\ a_{24} \end{Bmatrix}$$

where

$$t = t^{(2)}\left[I - r'^{(1)} r^{(2)}\right]^{-1} t^{(1)}, \quad r = r^{(1)} + t'^{(1)} r^{(2)}\left[I - r'^{(1)} r^{(2)}\right]^{-1} t^{(1)}$$

$$t' = t'^{(1)}\left[I - r^{(2)} r'^{(1)}\right]^{-1} t'^{(2)}, \quad r' = r'^{(2)} + t^{(2)}\left[I - r'^{(1)} r^{(2)}\right]^{-1} r'^{(1)} t'^{(2)} \tag{3.2.1}$$

Feynman paths

We can get some insight into this result by expanding the expressions in Eq.(3.2.1) in a geometric series as follows:

$$t = t^{(2)}\left[I - r'^{(1)} r^{(2)}\right]^{-1} t^{(1)}$$

$$= t^{(2)} t^{(1)} + t^{(2)}\left[r'^{(1)} r^{(2)}\right] t^{(1)} + t^{(2)}\left[r'^{(1)} r^{(2)}\right]\left[r'^{(1)} r^{(2)}\right] t^{(1)} + \ldots$$

The successive terms in this series have a simple physical interpretation (see Fig. 3.2.2). The first term is the amplitude for transmission through the two obstacles without any reflection, the second term for transmission with two reflections, the third term for transmission with four reflections and so on.

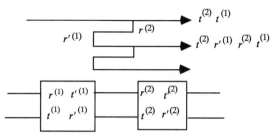

Fig. 3.2.2. Two obstacles with scattering matrices as shown are placed in series. The problem is to find the S-matrix of the composite structure.

Consider now the element (m,n) of one of these terms (note that the quantities t and r are all matrices)

$$\left(t^{(2)}\left[r'^{(1)} r^{(2)}\right]t^{(1)}\right)_{mn} \rightarrow \sum_{m_1}\sum_{m_2}\sum_{m_3}\left(t^{(2)}\right)_{m,m_3}\left(r'^{(1)}\right)_{m_3,m_2}\left(r^{(2)}\right)_{m_2,m_1}\left(t^{(1)}\right)_{m_1,n}$$

We could depict each term in this summation by a 'Feynman path' as shown in Fig. 3.2.3. A path starts out in mode n, transmits to mode m_1,

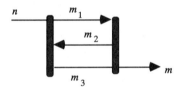

Fig. 3.2.3. A typical Feynman path leading from an input mode n to an output mode m.

reflects to mode m_2, reflects again to mode m_3 and then transmits to mode m.

The overall transmission amplitude from mode n to mode m can be expressed as

$$(t)_{mn} = \sum_{P} A_P, \, P \in \text{all paths starting in mode } n \text{ on the left}$$

$$(3.2.2)$$

$$\text{and ending in mode } m \text{ on the right}$$

where A_P is the amplitude associated with a particular path. As a practical computational tool this viewpoint may not be very useful because the number of paths involved is enormous and the phase of A_p often changes quite rapidly from one path to the next. But it provides a very appealing conceptual picture that is often useful in 'understanding' the physics which would otherwise be buried in the algebra of S-matrices.

Combining successive sections incoherently

For coherent transport we can combine the S-matrices of individual sections as described above. The transmission probability is obtained by squaring the transmission amplitude:

$$T_{mn} = t_{mn}{}^* t_{mn} = \sum_P A_P{}^* A_P + \sum_{P \neq P'} \sum A_P{}^* A_{P'} \qquad (3.2.3)$$

The first term represents the sum of the probabilities of all the paths, while the second term is due to interference among the different paths. If the length of a section is much greater than the phase-relaxation length then successive sections could be treated as incoherent. This means that instead of adding the amplitudes of the different paths shown in Fig. 3.2.2 we should add their probabilities; that is, we should ignore the second term in Eq.(3.2.3).

The simplest way to do this is to calculate the transmission probability by combining probability matrices instead of S-matrices; that is, instead of $s = s_1 \otimes s_2$ we calculate $S = S_1 \otimes S_2$ where the probability matrix S is obtained simply by squaring the corresponding elements of the amplitude matrix s:

$$S(m,n) = |s(m,n)|^2$$

Recently this approach has been used to describe semiclassical transport in electronic devices (see M. A. Alam *et al.* (1993), *Solid State Electronics*, **36**, 263). In device applications one is often interested in non-degenerate systems so that even vertical flow is readily included in this approach. Degenerate systems too can be handled straightforwardly using an iterative approach since quantum interference effects are absent.

One can gain insight into the effect of quantum interference by comparing the results obtained by combining successive scatterers coherently and incoherently (see, for example, the results shown in Fig. 5.2.1). These two methods for combining S-matrices are easily appreciated with a simple example involving two scatterers in a single-moded conductor whose S-matrices are given by

$$s_1 = \begin{bmatrix} r_1 & t_1' \\ t_1 & r_1' \end{bmatrix} \quad \text{and} \quad s_2 = \begin{bmatrix} r_2 & t_2' \\ t_2 & r_2' \end{bmatrix}$$

Since we are considering just one mode, the quantities r, t etc. are just numbers and not matrices. The probability matrices are given by

$$S_1 = \begin{bmatrix} R_1 & T_1 \\ T_1 & R_1 \end{bmatrix} \quad \text{and} \quad S_2 = \begin{bmatrix} R_2 & T_2 \\ T_2 & R_2 \end{bmatrix}$$

where

$$T_1 = |t_1|^2 = |t_1'|^2 \quad R_1 = |r_1|^2 = |r_1'|^2 \quad T_1 + R_1 = 1$$

$$T_2 = |t_2|^2 = |t_2'|^2 \quad R_2 = |r_2|^2 = |r_2'|^2 \quad T_2 + R_2 = 1$$

If we wish to combine the scatterers coherently then we should combine the individual amplitude matrices to obtain the composite transmission amplitude (according to the rules given in Eq.(3.2.1))

$$t = \frac{t_1 t_2}{1 - r_1' r_2}$$

Squaring we obtain the transmission probability:

$$T = |t|^2 = \frac{T_1 T_2}{1 - 2\sqrt{R_1 R_2}\,\cos\theta + R_1 R_2} \quad \text{(coherent)} \qquad (3.2.4)$$

where θ is the phase shift acquired in one round-trip between the scatterers: $\theta = \text{phase}(r_1') + \text{phase}(r_2)$.

If we wish to combine the scatterers incoherently then we should combine the individual probability matrices to obtain the composite transmission probability (according to the same rules):

$$T = \frac{T_1 T_2}{1 - R_1 R_2} \quad \text{(incoherent)} \qquad (3.2.5)$$

This result is independent of the phase θ as we would expect.

Combining successive sections with partial coherence

We have seen how successive sections can be combined coherently or incoherently depending on whether we combine the amplitude or the probability matrices. But can we combine two sections with partial coherence? One simple way to achieve this is to interpose a voltage probe between two scatterers as explained earlier (see Fig. 2.6.2). As shown in Fig. 3.2.4, a fraction of the electrons travel coherently from terminal '1' to '2'. The remainder travel coherently from terminal '1' to the probe 'P', are reinjected after phase-randomization and then reach terminal '2'. The

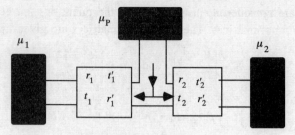

Fig. 3.2.4. A conceptual voltage probe introduced between two scatterers to simulate the effects of phase-breaking. Unfortunately this also introduces momentum relaxation.

electron flux reaching terminal '2' thus has a coherent component and an incoherent component.

One difficulty with this arrangement is that the probe introduces an additional resistance. This is because the electrons reinjected from the probe have equal probability of going to the left or to the right. The probe thus randomizes the momentum as well as the phase. It would be much better if we could adjust the two effects independently since it is well-known experimentally that the phase-relaxation time and the momentum relaxation time can be very different at low temperatures (see Section 1.3). We can introduce phase-relaxation without introducing any momentum randomization by using a pair of unidirectional probes as shown in Fig. 3.2.5. An electron coming in at 'A' has a certain probability of going onto 'B' and a certain probability of being deflected into 'P1'. If it goes into 'P1', it is reinjected after its phase has been randomized. When it is reinjected it can go on to 'B' but not back to 'A'. Thus the probe preserves the sense of current flow so that momentum is not relaxed though phase is randomized. The same is true of the other probe (P2).

The S-matrix for this twin probe configuration can be expressed in terms of a single parameter α (M. Büttiker (1988), *IBM J. Res. Dev.*, **32**, 72):

$$
\begin{array}{c|cccc}
m = & A & B & P1 & P2 \\
\hline
n = A & 0 & \sqrt{1-\alpha} & 0 & -\sqrt{\alpha} \\
n = B & \sqrt{1-\alpha} & 0 & -\sqrt{\alpha} & 0 \\
n = P1 & \sqrt{\alpha} & 0 & \sqrt{1-\alpha} & 0 \\
n = P2 & 0 & \sqrt{\alpha} & 0 & \sqrt{1-\alpha}
\end{array}
\qquad (3.2.6)
$$

where we have assumed each lead to be single-moded. The degree of phase-breaking can be adjusted through the parameter α which is a num-

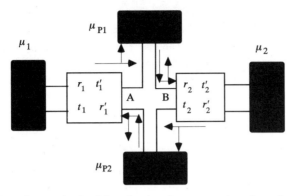

Fig. 3.2.5. By using a pair of unidirectional probes, we can introduce phase-relaxation without introducing momentum relaxation.

ber between 0 and 1. Note that all the elements are purely real numbers and that we have chosen the signs of the different elements so as to ensure that the matrix is unitary (this is essential to ensure current conservation).

The structure in Fig. 3.2.5 has to be treated as a four-terminal conductor whose conductance is obtained from the Büttiker formula (see Eq.(2.5.8)) by setting the currents at the floating probes to zero. To use Eq.(2.5.8) we first need to combine the individual scattering matrices to obtain a (4×4) scattering matrix relating the terminal quantities at the four terminals 1, 2, P1 and P2. In principle, the procedure is quite straightforward. We have one S-matrix relating terminal quantities at '1' and at 'A', one relating terminals 'B' and '2' and one relating 'A', 'B', 'P1' and 'P2'. We can eliminate the internal terminals 'A' and 'B' using straightforward algebra to obtain our desired S-matrix for the composite four-terminal structure. Alternatively we can enumerate all the 'Feynman paths' between each pair of terminals and add up their amplitudes.

This approach of simulating the effects of phase-breaking with floating probes provides a rather simple method for modeling partially coherent transport using the Landauer–Büttiker formalism and has been used by other authors since the initial work of Büttiker (see for example J. L. D'Amato and H. M. Pastawski (1990), *Phys. Rev. B*, **41**, 7411 and K. Maschke and M. Schreiber (1994), *Phys. Rev. B*, **49**, 2295). Although the approach appears to be purely phenomenological, we will see in Section 8.7 that it can be justified from a detailed microscopic theory with appropriate approximations (see also M. J. McLennan *et al.* (1991), *Phys. Rev. B*, **43**, 13846 and S. Hershfield (1991). *Phys. Rev. B*, **43**, 11586). In a microscopic theory, electron–phonon and electron–electron interactions are usually described by a 'self-energy function'. Later in this chapter we will show that the effect of a lead too can be described by a self-energy function, suggesting that the similarity between leads and real interactions is more than superficial.

3.3 Green's functions: a brief introduction

The S-matrix tells us the response at one lead due to an excitation at another. The Green's function is a more powerful concept that gives us the response at any point (inside or outside the conductor) due to an excitation at any other. For non-interacting transport, the only excitations we need to worry about are those due to waves incident from the leads. For such excitations, the Green's function and the S-matrix are related concepts and what we use is largely a matter of taste (we will derive their relationship in Section 3.4). The real power of Green's functions is evident when we try to include the effect of interactions (electron–electron or electron–phonon), as we will do in Chapter 8. Such interactions give rise to excitations within the conductor, and cannot be described by simple S-matrices.

In this chapter we will restrict our discussion to non-interacting transport. For non-interacting transport, the language of Green's function is really not necessary and our main purpose in introducing it is that it provides a useful practical tool for computing the S-matrix of arbitrarily shaped conductors (Section 3.5). Besides it is useful in relating the scattering viewpoint to other viewpoints that are widely used in the literature (Sections 3.6–3.8).

In this section we will briefly summarize some properties of the Green's function that we will need for our discussion. The concept of Green's functions appears in many physical contexts including circuit theory, electrostatics and electromagnetics. Whenever the response R is related to the excitation S by a differential operator D_{op}

$$D_{op}R = S$$

we can define a Green's function and express the response in the form

$$R = D_{op}^{-1}S = GS \quad \text{where} \quad G \equiv D_{op}^{-1}$$

Our problem can be expressed in the form

$$[E - H_{op}]\Psi = S$$

where Ψ is the wavefunction and S is an equivalent excitation term due to a wave incident from one of the leads. The corresponding Green's function can be written as

$$G = \left[E - H_{op}\right]^{-1} \qquad (3.3.1)$$

where H_{op} is the Hamiltonian operator (see Eq.(1.2.2), the subband energy E_s has been included as part of the potential $U(x,y)$):

$$H_{op} \equiv \frac{\left(i\hbar\nabla + e\mathbf{A}\right)^2}{2m} + U(\mathbf{r}) \qquad (3.3.2)$$

Retarded and advanced Green's functions

The inverse of a differential operator is not uniquely specified till we specify the boundary conditions. It is common to define two different Green's functions (retarded and advanced) corresponding to two different boundary conditions. The difference is best appreciated with a simple example.

Consider a simple one-dimensional wire with a constant potential energy U_0 and zero vector potential. From Eqs.(3.3.1) and (3.3.2) we can write

$$G = \left[E - U_0 + \frac{\hbar^2}{2m}\frac{\partial^2}{\partial x^2}\right]^{-1}$$

that is, $$\left(E - U_0 + \frac{\hbar^2}{2m}\frac{\partial^2}{\partial x^2}\right)G(x,x') = \delta(x - x') \qquad (3.3.3)$$

This looks just like the Schrödinger equation

$$\left(E - U_0 + \frac{\hbar^2}{2m}\frac{\partial^2}{\partial x^2}\right)\Psi(x) = 0$$

except for the source term $\delta(x - x')$ on the right. We could view the Green's function $G(x,x')$ as the wavefunction at x resulting from a unit excitation applied at x'. Physically we expect such an excitation to give rise to two waves traveling outwards from the point of excitation, with amplitudes A^+ and A^- as shown in Fig. 3.3.1.

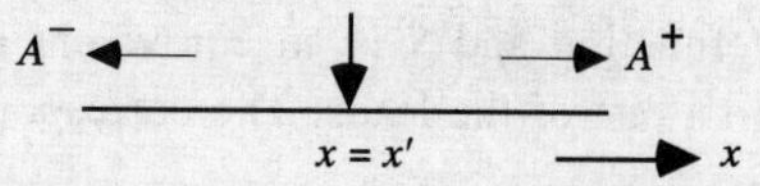

Fig. 3.3.1. Retarded Green's function for an infinite 1-D wire.

We can write

$$G(x,x') = A^+ \exp\left[ik(x - x')\right], \quad x > x'$$
$$G(x,x') = A^- \exp\left[-ik(x - x')\right], \quad x < x' \tag{3.3.4}$$

where $k = [2m(E - U_0)]^{1/2}/\hbar$. Regardless of what A^+ and A^- might be, this solution satisfies Eq.(3.3.3) at all points other than $x = x'$. In order to satisfy Eq.(3.3.3) at $x = x'$, the Green's function must be continuous

$$\left[G(x,x')\right]_{x=x'^+} = \left[G(x,x')\right]_{x=x'^-} \tag{3.3.5a}$$

while the derivative must be discontinuous by $2m/\hbar^2$.

$$\left[\frac{\partial G(x,x')}{\partial x}\right]_{x=x'^+} - \left[\frac{\partial G(x,x')}{\partial x}\right]_{x=x'^-} = \frac{2m}{\hbar^2} \tag{3.3.5b}$$

Substituting for $G(x,x')$ from Eq.(3.3.4) into Eq.(3.3.5), we obtain

$$A^+ = A^- \quad \text{and} \quad ik\left[A^+ + A^-\right] = \frac{2m}{\hbar^2}$$

Hence $A^+ = A^- = -\dfrac{i}{\hbar v}$ where $v \equiv \dfrac{\hbar k}{m}$ (3.3.6)

and the Green's function is given by

$$G(x,x') = -\frac{i}{\hbar v} \exp\left[ik|x - x'|\right]$$

It is important to note that there is another solution

$$G(x,x') = +\frac{i}{\hbar v} \exp\left[-ik|x - x'|\right]$$

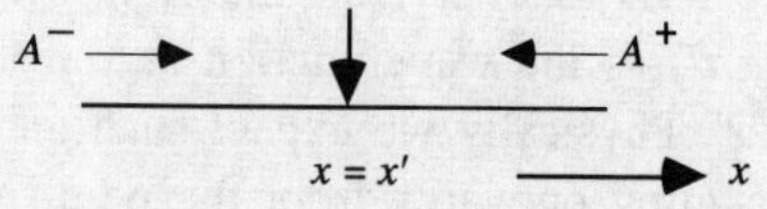

Fig. 3.3.2. Advanced Green's function for an infinite 1-D wire.

which satisfies Eq.(3.3.3) just as well. This solution consists of incoming waves that disappear at the point of excitation (Fig. 3.3.2) rather than outgoing waves that originate at the point of excitation (Fig. 3.3.1). The two solutions are referred to as the advanced Green's function (G^A) and the retarded Green's function (G^R):

$$G^R(x,x') = -\frac{i}{\hbar v}\exp\left[ik|x-x'|\right] \qquad (3.3.7)$$

and

$$G^A(x,x') = +\frac{i}{\hbar v}\exp\left[-ik|x-x'|\right]$$

where

$$k \equiv \frac{\sqrt{2m(E-U_0)}}{\hbar} \quad \text{and} \quad v \equiv \frac{\hbar k}{m}$$

Both these solutions satisfy the same equation (Eq.(3.3.3)) but they correspond to different *boundary conditions*: the retarded function corresponds to outgoing waves while the advanced function corresponds to incoming waves (far away from the source).

The infinitesimal η

One way to incorporate the boundary conditions into the equation itself is to add an infinitesimal imaginary part to the energy. Instead of Eq.(3.3.3) we write ($\eta > 0$)

$$\left(E - U_0 + \frac{\hbar^2}{2m}\frac{\partial^2}{\partial x^2} + i\eta\right)G^R(x,x') = \delta(x-x')$$

for the retarded function. The small imaginary part of the energy introduces a positive imaginary component to the wavenumber.

$$k' = \frac{\sqrt{2m(E+i\eta-U_0)}}{\hbar} = \frac{\sqrt{2m(E-U_0)}}{\hbar}\sqrt{1+\frac{i\eta}{E-U_0}}$$

$$\approx \frac{\sqrt{2m(E-U_0)}}{\hbar}\left[1+\frac{i\eta}{2(E-U_0)}\right] \equiv k(1+i\delta)$$

This imaginary part makes the advanced function grow indefinitely as we move away from the point of excitation. This makes the retarded function the only acceptable solution, since a proper solution must be bounded.

Similarly the advanced function is the only acceptable solution of the equation ($\eta > 0$)

$$\left(E - U_0 + \frac{\hbar^2}{2m}\frac{\partial^2}{\partial x^2} - i\eta\right)G^A(x,x') = \delta(x - x')$$

In general the retarded Green's function is defined as (cf.Eq.(3.3.1))

$$G^R = \left[E - H_{op} + i\eta\right]^{-1} \quad (\eta \to 0^+) \tag{3.3.8a}$$

while the advanced Green's function is defined as

$$G^A = \left[E - H_{op} - i\eta\right]^{-1} \quad (\eta \to 0^+) \tag{3.3.8b}$$

From hereon we will generally refer to the retarded Green's function as just the 'Green's function'.

Green's function for a multi-moded wire

Next let us look at the Green's function for an infinite multi-moded wire (Fig. 3.3.3). The Green's function $G^R(x,y;x',y')$ represents the wavefunction

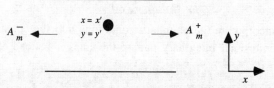

Fig. 3.3.3. Green's function for an infinite multi-moded wire.

at (x,y) due to an excitation at $x = x'$, $y = y'$. We would expect such an excitation to give rise to outgoing waves in different modes as shown. We could write the Green's function in the form

$$G^R(x,x') = \sum_m A_m^\pm \chi_m(y)\exp\left[ik_m|x - x'|\right] \tag{3.3.9}$$

where the A_m^+ and A_m^- are the amplitudes of the different modes that propagate away from the source. The transverse mode wavefunctions, $\chi_m(y)$ satisfy the equation (see Eq.(1.6.3) with B = 0)

$$\left[-\frac{\hbar^2}{2m}\frac{\partial^2}{\partial y^2} + U(y)\right]\chi_m(y) = \varepsilon_{m,0}\chi_m(y) \tag{3.3.10}$$

where $U(y)$ is the transverse confining potential in the y-direction. These functions are orthogonal

$$\int \chi_n(y)\chi_m(y)dy = \delta_{nm} \qquad (3.3.11)$$

since they satisfy the same equation with different eigenvalues. We will assume that these functions $\chi_m(y)$ are real.

To calculate the mode amplitudes A_m^+ and A_m^-, we proceed as we did for the 1-D wire and obtain (cf.Eq.(3.3.5))

$$\left[G^R(x,x')\right]_{x=x'+} = \left[G^R(x,x')\right]_{x=x'-}$$

$$\left[\frac{\partial G^R(x,x')}{\partial x}\right]_{x=x'+} - \left[\frac{\partial G^R(x,x')}{\partial x}\right]_{x=x'-} = \frac{2m}{\hbar^2}\delta(y-y') \qquad (3.3.12)$$

Substituting from Eq.(3.3.9) into (3.3.12) we obtain

$$\sum_m A_m^+ \chi_m(y) = \sum_m A_m^- \chi_m(y)$$

$$\sum_m ik_m\left[A_m^+ + A_m^-\right]\chi_m(y) = \frac{2m}{\hbar^2}\delta(y-y') \qquad (3.3.13)$$

Multiplying Eq.(3.3.13) by $\chi_n(y)$, integrating over y and using the orthogonality relation (Eq.(3.3.11)) we obtain

$$A_m^+ = A_m^- \quad \text{and} \quad ik_m\left[A_m^+ + A_m^-\right] = \frac{2m}{\hbar^2}\chi_m(y')$$

Hence the mode amplitudes are given by

$$A_m^+ = A_m^- = -\frac{i}{\hbar v_m}\chi_m(y') \qquad (3.3.14)$$

As we might expect, the amplitude A_m of mode m is proportional to the transverse wavefunction at the point of excitation, $\chi_m(y')$. Substituting Eq.(3.3.14) into Eq.(3.3.9) we obtain the Green's function:

$$G^R(x,y;x',y') = \sum_m -\frac{i}{\hbar v_m}\chi_m(y)\chi_m(y')\exp\left[ik_m|x-x'|\right] \qquad (3.3.15)$$

where

$$k_m \equiv \frac{\sqrt{2m(E-\varepsilon_{m,0})}}{\hbar} \quad \text{and} \quad v_m \equiv \frac{\hbar k_m}{m}$$

Eigenfunction expansion

We end this section by deriving a result that is often used to calculate Green's functions. The basic idea is that for any structure, if we know the eigenfunctions of the Hamiltonian operator

$$H_{op}\psi_\alpha(\mathbf{r}) = \varepsilon_\alpha\psi_\alpha(\mathbf{r}) \qquad (3.3.16)$$

then we can calculate the Green's function by performing the following summation:

$$G^R(\mathbf{r},\mathbf{r}') = \sum_\alpha \frac{\psi_\alpha(\mathbf{r})\psi_\alpha^*(\mathbf{r}')}{E - \varepsilon_\alpha + i\eta} \qquad (3.3.17)$$

We could have used this result to calculate the Green's function that we just obtained (Eq.(3.3.15)). However, the mathematics involves contour integration and is less transparent (see Exercise E.3.2 at the end of this chapter).

To derive Eq.(3.3.17), we first note that the eigenfunctions form a complete orthonormal set

$$\int \psi_\beta^*(\mathbf{r})\psi_\alpha(\mathbf{r})d\mathbf{r} = \delta_{\beta\alpha} \qquad (3.3.18)$$

so that we can expand the Green's function in the form

$$G^R(\mathbf{r},\mathbf{r}') = \sum_\alpha C_\alpha(\mathbf{r}')\psi_\alpha(\mathbf{r}) \qquad (3.3.19)$$

where the coefficients C_α have to be determined appropriately. Next we substitute Eq.(3.3.19) into the equation for the Green's function

$$(E - H_{op} + i\eta)G^R(\mathbf{r},\mathbf{r}') = \delta(\mathbf{r} - \mathbf{r}')$$

and make use of Eq.(3.3.16) to obtain (note that H_{op} acts only on $\mathbf{r}$, not $\mathbf{r}'$)

$$\sum_\alpha (E - \varepsilon_\alpha + i\eta)C_\alpha\psi_\alpha(\mathbf{r}) = \delta(\mathbf{r} - \mathbf{r}')$$

Multiplying by $\psi_\alpha^*(\mathbf{r})$, integrating over $\mathbf{r}$ and making use of the orthogonality relation (Eq.(3.3.18)) we obtain the coefficients C_α:

$$C_\alpha = \frac{\psi_\alpha^*(\mathbf{r}')}{E - \varepsilon_\alpha + i\eta}$$

Substituting back into Eq.(3.3.19) we obtain the result stated earlier (Eq.(3.3.17)). Similarly we can show that

$$G^A(\mathbf{r},\mathbf{r}') = \sum_\alpha \frac{\psi_\alpha(\mathbf{r})\psi_\alpha^*(\mathbf{r}')}{E-\varepsilon_\alpha-i\eta} \qquad (3.3.20)$$

From Eqs.(3.3.17) and (3.3.20) it is straightforward to show that

$$G^A(\mathbf{r},\mathbf{r}') = \left[G^R(\mathbf{r}',\mathbf{r})\right]^* \quad \to \quad G^A = \left[G^R\right]^+ \qquad (3.3.21)$$

so that the advanced function is the Hermitian conjugate of the retarded function.

3.4 S-matrix and the Green's function

With this brief introduction to Green's functions, we are ready to discuss the Fisher–Lee relation which expresses the elements of the S-matrix in terms of the Green's function (see D. S. Fisher and P. A. Lee (1981), *Phys. Rev. B*, **23**, 6851). Consider a conductor connected to a set of leads. For convenience, we use a different coordinate system in each lead as shown in Fig. 3.4.1. The interface between lead p and the conductor is

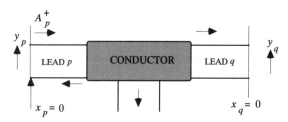

Fig. 3.4.1. A unit impulse in lead p generates an incident wave which is partially transmitted to each of the other leads.

defined by $x_p = 0$. We will use the symbol G_{qp}^R to denote the Green's function between a point lying on the plane $x_p = 0$ and another point lying on $x_q = 0$:

$$G_{qp}^R(y_q;y_p) \equiv G^R(x_q = 0, y_q; x_p = 0, y_p) \qquad (3.4.1)$$

Let us try to write this quantity in terms of the S-matrix element connecting the two leads. This is easy to do if we neglect the transverse dimension (y) of the leads and treat them as one-dimensional. We know that the unit excitation at $x_p = 0$ gives rise to a wave of amplitude A_p^- away from the conductor (not shown in the Figure) and a wave of amplitude A_p^+ toward the conductor. The wave traveling toward the conductor is scattered by the conductor into different leads. Hence we can write

$$G_{qp}^R = \delta_{qp}A_p^- + s'_{qp}A_p^+ \qquad (3.4.2)$$

But we know that (see Eq.(3.3.6))

$$A_p^+ = A_p^- = -\frac{i}{\hbar v_p}$$

Also, $\qquad\qquad s'_{qp} = s_{qp}\sqrt{v_p/v_q}$ $\qquad\qquad$ (see Eq.(3.1.4))

Hence from Eq.(3.4.2) we obtain

$$s_{qp} = -\delta_{qp} + i\hbar\sqrt{v_q v_p}\, G_{qp}^R \qquad (3.4.3)$$

This is the desired relation expressing the *S*-matrix in terms of the Green's function.

Multi-moded leads

The details are slightly more complicated if we take multiple modes in the leads into account. Instead of Eq.(3.4.2) we now have

$$G_{qp}^R(y_q; y_p) = \sum_{m\in p}\sum_{n\in q}\big[\delta_{nm}A_m^- + s'_{nm}A_m^+\big]\chi_n(y_q) \qquad (3.4.4)$$

From Eq.(3.3.14) we know that

$$A_m^+ = A_m^- = -\frac{i}{\hbar v_m}\chi_m(y_p)$$

Also, $\qquad\qquad s'_{nm} = s_{nm}\sqrt{v_m/v_n}$ $\qquad\qquad$ (see Eq.(3.1.4))

Hence from Eq.(3.4.4)

$$G_{qp}^R(y_q; y_p) = \sum_{m\in p}\sum_{n\in q} -\frac{i}{\hbar\sqrt{v_n v_m}}\chi_n(y_q)\big[\delta_{nm} + s_{nm}\big]\chi_m(y_p) \qquad (3.4.5)$$

In order to obtain an expression for an individual *S*-matrix element, we multiply Eq.(3.4.5) by $\chi_m(y_p)\chi_n(y_q)$, integrate over y_p and y_q and make use of the orthogonality relation (see Eq.(3.3.11)):

$$s_{nm} = -\delta_{nm} + i\hbar\sqrt{v_n v_m}\iint\chi_n(y_q)\big[G_{qp}^R(y_q; y_p)\big]\chi_m(y_p)\mathrm{d}y_q\mathrm{d}y_p \qquad (3.4.6)$$

Magnetic field in the leads

In general there is a non-zero magnetic field present in the leads which complicates the discussion considerably. Although we can calculate transverse mode wavefunctions even when a magnetic field is present (see Section 1.6) these wavefunctions do not satisfy the orthogonality relation stated in Eq.(3.3.11) (see Exercise E.2.6 at the end of Chapter 2). Consequently the derivation of a Fisher–Lee relation gets more complicated (see for example Eq.(88) of H. U. Baranger and A. D. Stone (1989), *Phys. Rev. B*, **40**, 8169). To simplify our discussion, we will assume that any magnetic field is present only inside the conductor and not in the leads. In an actual calculation we can include a length of the lead as part of the conductor and reduce the vector potential to zero over this length. If the vector potential were to have a transverse component (along y) then this would introduce a spurious magnetic field due to the non-zero $\partial A_y / \partial x$. But as we explained in Section 2.6, the vector potential must be chosen to be purely longitudinal (along x) in every lead, in order to permit us to define scattering states. Thus it can be reduced to zero without introducing spurious effects. This approach has been used by several authors to give sensible results both at low and high magnetic fields (see for example, H. U. Baranger and A. D. Stone (1991), *Phys. Rev. B*, **44**, 10637 and M. J. McLennan *et al.* (1991), *Phys. Rev. B*, **43**, 13846, 14333). However, this is not a necessary assumption. Several authors have reported numerical calculations taking a non-zero magnetic field in the leads into account (see for example, H. Tamura and T. Ando (1991), *Phys. Rev. B*, **44**, 1792, M. Leng and C. S. Lent (1993), *Phys. Rev. Lett.*, **71**, 137 and Y. Wang *et al.* (1994), *Phys. Rev. B*, **49**, 1928).

3.5 Tight-binding model (or the method of finite differences)

Next we address the question of how we can calculate the Green's function (and hence the S-matrix via the Fisher–Lee relation), for an arbitrarily shaped conductor. Basically we need to solve the differential equation for the Green's function (see Eqs.(3.3.8a), (3.3.2)):

$$\left[E - H_{\mathrm{op}}(\mathbf{r}) + i\eta \right] G^{\mathrm{R}}(\mathbf{r};\mathbf{r}') = \delta(\mathbf{r} - \mathbf{r}') \tag{3.5.1}$$

$$H_{\mathrm{op}}(\mathbf{r}) = \frac{(i\hbar\nabla + e\mathbf{A})^2}{2m} + U(\mathbf{r}) \tag{3.5.2}$$

for arbitrary $U(\mathbf{r})$ and $\mathbf{A}(\mathbf{r})$. We will restrict the discussion to two sional conductors, but the approach can be applied to three-dimensional structures in a straightforward manner.

A common approach for solving a differential equation like Eq.(3.5.1) is to discretize the spatial coordinate so that the Green's function becomes a matrix:

$$G^R(\mathbf{r}, \mathbf{r}') \quad \rightarrow \quad G^R(i, j)$$

where the indices i, j denote points on a discrete lattice. The differential equation becomes a matrix equation

$$[(E + i\eta)I - H]G^R = I \tag{3.5.3}$$

where $[I]$ is the identity matrix and $[H]$ is the matrix representation of the Hamiltonian operator H_{op}. We can then calculate G^R by inverting the matrix $[(E + i\eta)I - H]$ numerically. How do we write down the matrix $[H]$? The basic idea is best appreciated with a simple 1-D example without any magnetic field. Later we will generalize it to 2-D conductors in arbitrary magnetic fields.

Matrix representation for H_{op} in 1-D

In 1-D, with the vector potential set to zero, the Hamiltonian operator (Eq.(3.5.2)) simplifies to

$$H_{op} = -\frac{\hbar^2}{2m}\frac{d^2}{dx^2} + U(x)$$

To obtain a matrix representation for this operator, we consider the quantity $H_{op}F(x)$ where $F(x)$ is *any* function of x. Now we choose a discrete

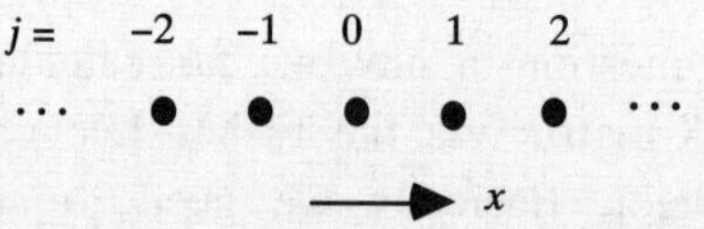

$$j = \quad -2 \quad -1 \quad 0 \quad 1 \quad 2$$
$$\cdots \quad \bullet \quad \bullet \quad \bullet \quad \bullet \quad \bullet \quad \cdots$$
$$\longrightarrow x$$

Fig. 3.5.1. An infinite linear chain discretized into a lattice.

lattice whose points are located at $x = ja$, j being an integer (Fig. 3.5.1) and write

$$[H_{op}F]_{x=ja} = \left[-\frac{\hbar^2}{2m}\frac{d^2F}{dx^2}\right]_{x=ja} + U_jF_j \tag{3.5.4}$$

where $\qquad F_j \rightarrow F(x = ja) \quad U_j \rightarrow U(x = ja)$

We now use the method of finite differences to approximate the derivative operators. Assuming a is small we can approximate the first derivative by

$$\left[\frac{dF}{dx}\right]_{x=(j+\frac{1}{2})a} \rightarrow \frac{1}{a}\left[F_{j+1} - F_j\right]$$

and the second derivative by

$$\left[\frac{d^2F}{dx^2}\right]_{x=ja} \rightarrow \frac{1}{a}\left\{\left[\frac{dF}{dx}\right]_{x=(j+\frac{1}{2})a} - \left[\frac{dF}{dx}\right]_{x=(j-\frac{1}{2})a}\right\}$$

$$\rightarrow \frac{1}{a^2}\left\{F_{j+1} - 2F_j + F_{j-1}\right\}$$

With this approximation we can write from Eq.(3.5.4),

$$\left[H_{op}F\right]_{x=ja} = (U_j + 2t)F_j - tF_{j-1} - tF_{j+1} \qquad (3.5.5)$$

where $\qquad\qquad t \equiv \hbar^2 / 2ma^2 \qquad\qquad (3.5.6)$

We can rewrite Eq.(3.5.5) in the form

$$\left[H_{op}F(x)\right]_{x=ja} = \sum_i H(j,i)F_i \qquad (3.5.7)$$

where $\qquad H(j,i) = U_i + 2t \quad$ if $i = j$

$\qquad\qquad\qquad = -t \qquad\quad$ if i and j are nearest neighbors

$\qquad\qquad\qquad = 0 \qquad\quad\ $ otherwise

Eq.(3.5.7) gives us the desired matrix representation for the Hamiltonian operator for a 1-D linear chain (Fig. 3.5.1).

$$H = \begin{bmatrix} \cdots & -t & 0 & 0 & 0 \\ -t & U_{-1}+2t & -t & 0 & 0 \\ 0 & -t & U_0+2t & -t & 0 \\ 0 & 0 & -t & U_1+2t & -t \\ 0 & 0 & 0 & -t & \cdots \end{bmatrix}$$

Each site is linked to its nearest neighbor by the element t, while the diagonal elements are given by the potential energy plus $2t$.

It is interesting to note the similarity of this discretized Hamiltonian to the tight-binding Hamiltonian which is widely used to model electronic transfer in molecules and condensed matter. In the tight-binding model the wavefunction is expressed in terms of localized atomic orbitals, one at each site. Orbitals on neighboring sites are connected by what is referred to as a 'hopping matrix element' or an 'overlap integral'. The local potential U_j (+ $2t$) in our model plays the role of the energy of the orbital localized at site j while t ($\equiv \hbar^2/2ma^2$) plays the role of the overlap integral between orbitals on neighboring sites.

Dispersion relation for a discrete lattice

We know that for a uniform 1-D wire with a constant potential $U(x) = U_0$, the eigenfunctions are plane waves with a parabolic dispersion relation:

$$\psi_k(x) = \exp[ikx] \quad \text{where} \quad E = U_0 + \frac{\hbar^2 k^2}{2m}$$

How is this result modified for a discrete lattice? For a discrete wire the Schrödinger equation can be written as (see Eq.(3.5.5))

$$E\psi_j = (U_0 + 2t)\psi_j - t\psi_{j-1} - t\psi_{j+1}$$

which is satisfied by a solution of the form

$$\psi_j = \exp[ikx_j] \quad \text{where} \quad x_j = ja$$

provided $\qquad\qquad E = U_0 + 2t(1 - \cos(ka)) \qquad\qquad$ (3.5.8a)

This is the dispersion relation for a discrete lattice. It is easy to show that as we let the lattice constant a tend to zero we recover the usual parabolic relationship. Also, the velocity is given by

$$\hbar v = \frac{\partial E}{\partial k} = 2at\sin(ka) \qquad\qquad (3.5.8b)$$

Again as we let a tend to zero, we recover the usual result $v = \hbar k/m$.

Matrix representation for H_{op} in 2-D

It is straightforward to extend Eq.(3.5.7) to two or more dimensions. In general the matrix elements of $[H]$ are given by

$$[H]_{ij} = U(\mathbf{r}_i) + zt \quad \text{if } i = j$$

$$= -\tilde{t}_{ij} \qquad \text{if } i \text{ and } j \text{ are nearest neighbors} \qquad (3.5.9a)$$

$$= 0 \qquad \text{otherwise}$$

where (1) z is the number of nearest neighbors ($z = 2$ for a linear chain and $z = 4$ for a square lattice), (2) $\mathbf{r}_i$ is the position vector for lattice site i. If the vector potential is zero, then the nearest neighbor coupling is equal to $-t$ as in the 1-D example. With a non-zero vector potential it is modified to

$$\tilde{t}_{ij} = t \exp\left[i e \mathbf{A} . (\mathbf{r}_i - \mathbf{r}_j)/\hbar \right] \qquad (3.5.9b)$$

The vector potential $\mathbf{A}$ is evaluated at a point halfway between sites i and j, that is, at $(\mathbf{r}_i + \mathbf{r}_j)/2$. For a discussion of Eq.(3.5.9b) see R. P. Feynman (1965), *Lectures on Physics*, Vol.III, 21–2, (New York, Addison–Wesley).

Truncating the matrix

Now that we have a matrix representation for the Hamiltonian operator it may seem straightforward to set up the matrix $[(E + i\eta)I - H]$ and invert it (see Eq.(3.5.3)):

$$G^{\mathrm{R}} = \left[(E + i\eta)I - H \right]^{-1} \qquad (3.5.10)$$

The only problem is that the matrix is infinite dimensional! This is because we are dealing with an open system connected to leads that stretch out to infinity. If we simply truncate the matrices at some point, then we would effectively be describing a closed system with fully reflecting boundaries and not the open system with non-reflecting boundaries that we would like to describe. The truncation has to be done more carefully as we will now describe.

Consider a conductor connected to the lead p as shown in Fig. 3.5.2.

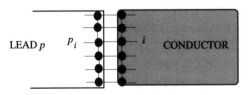

Fig. 3.5.2. A conductor described by a Hamiltonian H_{C}, connected to lead p described by H_p, through the coupling matrix τ_p (see Eq.(3.5.11b)). A point in lead p is labeled p_i if it is adjacent to point i inside the conductor

We can partition the overall Green's function in Eq.(3.5.10) into submatrices as follows:

$$\begin{bmatrix} G_p & G_{pC} \\ G_{Cp} & G_C \end{bmatrix} = \begin{bmatrix} (E+i\eta)I - H_p & \tau_p \\ \tau_p^+ & EI - H_C \end{bmatrix}^{-1} \tag{3.5.11a}$$

where the matrix $[(E + i\eta)I - H_p]$ represents the isolated lead, while the matrix $[EI - H_C]$ represents the isolated conductor. We could add an infinitesimal imaginary term $(i\eta)$ to the conductor matrix as well but it is not necessary. As we will see, the coupling to the leads effectively gives rise to a finite imaginary term that will swamp it. The coupling matrix is non-zero only for adjacent points i and p_i (see Fig. 3.5.2):

$$\tau_p(p_i, i) = t \tag{3.5.11b}$$

We will now derive an explicit expression for the sub-matrix G_C, since this is the part that we are really interested in. From Eq.(3.5.11) we can write

$$[(E+i\eta)I - H_p]G_{pC} + [\tau_p]G_C = 0 \tag{3.5.12a}$$

$$[EI - H_C]G_C + [\tau_p^+]G_{pC} = I \tag{3.5.12b}$$

From Eq.(3.5.12a) we obtain

$$G_{pC} = -g_p^R \tau_p G_C \tag{3.5.13}$$

where

$$g_p^R = [(E+i\eta)I - H_p]^{-1} \tag{3.5.14}$$

is the Green's function for the isolated semi-infinite lead. Substituting Eq.(3.5.13) into Eq.(3.5.12b) we obtain

$$G_C = [EI - H_C - \tau_p^+ g_p^R \tau_p]^{-1} \tag{3.5.15}$$

Note that the matrices in this expression are finite matrices of dimensions $(C \times C)$, C being the number of points inside the conductor. But the infinite lead is taken into account exactly through the term $\tau_p^+ g_p^R \tau_p$. It might seem that we have not gained much since we have to invert an infinite matrix to obtain g_p^R from Eq.(3.5.14). However, this is not necessary. Since g_p^R represents the Green's function for an isolated lead it can usually be determined analytically. From Eqs.(3.5.15) and (3.5.11b) we can write

$$\left[\tau_p^+ g_p^R \tau_p\right]_{ij} = t^2 g_p^R(p_i, p_j) \qquad (3.5.16)$$

since the coupling matrix τ_p is zero for all points in the lead except those (p_i, p_j) that are adjacent to points (i, j) inside the conductor.

Assuming that different leads are independent so that their effects are additive, we can write:

$$G^R = \left[EI - H_C - \Sigma^R\right]^{-1} \qquad (3.5.17a)$$

where
$$\Sigma^R = \sum_p \Sigma_p^R \quad \text{and} \quad \Sigma_p^R(i, j) = t^2 g_p^R(p_i, p_j) \qquad (3.5.17b)$$

From hereon we will use G^R to denote what we were writing as G_C, since this is the only component of the overall Green's function that we will be using in our discussion. It represents the propagation of electrons between two points inside the conductor, taking the effect of the leads into account through the term Σ^R.

The term Σ^R can be viewed as an effective Hamiltonian arising from the interaction of the conductor with the leads. A similar term is often used to describe the interaction of the electrons with phonons and other electrons (which we are neglecting) and is called the self-energy. By analogy we will refer to Σ^R as the 'self-energy' due to the leads. However, it should be noted that the self-energy usually provides only an approximate description of the electron–phonon and electron–electron interactions which involve complicated systems with internal degrees of freedom. Leads, on the other hand, are simple inert objects whose effect is described *exactly* by Σ^R. We have made no approximations in arriving at Eq.(3.5.17) from Eq.(3.5.10).

Self-energy

To use Eq.(3.5.17) we need the self-energy Σ^R. For this we need to calculate the Green's function g_p^R for an isolated lead. Earlier we had calculated the Green's function for an infinite wire (see Eq.(3.3.15)), but we cannot use it here because what we need is the Green's function for a *semi-infinite* wire that terminates on one side. For a semi-infinite discrete wire it can be shown that the Green's function between two points along the edge is given by (see Exercise E.3.3 at the end of this chapter)

$$g_p^R(p_i, p_j) = -\frac{1}{t} \sum_m \chi_m(p_i) \exp[ik_m a] \chi_m(p_j) \qquad (3.5.18)$$

where the (real) function $\chi_m(y_p)$ describes the transverse profile of mode m in lead p and v_m is its velocity. Substituting from Eq.(3.5.18) into Eq.(3.5.17b) we obtain the desired expression for the self-energy:

$$\Sigma_p^R(i, j) = -t \sum_{m \in p} \chi_m(p_i) \exp[+ik_m a] \chi_m(p_j) \qquad (3.5.19)$$

Note that it is not necessary to assume the leads to have regular shapes with well-defined transverse modes, though we have done so for convenience. Even for irregularly shaped leads we can in principle calculate the Green's function and obtain the appropriate self-energy.

Transmission function

Once we have calculated the Green's function, we can use the Fisher–Lee relation (see Eq.(3.4.6)) to obtain the elements of the S-matrix, from which the transmission function can be calculated. Using this procedure we can express the transmission function in a rather compact form (see Exercise E.3.4 at the end of this chapter):

$$\bar{T}_{pq} = \text{Tr}\left[\Gamma_p G^R \Gamma_q G^A\right] \qquad (3.5.20)$$

where 'Tr' represents the trace and the elements of the matrix Γ_p are given by

$$\Gamma_p(i, j) = \sum_m \chi_m(p_i) \frac{\hbar v_m}{a} \chi_m(p_j) \qquad (3.5.21)$$

Using the expression for the self-energy (Eq.(3.5.19)) and the relation between velocity and wavenumber (Eq.(3.5.8b)), it is straightforward to show that

$$\Gamma_p = i\left[\Sigma_p^R - \Sigma_p^A\right] \qquad (3.5.22)$$

where the advanced self-energy (Σ_p^A) is the Hermitian conjugate of the retarded self-energy (Σ_p^R). It should be mentioned that Eq.(3.5.20) for the transmission function follows from the Fisher–Lee relation only for $p \neq q$. But this is not very important because the elements of the transmission function with $p = q$ have no effect on the current (see Eq.(2.5.7)).

The physical meaning of Eq.(3.5.20) is not very transparent, but it provides a compact expression for calculating the transmission function. The Green's function G^R describes the dynamics of the electrons inside the conductor, taking the effect of the leads into account through the self-energy Σ^R. The function Γ describes the coupling of the conductor to the leads. We will discuss the meaning of these functions further in Section 3.6.

Sum rule

Earlier in this chapter we derived the 'sum rule' for the transmission function (see Eq.(3.1.5))

$$\sum_q \overline{T}_{pq} = \sum_q \overline{T}_{qp} = M_p$$

where M_p is the number of modes in lead p. This sum rule is very important because it ensures current conservation. We can derive a similar sum rule in terms of the spectral function. Starting from our expression for the transmission function (Eq.(3.5.20)) we can write

$$\sum_q \overline{T}_{pq} = \text{Tr}\left[\Gamma_p G^R \Gamma G^A\right] \quad \text{and} \quad \sum_q \overline{T}_{qp} = \text{Tr}\left[\Gamma_p G^A \Gamma G^R\right] \quad (3.5.23)$$

where

$$\Gamma \equiv \sum_p \Gamma_p = \sum_p i\left[\Sigma_p^R - \Sigma_p^A\right] = i\left[\Sigma^R - \Sigma^A\right] \quad (3.5.24)$$

From Eq.(3.5.23) we obtain the desired 'sum rule'

$$\sum_q \overline{T}_{pq} = \sum_q \overline{T}_{qp} = \text{Tr}\left[\Gamma_p A\right] \quad (3.5.25)$$

if we make use of the following identity (see Eq.(3.6.4) in the next section)

$$A \equiv i\left[G^R - G^A\right] = G^R \Gamma G^A = G^A \Gamma G^R$$

The quantity A is known as the spectral function which plays the role of a generalized density of states inside the conductor taking the leads into account. We will discuss this concept in the next section. For the moment we just wish to point out that in the language of Green's functions, the quantity $\text{Tr}\left[\Gamma_p A\right]$ plays the role of the number of modes M_p.

A few remarks

Before we leave this section let us take a moment to recognize what we have accomplished. We have managed to *eliminate the infinite leads* from the formulation. The transmission function is expressed completely in terms of quantities that are defined inside the finite-sized conductor (see Eq.(3.5.20)). Even the function $\Sigma_p^R(i,j)$ describing the effect of the lead p is defined for points i and j located inside the conductor. Thus if we discretize the conductor into N lattice sites then (at each energy E) each of the matrices in the formulation has dimensions $(N \times N)$. The numerical procedure for calculating the transmission function is thus quite straightforward. The only practical limitation is that there is an upper limit to the size of matrices that can be handled, and this limits the number of lattice sites that we can use.

In arriving at this formulation, the key concept is that a lead can be replaced by a 'self-energy function' given by Eq.(3.5.19). As we let the lattice constant a tend to zero and go to the continuum limit, the self-energy reduces to

$$\left[\Sigma_p^R(i,j)\right]_{a\to 0} = -t\delta_{ij} - \frac{i}{2}\delta(x_p)\sum_m \chi_m(p_i)\hbar v_m \chi_m(p_j) \qquad (3.5.26)$$

where we have replaced $(1/a)$ by a delta function at the interface, made use of Eq.(3.5.8b), and also assumed that the transverse mode functions form a complete set:

$$\sum_m \chi_m(p_i)\chi_m(p_j) = \delta_{p_i,p_j}$$

The real part of the self-energy is given by the first term on the right of Eq.(3.5.26). It simply changes the diagonal element (see Eq.(3.5.9a)) from $(U_i + zt)$ to $(U_i + (z-1)t)$, as if there were one less nearest neighbor. The imaginary part of the self-energy is given by the second term on the right. As we will discuss in the next section, it represents the rate at which particles can escape through the lead, somewhat like the concept of a 'surface recombination velocity' used by device engineers.

To get an intuitive feeling for the method described here, we *strongly recommend* that the reader go through the simple example described in *Exercise E.3.5.* The reader may also find the discussion in M. J. McLennan *et al.* (1991), *Phys. Rev. B*, **43**, 13 846 useful (see Appendix). The approach we have described here is basically an extension of that described by Caroli *et al.* (1971), *J. Phys. C: Solid State Physics*, **4**, 916. Results similar to Eq.(3.5.20) have been obtained by several authors (see for example, Eq.(7) of Y. Meir and N. S. Wingreen (1992). *Phys. Rev. Lett.*, **68**, 2512 or Eq.(4) of M. Sumetskii (1991), *J. Phys. Cond. Matter*, **3**, 2651). Indeed many different approaches have been described in the literature for handling the open boundary conditions associated with the leads, although many authors may not use the word 'self-energy' in this context.

3.6 'Self-energy'

We have seen in the last section that the Green's function which is defined as (Eq.(3.3.8a))

$$G^R = \left[E - H + i\eta\right]^{-1}$$

can be written in the form (Eq.(3.5.17a))

$$G^R = \left[E - H_C - \Sigma^R\right]^{-1} \tag{3.6.1}$$

where H_C is the Hamiltonian for the isolated conductor, and the self-energy Σ^R describes the effect of the leads on the conductor. This is an important conceptual step, for it allows us to replace an infinite open system with a finite one (Fig.3.6.1). In this section we will discuss two important concepts defined by the relations

$$\Gamma \equiv i\left[\Sigma^R - \Sigma^A\right] \tag{3.6.2}$$

$$A \equiv i\left[G^R - G^A\right] \tag{3.6.3}$$

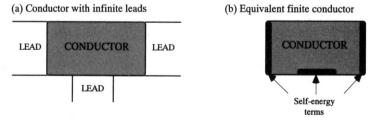

Fig. 3.6.1. A conductor connected to infinite leads can be replaced conceptually by an equivalent finite conductor with self-energy terms.

and derive a useful relationship between them:

$$A = G^R \Gamma G^A = G^A \Gamma G^R \tag{3.6.4}$$

We have mentioned earlier that the Green's function is like a generalized S-matrix that tells us the response at any point due to an excitation at any other. From this point of view Eq.(3.6.4) is the analog of the unitarity relation

$$I = SS^+ = S^+S \qquad \text{(same as Eq.(3.1.3a))}$$

satisfied by the S-matrix. Indeed in the last section we made use of Eq.(3.6.4) in deriving the sum rule for the transmission function (just as we used the unitarity of the S-matrix in Section 3.1).

Eigenstate lifetime

Usually in quantum mechanics one deals with isolated or closed systems whose eigenstates are found by diagonalizing the Hamiltonian H_C:

$$H_C \psi_{\alpha 0} = \varepsilon_{\alpha 0} \psi_{\alpha 0} \tag{3.6.5}$$

However, our interest is often in open systems with strong coupling to the leads. As we have seen this gives rise to a self-energy Σ^R (see Eq.(3.5.19)) so that we have an effective Hamiltonian $[H_C + \Sigma^R]$ whose eigenstates can be written as

$$\left[H_C + \Sigma^R \right] \psi_\alpha = \varepsilon_\alpha \psi_\alpha \tag{3.6.6}$$

This looks just like Eq.(3.6.5) but there is an important difference: *the eigenenergies are complex* since the self-energy is in general not Hermitian. If it were Hermitian then the quantity Γ defined by Eq.(3.6.3) would be zero.

We could write the new eigenenergies as

$$\varepsilon_\alpha = \varepsilon_{\alpha 0} - \Delta_\alpha - i(\gamma_\alpha/2) \tag{3.6.7}$$

where $\varepsilon_{\alpha 0}$ is the eigenenergy of the isolated conductor described by $[H_C]$ (see Eq.(3.6.5)). The time dependence of an eigenstate of $[H_C + \Sigma^R]$ has the form

$$\exp\left[-i\varepsilon_\alpha t/\hbar\right] \rightarrow \exp\left[-i(\varepsilon_{\alpha 0} - \Delta_\alpha)t/\hbar\right] \exp\left[-\gamma_\alpha t/\hbar\right]$$

Δ_α represents the shift in the energy due to the modification of the dynamics of an electron inside the conductor by the interaction with the leads. On the other hand, the imaginary part of the energy γ_α reflects the fact that an electron injected anywhere in the conductor will eventually disappear through one of the leads stretching out to infinity. Taking the squared magnitude of the wavefunction we obtain the probability:

$$\left|\psi_\alpha\right|^2 \exp\left[-2\gamma_\alpha t/\hbar\right]$$

The quantity $2\hbar/\gamma_\alpha$ thus represents the 'lifetime' or the average time an electron remains in state α before it escapes out into the leads. The lifetime is infinite if the self-energy is Hermitian, that is, if $\Gamma = 0$.

Eigenfunction expansion

To see the effect of the self-energy term on the Green's function it is convenient to express the Green's function in terms of the eigenstates of the Hamiltonian. It might seem that we can use the expression we derived earlier (see Eq.(3.3.17))

$$G^R(\mathbf{r},\mathbf{r}') = \sum_\alpha \frac{\psi_\alpha(\mathbf{r})\psi_\alpha^*(\mathbf{r}')}{E - \varepsilon_\alpha} \qquad \text{(WRONG)}$$

if we define ψ_α as the eigenstate of the effective Hamiltonian $[H_C + \Sigma^R]$:

$$\left[H_C + \Sigma^R\right]\psi_\alpha = E_\alpha\psi_\alpha \qquad (3.6.8a)$$

However, this is not correct. Since $[H_C + \Sigma^R]$ is not Hermitian its eigenstates ψ_α do not form an orthogonal set, unlike those of $[H_C]$. As a result we need to modify Eq.(3.3.17), making use of the eigenstates Φ_α of the adjoint operator as well.

$$\left[H_C + \Sigma^A\right]\Phi_\alpha = E_\alpha^*\Phi_\alpha \qquad (3.6.8b)$$

The Φs and ψs are identical if we are dealing with a Hermitian operator, but not with non-Hermitian operators. Together the Φs and ψs form what is known as a bi-orthonormal set having the property that

$$\int \Phi_\alpha(\mathbf{r})\psi_\beta^*(\mathbf{r})d\mathbf{r} = \delta_{\alpha\beta} \qquad (3.6.9)$$

(See p. 884 of P. M. Morse and H. Feshbach (1953), *Methods of Theoretical Physics*, (New York, McGraw–Hill)). The correct expression for the Green's function has the form (see for example, W. van Haeringen, B. Farid and D. Lenstra (1987), *Physica Scripta*, **T19**, 282):

$$G^R(\mathbf{r}, \mathbf{r}') = \sum_\alpha \frac{\psi_\alpha(\mathbf{r}) \Phi_\alpha^*(\mathbf{r}')}{E - \varepsilon_\alpha} \tag{3.6.10}$$

For an isolated conductor the eigenvalues are real and given by $\varepsilon_{\alpha 0}$. The individual terms in Eq.(3.6.10) then have singularities at $E \approx \varepsilon_{\alpha 0}$ and we need the infinitesimal η (see Eqs.(3.3.8)) to keep the Green's function from blowing up at these energies. The coupling to the leads takes care of this problem automatically by making the eigenenergies complex, so that there are no singularities for real values of E. The effect of the complex eigenenergies is best appreciated by looking at the spectral function.

Spectral function

A very important concept is that of the spectral function defined by Eq.(3.6.3). Using Eqs.(3.6.10) and (3.6.7) we can write the spectral function as

$$A(\mathbf{r}, \mathbf{r}') = \sum_\alpha \psi_\alpha(\mathbf{r}) \Phi_\alpha^*(\mathbf{r}') \frac{\gamma_\alpha}{(E - \varepsilon_{\alpha 0} + \Delta_\alpha)^2 + (\gamma_\alpha/2)^2} \tag{3.6.11}$$

Fig. 3.6.2 shows how the spectral function evolves as the coupling to the leads is increased. The self-energy Σ^R becomes larger and shifts the eigenenergies as indicated in Eq.(3.6.7). As a result, the peaks of the quantity

$$\frac{\gamma_\alpha}{(E - \varepsilon_{\alpha 0} + \Delta_\alpha)^2 + (\gamma_\alpha/2)^2}$$

(1) *shift* by Δ_α and (2) *broaden* by γ_α. When the coupling is very large the peaks all merge and the peaks corresponding to different eigenstates can no longer be distinguished.

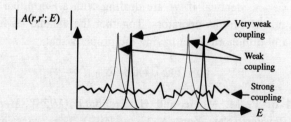

Fig. 3.6.2. The spectral function for a conductor weakly coupled to the leads shows peaks at energies corresponding to the eigenvalues of the isolated conductor. As the coupling is increased the peaks shift and broaden.

One small point. Eq.(3.6.11) conveys the impression that each of the peaks has the form of a Lorentzian function:

$$\frac{\gamma}{(E - \varepsilon)^2 + (\gamma/2)^2}$$

However, this is not necessarily true. This is because the self-energy itself is a function of E. Consequently the eigenfunctions Φ_α, ψ_α as well as the eigenvalues ε_α are functions of E and the peaks in the spectral function can have very non-Lorentzian shapes.

The trace of the spectral function represents the *density of states*:

$$N(E) = \frac{1}{2\pi} \mathrm{Tr}[A(E)] \tag{3.6.12}$$

If the lifetime of the eigenstates is large, then Eq.(3.6.12) leads to the conventional expression for the density of states. To see this, we make use of Eq.(3.6.9) to write

$$\mathrm{Tr}[A] \equiv \int A(\mathbf{r},\mathbf{r})d\mathbf{r} = \sum_\alpha \frac{\gamma_\alpha}{(E - \varepsilon_{\alpha 0} + \Delta_\alpha)^2 + (\gamma_\alpha/2)^2} \tag{3.6.13}$$

From Eq.(3.6.12) we can write

$$N(E) = \sum_\alpha \frac{1}{2\pi} \frac{\gamma_\alpha}{(E - \varepsilon_{\alpha 0} + \Delta_\alpha)^2 + (\gamma_\alpha/2)^2}$$
$$\rightarrow \sum_\alpha \delta(E - \varepsilon_{\alpha 0} + \Delta_\alpha) \quad \text{as} \quad \gamma_\alpha \rightarrow 0$$

in agreement with what we would expect for long-lived eigenstates with well-defined energies. Eq.(3.6.12) provides us with a general expression for the density of states that can be used even when eigenstates have finite lifetimes.

Local density of states

The diagonal elements of the spectral function give us the local density of states

$$\rho(\mathbf{r},E) = \frac{1}{2\pi} A(\mathbf{r},\mathbf{r};E) = -\frac{1}{\pi} \mathrm{Im}[G^R(\mathbf{r},\mathbf{r};E)] \tag{3.6.14}$$

From Eq.(3.6.11) we can write

$$\rho(\mathbf{r},E) \sim \sum_{\alpha} \frac{1}{2\pi} \frac{\gamma_{\alpha}}{(E - \varepsilon_{\alpha 0} + \Delta_{\alpha})^2 + (\gamma_{\alpha}/2)^2} \psi_{\alpha}(\mathbf{r})\Phi_{\alpha}^*(\mathbf{r})$$

$$\rightarrow \sum_{\alpha} \delta(E - \varepsilon_{\alpha 0})|\psi_{\alpha}(\mathbf{r})|^2 \quad \text{as} \quad \gamma_{\alpha} \rightarrow 0$$

again in agreement with what we would expect for long-lived eigenstates with well-defined energies. This function provides insight into the spatial variation of the states in a conductor as can be appreciated from the simple example (see Fig. 3.6.3) showing the variation in the local density of states across the width of a quantum wire in a magnetic field.

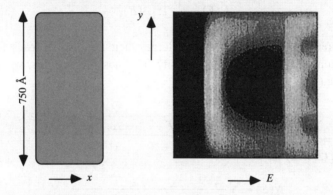

Fig. 3.6.3. Gray scale plot of the local density of states $\rho(y,E)$ for a quantum wire (assumed uniform in the x-direction) in a magnetic field of $B = 5.6$ T. Bright regions indicate a large density of states. Adapted with permission from Y. Lee *et al.* (1992), *Superlattices and Microstructures*, **11**, 137.

When the energy matches the energy of a bulk Landau level

$$E \sim (n + \frac{1}{2})\hbar\omega_c$$

the corresponding eigenstates are located in the interior of the sample giving rise to a large density of states inside the sample. But when the energy lies between two Landau levels the corresponding states are localized near the edges (see Section 1.6 and also Chapter 4) giving rise to a large density of states near the edges.

Mesoscopic conductors in general often display significant spatial variations in the local density of states that influence the nature of conduction through the sample. With the advent of scanning tunneling microscopy (STM) it has become feasible to probe the local density of states on an atomic scale thus making this concept very real from an experimental point of view (see for example, C. J. Chen (1993), *Introduction to Scanning Tunneling Microscopy*, Oxford University Press).

A useful identity

We will now prove the identity stated earlier:

$$A = G^{R}\,\Gamma\,G^{A} = G^{A}\,\Gamma\,G^{R} \qquad \text{(same as (Eq.(3.6.4))}$$

First we use Eq.(3.5.17a) and its Hermitian adjoint to write

$$\left[G^{R}\right]^{-1} - \left[G^{A}\right]^{-1} = \Sigma^{A} - \Sigma^{R} = i\Gamma \qquad (3.6.15)$$

Multiplying by G^{R} from the left and by G^{A} from the right we obtain

$$G^{A} - G^{R} = i\,G^{R}\,\Gamma\,G^{A} \quad \rightarrow \quad A = G^{R}\,\Gamma\,G^{A}$$

as stated above. The other half of the relation is obtained if we multiply Eq.(3.6.15) by G^{A} from the left and by G^{R} from the right.

3.7 Relation to other formalisms

In Section 3.5 we showed that the transmission function $\overline{T}_{pq}$ between two leads p and q can be written in terms of the Green's functions as

$$\overline{T}_{pq} = \text{Tr}\!\left[\Gamma_{p}\,G^{R}\,\Gamma_{q}\,G^{A}\right] \qquad \text{(same as Eq.(3.5.20))}$$

The Green's function G^{R} describes the dynamics of the electrons inside the conductor (taking the leads into account), while Γ_{p} describes the strength of the coupling of lead p to the conductor (see Eqs.(3.5.19)–(3.5.22) for the definitions of these quantities). Although we are using the language of Green's functions, these results are basically obtained from scattering theory. In this section we will relate these results to those obtained from other formalisms that are widely used, namely the Kubo formalism and the transfer Hamiltonian formalism. We will also use Eq.(3.5.20) later to relate to the non-equilibrium Green function formalism in Chapter 8.

Kubo formalism

The Kubo formalism is based on the fluctuation-dissipation theorem which relates the equilibrium noise (that is, fluctuations) to the linear response conductance (that is, dissipation). A special case of this general theorem is the Nyquist–Johnson formula mentioned earlier (see Eq.(2.7.7)):

$$G = \frac{1}{2k_{\mathrm{B}}T} \int_{-\infty}^{+\infty} \langle I(t + t_0)I(t_0) \rangle_{\mathrm{eq}} \, dt$$

This relationship allows us to calculate a *non-equilibrium* property like conductance indirectly by calculating the equilibrium noise using the methods of *equilibrium* statistical mechanics (see for example M. Büttiker *et al.* (1993), *Phys. Rev. Lett.*, **70**, 4114). There is a similar relationship between the non-local conductivity $\sigma(\mathbf{r}, \mathbf{r}')$ and the current density $\mathbf{J}(\mathbf{r})$ which is widely used in the literature (for a fairly extensive list of references see H. U. Baranger and A. D. Stone (1989), *Phys. Rev. B*, **40**, 8169).

The Kubo formalism is commonly used to derive the following expression for the zero temperature conductivity of a uniform rectangular conductor with all sides equal to L (d = number of dimensions and all quantities are evaluated at the Fermi energy)

$$\sigma = \frac{2e^2}{h} \frac{\hbar^2}{L^d} \sum_{\mathbf{k}, \mathbf{k}'} v_x(\mathbf{k}') v_x(\mathbf{k}) \left| G^{\mathrm{R}}(\mathbf{k}', \mathbf{k}) \right|^2 \qquad (3.7.1)$$

Eq.(3.7.1) is used as the starting point for calculating the ensemble-averaged conductivity of large conductors (see, for example, Eq.(2.20) of P. A. Lee and T. V. Ramakrishnan (1985), *Rev. Mod. Phys.*, **57**, 287). We will use this result in Section 5.5 to calculate the conductivity corrections due to interference effects. From Eq.(3.7.1) we can write the conductance as

$$G = \frac{\sigma L^{d-1}}{L} = \frac{2e^2}{h} \frac{\hbar^2}{L^2} \sum_{\mathbf{k}, \mathbf{k}'} v_x(\mathbf{k}') v_x(\mathbf{k}) \left| G^{\mathrm{R}}(\mathbf{k}', \mathbf{k}) \right|^2 \qquad (3.7.2)$$

We will now show that Eq.(3.7.2) follows from the Landauer formula $G = (2e^2/h)\overline{T}$, making use of Eq.(3.5.20) for the transmission function.

Although Eq.(3.5.20) was derived using a discrete lattice in real space, we can use a unitary transformation to transform it to any other convenient representation. For example in the k-representation we can write

$$G = \frac{2e^2}{h} \sum_{\mathbf{k}_1,\mathbf{k}_2,\mathbf{k},\mathbf{k'}} \Gamma_1(\mathbf{k}_1,\mathbf{k'}) G^R(\mathbf{k'},\mathbf{k}) \Gamma_2(\mathbf{k},\mathbf{k}_2) G^A(\mathbf{k}_2,\mathbf{k}_1)$$

Note that the terms in this summation with $\mathbf{k}_1 = \mathbf{k'}$ and $\mathbf{k}_2 = \mathbf{k}$

$$\Gamma_1(\mathbf{k'},\mathbf{k'}) G^R(\mathbf{k'},\mathbf{k}) \Gamma_2(\mathbf{k},\mathbf{k}) G^A(\mathbf{k},\mathbf{k'})$$

are all positive real quantities since $G^A(\mathbf{k},\mathbf{k'}) = G^R(\mathbf{k'},\mathbf{k})^*$ and the diagonal elements of Γ are real since Γ is Hermitian (see Eq.(3.6.2)). By contrast the other terms with $\mathbf{k}_1 \neq \mathbf{k}$ and $\mathbf{k}_2 \neq \mathbf{k'}$ are complex quantities with randomly varying phases that average out to zero, and can be omitted from the summation:

$$G = \frac{2e^2}{h} \sum_{\mathbf{k},\mathbf{k'}} \Gamma_1(\mathbf{k'},\mathbf{k'}) \left| G^R(\mathbf{k'},\mathbf{k}) \right|^2 \Gamma_2(\mathbf{k},\mathbf{k}) \tag{3.7.3}$$

This is identical to the result we are trying to prove (Eq.(3.7.2)) if

$$\Gamma_1(\mathbf{k},\mathbf{k}) = \Gamma_2(\mathbf{k},\mathbf{k}) = \frac{\hbar v_x(\mathbf{k})}{L} \tag{3.7.4}$$

We will now justify this expression for Γ for a uniform conductor with two identical leads.

In Section 3.6 we saw that the physical significance of the function $\Gamma_p(\mathbf{k},\mathbf{k})$ is that it represents $2\hbar$ times the rate at which an electron placed in a state k will escape into lead p. From this point of view Eq.(3.7.4) is quite reasonable. Since the electron wavefunction is spread out over a box of length L and the electron can only escape through the two surfaces with a velocity v_x, the rate of escape is given by $v_x/2L$, so that

$$\Gamma_p(\mathbf{k},\mathbf{k}) = 2\hbar \frac{v_x(\mathbf{k})}{2L} = \frac{\hbar v_x(\mathbf{k})}{L}$$

For a formal derivation of Eq.(3.7.4), consider a one-dimensional conductor with N sites as shown in Fig. 3.7.1. The only non-zero components of the self-energy involve the end-points. Assuming both leads to be identical we can write (see Eq.(3.5.21))

$$\Gamma_1(1,1) = \frac{\hbar v}{a} = \Gamma_2(N,N)$$

Now we transform from the site representation to a k-representation using the unitary transformation

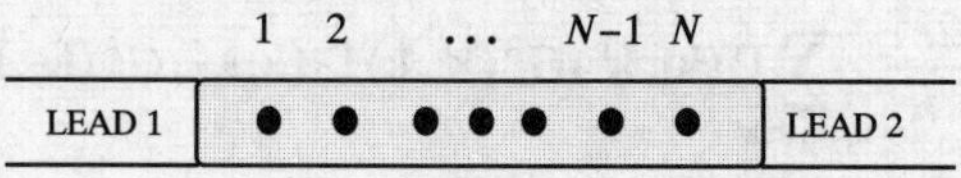

1   2   ...   N–1   N

| LEAD 1 | ● ● ● ● ● ● ● | LEAD 2 |

Fig. 3.7.1. A one-dimensional conductor with N discrete points connected to two leads.

$$|k\rangle \equiv \frac{1}{\sqrt{N}} \sum_j \exp(-ik.x_j)|j\rangle$$

Using the usual rules for matrix transformation

$$\Gamma_p(k,k) = \sum_{i,j} \langle k|i\rangle \Gamma_p(i,j)\langle j|k\rangle$$

we obtain (note $L = Na$)

$$\Gamma_1(k,k) = \frac{\hbar v}{L} = \Gamma_2(k,k)$$

It is straightforward to extend this result beyond the one-dimensional case considered here. For a conductor of uniform cross-section we could transform to a mode representation in the transverse direction and a plane wave representation in the longitudinal direction and show that

$$\Gamma_1(m,k;m,k) = \frac{\hbar v_m}{L} = \Gamma_2(m,k;m,k)$$

For wide conductors the boundaries have a negligible effect on the conductance so that we could replace the real boundary conditions in the transverse direction with simple periodic boundary conditions. The modes in the transverse direction would then be plane waves just like those in the longitudinal direction and we could replace the index (m,k) with the two-dimensional vector **k** and obtain Eq.(3.7.4).

It is interesting to note that Eq.(3.7.2) applied to a ballistic conductor yields a finite conductance equal to $(2e^2/h)M$, (see Exercise E.3.8 at the end of this chapter). As we have discussed in Chapter 2, this represents the contact resistance between the conductor and the reservoirs at the ends. Although we are not explicitly drawing these reservoirs, their presence is always implicit in our use of the Landauer formula. The usual derivations of Eq.(3.7.2) are not based on the Landauer formula and often make no explicit reference to reservoirs. Yet the influence of the reservoir creeps in through the boundary conditions giving rise to a contact resistance.

Transfer Hamiltonian formalism

The transfer Hamiltonian formalism is widely used to describe tunneling processes involving electron transfer between two leads separated by an

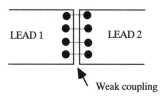

Fig. 3.7.2. Transfer Hamiltonian formalism is used widely to describe the transfer of electrons from one lead to another by tunneling through an insulator.

insulator (Fig. 3.7.2). In the transfer Hamiltonian formalism the current is related to the matrix elements, M, between lead 1 and lead 2:

$$I = \frac{4\pi e}{\hbar} \int [f_1(E) - f_2(E)] |M(E)|^2 \rho_1(E)\rho_2(E)\mathrm{d}E \qquad (3.7.5)$$

where $\rho_1(E)$ and $\rho_2(E)$ are the density of states in leads 1 and 2 respectively (see for example, Eq.(2.3.5) on p. 69 of C. J. Chen (1993), *Introduction to Scanning Tunneling Microscopy*, Oxford University Press). The change in the current in response to a change in the potential μ_2 (keeping μ_1 constant) can be written as

$$\frac{\partial I}{\partial \mu_2} = \frac{4\pi e}{\hbar} \Big[|M(E)|^2 \rho_1(E)\rho_2(E) \Big]_{E=\mu_2} \qquad (3.7.6)$$

where we have assumed low temperature (so that $f_2(E) \approx \vartheta(\mu_2 - E)$) and also neglected any change in M, ρ_1 and ρ_2 due to the applied bias.

Eq.(3.7.6) relates the slope of the current–voltage curve to the density of states in the leads. This allows one to use current–voltage measurements to deduce the density of states $\rho_2(E)$ in one lead, if the density of states $\rho_1(E)$ in the other lead is known (any energy dependence of M is commonly neglected). Indeed tunneling experiments have often been used as a probe for the density of states in unknown materials.

Could we apply Eq.(3.7.5) to our problem which involves the transfer of electrons from one lead to another through a conductor (see Fig. 3.7.3)? In the scattering formalism the current is given by (see Eq.(2.5.1)):

$$I = \frac{2e}{h} \int [f_1(E) - f_2(E)] \overline{T}(E)\mathrm{d}E \qquad (3.7.7)$$

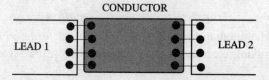

Fig. 3.7.3. Our problem involves the transfer of electrons from one lead to another via an intermediate state in the conductor.

where $\bar{T}$ is the transmission function. Eqs.(3.7.5) and (3.7.7) are consistent if the following relation is satisfied:

$$\bar{T} = 4\pi^2 \left| M \right|^2 \rho_1 \rho_2 \qquad (3.7.8)$$

We could use Eqs.(3.5.20), (3.5.22), (3.5.17b) and (3.6.3) to write the transmission function in the form $\bar{T} = \mathrm{Tr}[a_1 M a_2 M^+]$, where $M \equiv t^2 G^R$ and a_1, a_2 are the spectral functions for the isolated leads. This can be viewed as a generalized version of Eq.(3.7.8), since the spectral function is like a generalized density of states.

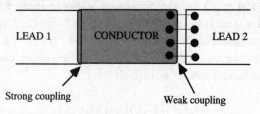

Fig. 3.7.4. If the conductor is very strongly coupled to one lead relative to the other then we can view the conductor as an extension of that lead.

However, there is an interesting situation where it may be physically meaningful to express the current in terms of the density of states. Suppose the conductor is very strongly coupled to lead '1' and only weakly to lead '2' (see Fig. 3.7.4). We could then view the conductor as a part of lead '1' that is always in equilibrium with it, regardless of the applied bias. We would then expect that the transmission function can be written in the form

$$\bar{T} = 4\pi^2 \left| M \right|^2 \rho_C \rho_2$$

where $\rho_C(E)$ is the density of states *in the conductor*. The matrix element M now represents just the coupling of lead '2' to the conductor and varies only weakly with energy. One should then be able to deduce the density of states in the conductor by measuring $\partial I/\partial\mu_2$ as discussed earlier (see Eq.(3.7.6)).

Starting from the general expression we derived for the transmission function (see Eq.(3.5.20)), we can easily obtain a simpler result that is appropriate in this limit. We simply note that

$$\Gamma \equiv \Gamma_1 + \Gamma_2 \approx \Gamma_1 \quad \text{since} \quad \Gamma_1 >> \Gamma_2$$

so that from Eq.(3.6.4) we can write

$$G^R \Gamma_1 G^A \approx A$$

Hence from Eq.(3.5.20) we can write the transmission function as

$$\bar{T}_{21} \approx \text{Tr}[\Gamma_2 A] \tag{3.7.9}$$

The quantity Γ_2 is non-zero only for points on the conductor adjacent to lead '2'. The transmission function is thus proportional to the spectral function at the edge of the conductor next to lead '2', and we could deduce this quantity from the slope of the current–voltage curve. This is in agreement with what we would have expected from the transfer Hamiltonian approach since the spectral function A is like a generalized density of states.

Eq.(3.7.9) is only valid under the special conditions depicted in Fig. 3.7.4 where the conductor is strongly coupled to lead '1' and weakly coupled to lead '2'. However, if these conditions are fulfilled, then the current–voltage curves will not be affected *even if there is strong inelastic scattering present inside the conductor*. This may seem surprising since we know that the scattering approach neglects inelastic scattering and 'vertical flow' (see Section 2.7). But the point is that even under high bias, the conductor remains essentially in equilibrium with lead '1' due to its close link to it. Since the conductor is already in equilibrium (without any scattering), inelastic scattering can do little to affect it!

3.8 Feynman paths

In Section 3.2 we have already seen how the concept of Feynman paths arises naturally when we combine the S-matrices of successive sections. However, it is difficult to extend that picture beyond a simple one-dimensional geometry. The Green's function on the other hand is like a generalized S-matrix that can be applied to arbitrarily shaped conductors using any convenient representation. In this section we will show how the Green's function can be visualized in terms of Feynman paths.

We know that the Green's function is given by the inverse of a matrix (see Eq.(3.5.17a)):

$$G^R = \left[EI - H_C - \Sigma^R \right]^{-1}$$

The basic idea is to write the matrix $[G^R]^{-1}$ as the sum of two parts

$$\left[G^R \right]^{-1} = \left[G_0 \right]^{-1} - \alpha \qquad (3.8.1)$$

one of which represents the unperturbed system (G_0^{-1}) and the other represents the perturbation (α). From Eq.(3.8.1) we can write

$$G^R = \left[G_0^{-1} - \alpha \right]^{-1} = \left[G_0^{-1} [I - G_0\alpha] \right]^{-1} = \left[I - G_0\alpha \right]^{-1} G_0$$

The inverse can be expanded in an infinite geometric series to obtain the following expression for G.

$$G^R = G_0 + G_0\alpha G_0 + G_0\alpha G_0\alpha G_0 + \ldots$$

Consider the element (i,j) of one of the terms in this series:

$$G_0(i,i)\,\alpha(i,A)\,G_0(A,A)\,\alpha(A,j)\,G_0(j,j)$$

We could visualize this term as the amplitude of a 'path' going from j to a point A and then on to i. The total Green's function $G^R(i,j)$ is equal to the sum of the amplitudes of all such paths leading from j to i (see Fig. 3.8.1):

$$G^R(i,j) = \sum_P A_P(i,j), \quad P \in \text{all paths starting in } j \text{ and ending in } i$$

Each path is composed of a series of segments which alternately involve unperturbed propagation (described by G_0) and a transition induced by the

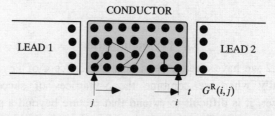

Fig. 3.8.1. Schematic representation of a Feynman path leading from a point j to another point i inside the conductor.

perturbation (described by α). The choice of G_0 and α is clearly not unique and has to be motivated by the physics of the problem at hand.

As an example, suppose we associate the unperturbed part with the diagonal elements of H_C (see Eq.(3.5.9))

$$\left[G_0^{-1}\right]_{ij} = (E - U_i - 2t)\delta_{ij}$$

then G_0 represents the Green's function for a set of isolated lattice sites:

$$\left[G_0\right]_{ij} = \frac{\delta_{ij}}{E - U_i - 2t} \tag{3.8.2a}$$

The perturbation consists of the hopping matrix element connecting adjacent sites:

$$\alpha_{ij} = t\exp\left[ie\mathbf{A}.(\mathbf{r}_i - \mathbf{r}_j)/\hbar\right], \quad \text{if } i \text{ and } j \text{ are nearest neighbors} \tag{3.8.2b}$$
$$= 0 \qquad\qquad\qquad \text{otherwise}$$

We are neglecting the effect of the leads which is represented by the self-energy term Σ^R. With this choice of G_0 and α, a Feynman path has the form shown in Fig. 3.8.1 consisting alternately of free propagation at an isolated lattice site and hopping to an adjacent site.

The expansion of the Green's function need not be carried out in the position representation as we have done here. It can be carried out in any convenient representation. For example in the k-representation the Green's function is diagonal in the absence of scattering. It is the scatterers that give rise to the off-diagonal terms. Later in Section 5.5 we will use an expansion in terms of Feynman paths in k-space to discuss the diagrammatic perturbation theory. For the moment let us illustrate the utility of this viewpoint by using Feynman paths in real space to understand the Aharonov–Bohm effect.

Aharonov–Bohm (A–B) effect

The concept of Feynman paths is often useful in understanding the physics underlying certain phenomena that would otherwise be obscured in the process of matrix inversion. The best example of this is the A–B effect. One of the pioneering experiments in mesoscopic physics was performed using a small ring 820 nm in diameter etched out of a high quality gold film (actual structure shown earlier in Fig. 0.2). Experimentally the conductance of the ring was observed to oscillate as a function of the magnetic flux enclosed by the ring (see, for example, the article by R. A. Webb and S. Washburn (1988) in *Physics Today*, **41**, 46):

$$G = G_0 + \hat{G}\cos\left(\frac{|e|BS}{\hbar} + \overline{\varphi}\right) \quad (S \equiv \text{area enclosed by ring})$$

The period B_p of the oscillations is obtained by setting the phase difference to 2π:

$$\frac{|e|B_pS}{\hbar} = 2\pi \quad \Rightarrow \quad B_pS = h/|e|$$

This is called the h/e effect since the period corresponds to a change in the enclosed flux by h/e.

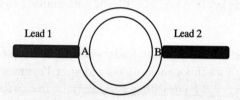

Fig. 3.8.2. A ring-shaped conductor exhibits periodic oscillations in its conductance as a function of the magnetic field. Actual measurements are made using voltage probes in a four-terminal configuration but we will not worry about it.

The oscillatory term arises from the interference between the waves traversing the two arms of the ring (see Fig. 3.8.2). The transmission from mode m in lead '1' to mode n in lead '2' can be written as

$$T(n \leftarrow m) \rightarrow |t_1 + t_2|^2$$

where t_1 is the total amplitude of all the Feynman paths going through the upper arm of the ring that start in mode m at A and end in mode n at B. Similarly t_2 is the total amplitude of all the paths going through the lower arm of the ring that start in mode m at A and end in mode n at B:

$$t_1 = \sum_{\substack{\text{all paths } P \text{ going} \\ \text{through upper arm}}} A_P \quad \text{and} \quad t_2 = \sum_{\substack{\text{all paths } P \text{ going} \\ \text{through lower arm}}} A_P$$

The oscillations in the conductance arise because a vector potential changes the phase relationship between t_1 and t_2. The conductance peaks whenever t_1 and t_2 are in phase. The effect of a vector potential **A** on t_1

and t_2 can be seen from Eq.(3.8.2b). The amplitude associated with a path connecting two nearest neighbors i and j is modified as follows:

$$\alpha_{ij} = t \exp\left[ie\mathbf{A}.(\mathbf{r}_i - \mathbf{r}_j)/\hbar\right] \quad \text{(same as Eq.(3.8.2b))}$$

Consequently the vector potential $\mathbf{A}$ introduces an extra phase to every path:

$$t_1(B) \sim t_1(0)\exp\left[i\varphi_1\right] \quad \text{and} \quad t_2(B) \sim t_2(0)\exp\left[i\varphi_2\right]$$

where $\displaystyle \varphi_1 = \exp\left(\frac{ie}{\hbar} \int_{\text{upper arm}} \mathbf{A}.\mathbf{dl}\right) \quad \varphi_2 = \exp\left(\frac{ie}{\hbar} \int_{\text{lower arm}} \mathbf{A}.\mathbf{dl}\right)$

We are assuming that the ring is thin enough that there is not much difference between the areas enclosed by the inner and the outer circumference. The difference between the two phases is proportional to the flux enclosed by the ring:

$$\varphi_1 - \varphi_2 = \frac{|e|}{\hbar} \oint \mathbf{A}.\mathbf{dl} = \frac{|e|}{\hbar} \int_S \mathbf{B}.\mathbf{dS} = \frac{|e|BS}{\hbar}$$

(B: magnetic field and S: area enclosed by ring). Hence

$$T(n \leftarrow m) = |t_1 + t_2|^2 = T_1 + T_2 + 2\sqrt{T_1 T_2}\cos\left(\frac{|e|BS}{\hbar} + \varphi\right)$$

where $T_1 = t_1 * t_1$; $T_2 = t_2 * t_2$; $\varphi = \text{Phase}(t_1 * t_2)$. Assuming $T_1 = T_2 \equiv T_0$,

$$T(n \leftarrow m) = 2T_0 + 2T_0 \cos\left(\frac{|e|BS}{\hbar} + \varphi\right) \qquad (3.8.3)$$

Thus the transmission from one mode in the left lead to one in the right lead oscillates as a function of the magnetic field. This leads to the oscillations in the conductance mentioned earlier. We will discuss this further in Section 5.4.

One small point. Actually t_1 and t_2 as defined above do not really exhaust all possible paths from m to n. Depending on the nature of the 'beam-splitters' at A and B there will be more complicated paths such as one that goes through the upper arm, transmits into the lower arm at B, gets reflected back into the lower arm at A and then exits into mode n at B. These paths would contribute to higher order oscillations that could be classified as h/Ne effects, N being an integer $(2, 3, ...)$. Experimentally the higher order effects are increasingly more difficult to observe because they involve longer path lengths and it is difficult to maintain phase coherence over the entire path. The h/Ne effect involves a path length of NL (L: length of one arm), so that its amplitude is reduced by a factor $\exp(-2NL/L_\varphi)$.

Regarding the name 'Aharonov–Bohm effect', it should be mentioned that Aharonov and Bohm originally proposed an experiment to show that observable effects could result from potentials (scalar or vector) even though the fields (electric or magnetic) were zero in the path of the electrons (see *Phys. Rev.*, **115**, 485 (1959)). The solid-state experiments which go under the name 'Aharonov–Bohm effect' (to be discussed further in Chapter 5) do not really demonstrate this since the magnetic fields used in these experiments are uniform everywhere. Their similarity to the original 'Aharonov–Bohm effect' lies in the use of a magnetic field to introduce a phase to the probability amplitudes associated with various paths.

Summary

In Section 3.1 we related the transmission function to the S-matrix and used the unitarity of the S-matrix to derive useful sum rules for the transmission function. In Section 3.2 we discussed how the S-matrix of a large conductor can be obtained by combining the S-matrices of the individual segments that comprise it, assuming complete coherence, partial coherence or complete incoherence among the segments. In Section 3.3 we introduced the concept of Green's functions. We then derived the following expression relating the S-matrix to the Green's function (Section 3.4):

$$S_{nm} = -\delta_{nm} + i\hbar\sqrt{v_n v_m} \iint \chi_n(y_q)\left[G_{qp}^R(y_q; y_p)\right]\chi_m(y_p)\,dy_q dy_p \qquad (3.4.6)$$

where the (real) function $\chi_m(y_p)$ describes the transverse profile of mode m in lead p and v_m is its velocity.

In Section 3.5 we showed how a discrete representation in real space could be used to calculate the Green's function G^R. Using this approach we showed that a conductor with infinite leads can be replaced by a finite conductor with the effects of the leads incorporated through a self-energy function Σ^R:

$$G^R = \left[EI - H_C - \Sigma^R\right]^{-1} \qquad (3.5.17a)$$

where H_C is the Hamiltonian for a finite-sized isolated conductor. The self-energy function is non-zero only for the points on the conductor that are adjacent to a lead. It can be written as

$$\Sigma^R = \sum_p \Sigma_p^R \quad \text{where} \quad \Sigma_p^R(i,j) = t^2 g_p^R(p_i, p_j) \qquad (3.5.17b)$$

where g_p^R is the Green's function for the isolated lead p given by

$$g_p^R(p_i, p_j) = -\frac{1}{t}\sum_m \chi_m(p_i)\exp[ik_m a]\chi_m(p_j) \qquad (3.5.18)$$

Finally we derived an expression for the transmission function in terms of the Green's function and the self-energies:

$$\overline{T}_{qp} = \text{Tr}\left[\Gamma_q G^R \Gamma_p G^A\right] \qquad (3.5.20)$$

$$\Gamma_p(i,j) \equiv i\left[\Sigma_p^R(i,j) - \Sigma_p^A(i,j)\right] = \sum_m \chi_m(p_i)\frac{\hbar v_m}{a}\chi_m(p_j) \qquad (3.5.21)$$

All quantities in the above equations are defined at points inside the conductor. They represent matrices of dimensions $(C \times C)$ if there are N lattice sites inside the conductor. The only exception is the function $g_p^R(p_i, p_j)$ which represents the Green's function for the isolated lead p, p_i

CONDUCTOR

LEADp LEADq

p_i i j q_j

A conductor with two or more leads p and q connected to it. A point in lead p is labeled p_i if it is adjacent to point i inside the conductor.

and p_j being points on this lead that are adjacent to points i and j inside the conductor as shown in the above figure.

This formulation is practically convenient because the infinite leads have been entirely eliminated. However, it also serves to introduce the very important concepts of 'self-energy' and spectral function which are widely used to describe the interaction with phonons and other electrons (Section 3.7). It also provides a convenient starting point for relating to other approaches that are widely used in the literature, namely the Kubo formalism and the transfer Hamiltonian formalism (Section 3.8). Finally we ended this chapter by discussing the concept of 'Feynman paths' which often provides useful physical insight into the factors that influence the Green's function and hence the S-matrix and the transmission function.

Exercises

E.3.1

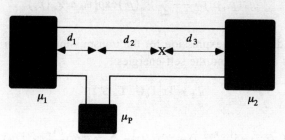

Fig. E.3.1. Single-moded waveguide with one scatterer as in Fig. 2.3.1, but with a probe attached to measure the potential before the scatterer.

In Section 2.3 we discussed the potential variation inside a conductor with a single scatterer. Instead one could ask the question: what is the potential measured by an external probe connected as shown in Fig. E.3.1 (see H. L. Engquist and P. W. Anderson, (1981), *Phys. Rev. B*, **24**, 1151 and M. Büttiker (1989), *Phys. Rev. B*, **40**, 3409)?

Assume that the scatterer is described by the (2×2) scattering matrix

$$\begin{bmatrix} i\sqrt{1-T} & \sqrt{T} \\ \sqrt{T} & i\sqrt{1-T} \end{bmatrix}$$

while the probe is described by the (3×3) scattering matrix

$$\begin{bmatrix} a & b & \sqrt{\varepsilon} \\ b & a & \sqrt{\varepsilon} \\ \sqrt{\varepsilon} & \sqrt{\varepsilon} & c \end{bmatrix} \quad \text{where} \quad c = \sqrt{1-2\varepsilon} \quad a = (1-c)/2 \quad b = (1+c)/2$$

This scattering matrix does not include the phase-shifts kd_1 and kd_2 due to propagation.

(a) Use Eq.(2.4.4) to show that the probe potential μ_P is given by

$$\mu_P = \frac{T_{P1}\mu_1 + T_{P2}\mu_2}{T_{P1} + T_{P2}}$$

in order that the probe current be zero.

(b) Sum the 'Feynman paths' as described in Section 3.2 to obtain the transmission probabilities T_{P1} and T_{P2}, with and without interference effects.

(c) Calculate the probe potential μ_P with and without interference effects, setting $\mu_1 = 1$ and $\mu_2 = 0$, assuming weak coupling ($\varepsilon \to 0$). How does this compare with what we have shown in Fig. 2.3.1?

E.3.2 Derive Eq.(3.3.15) for the Green's function of a semi-infinite wire using the eigenfunction expansion (Eq.(3.3.17)).

E.3.3 Show that the Green's function for a semi-infinite lead p can be written as

$$g_p^R(p_i, p_j) = -\frac{1}{t} \sum_m \chi_m(p_i) \exp[+ik_m a] \chi_m(p_j) \qquad (3.5.18)$$

where the index m represents the propagating modes in lead p.

E.3.4 Starting from the expression for the S-matrix elements

$$S_{nm} = -\delta_{nm} + \frac{i\hbar\sqrt{v_n v_m}}{a} \iint \chi_n(y_q) \left[G_{qp}^R(y_q; y_p) \right] \chi_m(y_p) dy_q dy_p$$

derive the expression for the transmission probability:

$$\overline{T}_{qp} = \mathrm{Tr}\left[\Gamma_q\, G^R\, \Gamma_p\, G^A \right] \qquad (3.5.20)$$

E.3.5 Consider a single-moded wire having a single scatterer with a scattering potential

$$U(x) = U_0\, \delta(x)$$

(a) Solve the Schrödinger equation directly to show that the S-matrix elements are given by

$$S_{11} = S_{22} = \frac{U_0}{i\hbar v - U_0} \quad \text{and} \quad S_{12} = S_{21} = \frac{i\hbar v}{i\hbar v - U_0}$$

where the velocity v ($= \hbar k/m = (2mE)^{1/2}/\hbar$) is the same in both leads, since the potential is assumed to be the same.

(b) Use the formulation developed in this chapter (namely, Eq.(3.5.17)) to calculate the Green's function and then use the Fisher–Lee relation

(Eq.(3.4.3)) to obtain the S-matrix elements. Compare with the results in part (a).

E.3.6 Consider a single-moded wire having two scatterers separated by a distance d such that the scattering potential can be written as

$$U(x) = U_0\big[\delta(x) + \delta(x - d)\big]$$

(a) Combine the individual S-matrices of the two scatterers (calculated in the last problem) to show that the transmission probability in this problem is given by

$$T(E) = \frac{T_1^2}{1 - 2R_1 \cos\theta + R_1^2}$$

where

$$T_1 = \frac{\hbar^2 v^2}{\hbar^2 v^2 + U_0^2} \qquad R_1 = \frac{U_0^2}{\hbar^2 v^2 + U_0^2}$$

and

$$\theta = 2\big[kd + \tan^{-1}(\hbar v/U_0)\big]$$

Plot this result in the energy range $0 < E < 250$ meV, assuming $U_0 = 9$ eV A and $d = 50$ Å.

(b) Use Eq.(3.5.20) to calculate the transmission numerically as the lattice constant a is reduced so that the number of lattice points in the region between the scatterers increases from 1 to 15. How does the result compare with the result in part (a)?

E.3.7 Consider a single-moded ring as shown in the figure.

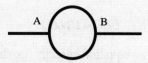

Fig. E.3.7. Single-moded ring connected to two leads.

(a) Assume that the 3-way junction at each end is described by an S-matrix of the form

$$\begin{bmatrix} c & \sqrt{\varepsilon} & \sqrt{\varepsilon} \\ \sqrt{\varepsilon} & a & b \\ \sqrt{\varepsilon} & b & a \end{bmatrix}$$

where a, b, c and ε are all real numbers. Show that in order to ensure the unitarity of the S-matrix the following relations must be satisfied:

$$c = \pm\sqrt{1-2\varepsilon} \quad a = (1-c)/2 \quad b = (1+c)/2$$

Thus the entire S-matrix can be specified by a single parameter ε, which determines the strength of the coupling of the leads to the ring $(0 < \varepsilon < 0.5)$.

(b) Combine S-matrices to show that the transmission probability from one lead to the other is given by

$$T = \frac{4\varepsilon^2}{1 - 2c^2 \cos(2\theta) + c^4}$$

where θ is the phase shift along one of the arms of the ring (assumed to be the same for both arms): $\theta = \sqrt{2mE}\,(\pi r/\hbar)$, where r is the radius of the ring. Plot the transmission as a function of the energy over the range $0 < E < 0.5$ meV for $\varepsilon = 0.025$, assuming $r = 1000$ Å.

(c) Use Eq.(3.5.20) to calculate the transmission numerically, taking approximately 140 lattice points along the ring (corresponding to a lattice constant $a \sim 45$ Å) and compare with the result in part (b).

(d) Plot the transmission as a function of the magnetic field $(0 < B < 0.3$ T) at an energy corresponding to a peak in part (c).

E.3.8 Starting from Eq.(3.7.2) (rewritten in terms of transverse modes)

$$G = \frac{2e^2}{h}\frac{\hbar^2}{L^2} \sum_{m,n,k,k'} v_m v_n \left| G^R(m,k;n,k') \right|^2$$

show that the conductance of a ballistic conductor is equal to $(2e^2/h)M$.

E.3.9 Reciprocity: Starting from the definition of the Green's function (Eq.(3.5.3)) show that

$$\left[G^R(r,r') \right]_{-B} = \left[G^R(r',r) \right]_{+B}$$

Hence use the Fisher–Lee relation (Eq.(3.4.3)) to prove the reciprocity relation for the S-matrix elements (Eq.(3.1.7)).

Further reading

A very large number of papers have appeared in the literature that calculate the transmission properties of small conductors. We have cited a few papers in the text that should provide a citation trail for the interested reader. The very brief introduction to Green's functions in Section 3.3 can be supplemented with standard texts such as

[3.1] Inkson, J. C. (1984), *Many-body Theory of Solids*, (New York, Plenum). See Chapter 2.

[3.2] Economou, E. N. (1983), *Green's Functions in Quantum Physics*, Springer Series in Solid-state Sciences, vol.7, (Heidelberg, Springer–Verlag).

[3.3] Schiff L. I. (1968). *Quantum Mechanics*, Chapter 9, Third Edition, (New York, McGraw–Hill).

4

Quantum Hall effect

4.1 Origin of 'zero' resistance
4.2 Effect of backscattering

One of the most significant discoveries of the 1980s is the quantum Hall effect (see K. von Klitzing, G. Dorda and M. Pepper (1980), *Phys. Rev. Lett.*, **45**, 494). Normally in solid state experiments, scattering processes introduce enough uncertainty that most results have an 'error bar' of plus or minus several per cent. For example, the conductance of a ballistic conductor has been shown (see Fig. 2.1.2) to be quantized in units of $(h/2e^2)$. But this is true as long as we are not bothered by deviations of a few per cent, since real conductors are usually not precisely ballistic. On the other hand, at high magnetic fields the Hall resistance has been observed to be quantized in units of $(h/2e^2)$ with an accuracy that is specified in parts per million. Indeed the accuracy of the quantum Hall effect is so impressive that the National Institute of Standards and Technology is interested in utilizing it as a resistance standard.

This impressive accuracy arises from the near complete suppression of momentum relaxation processes in the quantum Hall regime resulting in a truly ballistic conductor of incredibly high quality. Mean free paths of several millimeters have been observed. These unusually long mean free paths do not arise from any unusual purity of the samples. They arise because, at high magnetic fields, the electronic states carrying current in one direction are localized on one side of the sample while those carrying current in the other direction are localized on the other side of the sample. Due to the formation of this 'divided highway' there is hardly any overlap between the two groups of states and backscattering cannot take place even though impurities are present.

We will start out in Section 4.1 with a general discussion of the factors that lead to zero longitudinal resistance and consequently the quantization of the Hall resistance. We then discuss (Section 4.2) some of the surprising experiments in the quantum Hall regime reported in the late 1980s where controlled amounts of backscattering are deliberately introduced.

4.1 Origin of 'zero' resistance

We know that at high magnetic fields the longitudinal resistance (measured using a macroscopic Hall bridge) oscillates as a function of the magnetic field (see Fig. 1.4.2). As we discussed in Section 1.5 the density of states at high magnetic fields develops sharp peaks spaced by $\hbar\omega_c$ (see Fig. 1.5.1) and the resistivity oscillates as the position of these peaks is changed relative to the Fermi energy. This can be done either by changing the magnetic field as shown in Fig. 1.4.2 or by keeping the magnetic field fixed and changing the electron density (and hence the Fermi energy) by means of a gate voltage. Indeed the first experiment reporting the quantum Hall effect was performed on a silicon inversion layer as a function of the gate voltage at a magnetic field of $B = 18$ T.

Intuitively it might appear that the resistance should be a minimum whenever the Fermi energy coincides with a peak in the density of states, that is, with a Landau level. However, the correct answer is just the opposite. The resistance is a minimum when the Fermi energy lies between two Landau levels so that the density of states at the Fermi energy is a minimum! But how does a sample carry any current unless there are states at the Fermi energy? The answer is that there are states at the Fermi energy which are located near the edges of the sample. Normally in wide conductors we tend to ignore the edges since they form an insignificant fraction of the entire conductor. But these *edge states* play a very important role in carrying the current at the resistance minimum as discussed by several authors (see, for example, B. I. Halperin (1982), *Phys. Rev. B*, **25**, 2185, and A. H. MacDonald and P. Streda (1984), *Phys. Rev. B*, **29**, 1616). This is reminiscent of the boundary problems encountered in calculating the diamagnetism of a free electron gas (see Section 4.3 of R. Peierls (1979), *Surprises in Theoretical Physics*, Princeton University Press).

An important point to note from Fig. 1.4.2 is that at the minima the resistance is very nearly zero. This is particularly surprising since the voltage probes are typically spaced by hundreds of microns. The fact that the resistance is so close to zero shows that the electrons are able to travel such huge distances without losing their momentum. Clearly something rather special must be happening at the microscopic level leading to this fantastic suppression of momentum relaxation processes.

We have already seen in Section 1.6 that as we increase the magnetic field in a finite-width conductor, the states carrying current in one direction get spatially separated from the states carrying current in the opposite direction. The result is a significant reduction in the spatial overlap between the forward and the backward propagating states which leads to a suppression of backscattering (and hence momentum relaxation). Indeed at high magnetic fields the forward and backward propagating states are spatially separated by the width of the conductor and thus have practically zero overlap in wide conductors.

In Section 1.6 we assumed a parabolic confining potential $U(y) = m\omega_0^2 y^2/2$. This allowed us to obtain the eigenfunctions of the Schrödinger equation (see Eq.(1.2.2))

$$\left[E_{\mathrm{s}} + \frac{(i\hbar\nabla + e\mathbf{A})^2}{2m} + U(y) \right] \Psi(x,y) = E\Psi(x,y) \qquad (4.1.1)$$

analytically. A parabolic potential often provides a good description of narrow quantum wires. But for wide conductors, the transverse confining potential usually looks more like that shown in Fig. 4.1.1a. In general, analytical solutions are not available for arbitrary confining potentials. However, there is an approximate solution that we can use at high magnetic fields. It is quite accurate if the cyclotron radius is small enough that the confining potential can be assumed to be nearly constant on this scale. Let us start by deriving this approximate result.

Magneto-electric subbands at high magnetic fields

We know that if the confining potential were absent ($U(y) = 0$) then the solutions to Eq.(4.1.1) would be given by (see Eqs.(1.6.8a,b))

$$\Psi_{n,k}(x,y) = \frac{1}{\sqrt{L}} \exp[ikx] u_n(q - q_k) \equiv |n,k\rangle \qquad (4.1.2)$$

$$E(n,k) = E_{\mathrm{s}} + \left(n + \tfrac{1}{2}\right)\hbar\omega_{\mathrm{c}}, \quad n = 0,1,2,\ldots$$

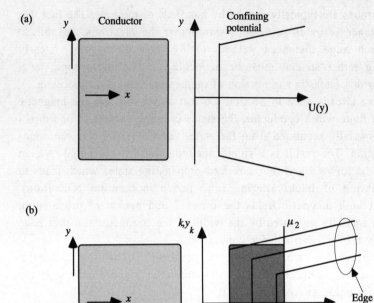

Fig. 4.1.1. A rectangular conductor assumed to be uniform in the *x*-direction. (*a*) Sketch of confining potential $U(y)$ versus *y*. (*b*) Sketch of the approximate dispersion relation assuming that the confining potential varies slowly over a cyclotron radius.

where
$$u_n(q) = \exp\left[-q^2/2\right] H_n(q)$$

$$q = \sqrt{m\omega_c/\hbar}\, y \quad \text{and} \quad q_k = \sqrt{m\omega_c/\hbar}\, y_k$$

$$y_k \equiv \frac{\hbar k}{eB} \quad \text{and} \quad \omega_c \equiv \frac{|e|B}{m}$$

$H_n(q)$ is the *n*th Hermite polynomial.

Now let us use lowest order perturbation theory to include the effect of the confining potential $U(y)$:

$$E(n,k) \approx E_s + \left(n + \tfrac{1}{2}\right)\hbar\omega_c + \langle n,k|U(y)|n,k\rangle$$

Note that each state (n,k) is centered around a different location $y = y_k$ in the transverse direction and has a spatial extent of $\sim (\hbar/m\omega_c)^{1/2}$. Assuming that the potential $U(y)$ is nearly constant over the extent of each state, we can write

$$E(n,k) \approx E_s + \left(n + \tfrac{1}{2}\right)\hbar\omega_c + U(y_k) \quad \text{where} \quad y_k = \hbar k/eB \qquad (4.1.3)$$

Figure 4.1.1b shows a sketch of the dispersion relation $E(n,k)$ vs. k. It looks just like the confining potential $U(y)$, with the coordinate y mapped onto the wavenumber k by the relation $y_k = \hbar k/eB$. In the middle of the sample the states look just like the Landau levels of an unconfined 2-D conductor spaced by $\hbar\omega_c$. Near the edges there are allowed states with a continuous distribution of energies. These are referred to as the edge states and they play a very important role in carrying the current at the resistance minimum.

What is the current carried by an edge state? From Eq.(4.1.3) we can calculate the velocity:

$$v(n,k) = \frac{1}{\hbar} \frac{\partial E(n,k)}{\partial k} = \frac{1}{\hbar} \frac{\partial U(y_k)}{\partial k} = \frac{1}{\hbar} \frac{\partial U(y)}{\partial y} \frac{\partial y_k}{\partial k} = \frac{1}{eB} \frac{\partial U(y)}{\partial y}$$

The edge states located at the two edges of the sample carry currents in opposite directions, since the quantity $\partial U(y)/\partial y$ changes sign. The bulk states too could carry current if there are electric fields in the interior of the sample due to, say, the Hall voltage. If $\mu_1 > \mu_2$ (as shown in Fig. 4.1.1) then the states below μ_2 are all filled (assuming 'zero' temperature) and essentially in equilibrium, so that they do not carry any net current. Any net current arises from the filled states between μ_1 and μ_2 (see Fig. 4.1.1b). The resistance of the sample is determined by the rate at which the electrons in these states can relax their momentum.

The situation is quite similar to that in an ordinary conductor carrying current. The positive k-states are occupied to a higher quasi-Fermi level than the negative k-states (see Fig. 1.7.2). The resistance at low temperatures is determined by the momentum relaxation time of the excess carriers in the positive k-states. What is unusual here is that the states carrying current in one direction are spatially separated from those carrying current in the opposite direction. To relax momentum an electron has to be scattered from the *left* of the sample to the *right* of the sample. This is all but impossible since the overlap between the wavefunctions is exponentially small and there are no allowed states in the interior of the sample in this energy range ($\mu_1 > E > \mu_2$).

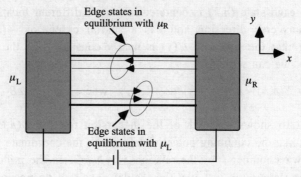

Fig. 4.1.2. A conductor in the quantum Hall regime. The edge states (two shown in the figure) carrying current to the right are in equilibrium with the left contact while those carrying current to the left are in equilibrium with the right contact.

As a result of this complete suppression of backscattering, electrons originating in the left contact enter the edge states carrying current to the right and empty into the right contact, while electrons in the right contact enter the edge states carrying current to the left and empty out into the left contact. Consequently, the edge states carrying current to the right are completely in equilibrium with the left contact and have a quasi-Fermi energy equal to μ_L. They are unaffected by μ_R since no electron originating in the right contact ever makes it to these states. Similarly we can argue that the edge states carrying current to the left all originate from the right contact and have a quasi-Fermi energy equal to μ_R (see Fig. 4.1.2):

$$\mu_1 = \mu_L \quad \text{and} \quad \mu_2 = \mu_R$$

Clearly the longitudinal voltage drop V_L as measured by two voltage probes located anywhere on the *same* side of the sample is zero, while the transverse (or Hall) voltage V_H measured by two probes located any-where on *opposite* sides of the sample is equal to the applied voltage:

$$V_L = 0 \quad \text{and} \quad eV_H = \mu_L - \mu_R \tag{4.1.4a}$$

Note that this situation arises only when the electrochemical potentials lie between two bulk Landau levels. If the electrochemical potentials lie on a bulk Landau level then there is a continuous distribution of allowed states from one edge to the other. Electrons can scatter from the left of the sample to the right of the sample through the allowed energy states in the interior of the sample. This backscattering gives rise to a maximum in the longitudinal resistance every time the Fermi energy lies on a bulk Landau level.

What is the current?

The current can be written down very simply by noting that the situation is very similar to what we had argued in Chapter 2 for a ballistic conductor (see Fig. 2.1.1). The number of edge states (which is equal to the number of filled Landau levels in the bulk) plays the role played by the number of modes in a ballistic conductor so that we can write (cf. Eq.(2.1.3))

$$I = \frac{2e}{h} M(\mu_L - \mu_R) \tag{4.1.4b}$$

We could derive this formally as follows (k_L and k_R are the wavenumbers corresponding to $E = \mu_L$ and $E = \mu_R$ respectively)

$$I = 2e \sum_n \int_{\mu_R}^{\mu_L} \frac{1}{2\pi} v(n,k) dk = 2e \sum_n \int_{\mu_R}^{\mu_L} \frac{1}{2\pi} \frac{1}{\hbar} \frac{\partial E(n,k)}{\partial k} dk$$

$$= \frac{2e}{h} \sum_n \int_{\mu_R}^{\mu_L} dE = \frac{2e}{h} M[\mu_L - \mu_R]$$

Hence from Eqs.(4.1.4a, b) we can write down the longitudinal and Hall resistances:

$$R_L = \frac{V_L}{I} = 0 \quad \text{and} \quad R_H = \frac{V_H}{I} = \frac{h}{2e^2 M} \tag{4.1.5}$$

Note that a two-terminal resistance measurement would yield the Hall resistance. Only a four-terminal measurement with voltage probes located on the same side of the sample yields zero resistance.

Thus whenever the Fermi energy lies between two bulk Landau levels, the longitudinal resistance is very nearly zero and corresponding to the zeros in the longitudinal resistance, there appear plateaus in the Hall resistance (see Fig. 1.4.2). At these plateaus the Hall resistance has the value

$$R_H = \frac{h}{2e^2 M} = \frac{25.8128 \, k\Omega}{2M}$$

where M = number of edge states at the Fermi energy = number of bulk Landau levels below the Fermi energy. M takes on integer values that decrease as the magnetic field is increased. What is impressive is the striking accuracy (better than one part per million) of this quantization of R_H that is obtained at high magnetic fields. This phenomenon is known as the quantum Hall effect (or QHE) and was discovered in 1980. It is characteristic of 2-D semiconducting films and is not observed in bulk materials.

Note that the quantized Hall resistance has the same form as the quantized resistance of ballistic conductors (see Section 2.1) with the number of edge states playing the role of the number of modes. In ordinary ballistic conductors the quantization is not very precise because backscattering processes are not completely eliminated. But in the quantum Hall regime we have a ballistic conductor of incredibly high quality due to the spatial separation of the forward and the backward propagating states. As a result the quantization is extremely precise.

Application of the Büttiker formula

So far we have not worried explicitly about the voltage probes used to measure the longitudinal or the transverse voltage drops. We have assumed that such probes would measure the local quasi-Fermi energy for the corresponding edge states. The Büttiker formula discussed in Section 2.4 (see Eq.(2.4.4) or (2.5.8)) provides a natural framework for the analysis of multi-terminal conductors taking the probes explicitly into account. Both the zero longitudinal resistance and the quantized Hall resistance follow readily from the Büttiker formula, if we postulate that electrons can travel from one terminal to another without scattering. The transmission functions can be written down by inspection without any messy calculations of the type discussed in the last chapter. We will assume the bias and temperature to be low enough that the transmission function is essentially constant over the energy range where transport occurs. This allows us to use the linear response formula (Eq.(2.5.8)) without worrying about vertical flow (see Section 2.7).

Consider an ordinary macroscopic Hall bridge hundreds of microns in length and in width. We assume that electrons can travel from one terminal to another without momentum relaxation due to the formation of edge states. Since there is no backscattering, the transmission function $\bar{T}_{pq}$ is very easy to evaluate. We just have to count the number of current carrying channels that start from terminal p and end in terminal q. For the Hall bridge depicted in Fig. 4.1.3 having M (= 2 shown in the figure) edge states that carry current around the sample, it is evident that $\bar{T}_{pq} = M$

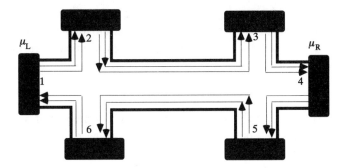

Fig. 4.1.3. Hall bridge at high magnetic fields showing two edge states at each edge.

only if $(p \leftarrow q)$ is equal to $(1 \leftarrow 6)$, $(2 \leftarrow 1)$, $(3 \leftarrow 2)$, $(4 \leftarrow 3)$, $(5 \leftarrow 4)$ or $(6 \leftarrow 5)$. All other transmission coefficients are zero. Neglecting any backscattering we can write down the conductance matrix (which is proportional to the transmission function) by inspection:

G_{pq}:	$q = 1$	$q = 2$	$q = 3$	$q = 4$	$q = 5$	$q = 6$
$p = 1$	0	0	0	0	0	G_C
$p = 2$	G_C	0	0	0	0	0
$p = 3$	0	G_C	0	0	0	0
$p = 4$	0	0	G_C	0	0	0
$p = 5$	0	0	0	G_C	0	0
$p = 6$	0	0	0	0	G_C	0

where

$$G_C = \frac{2e^2 M}{h}$$

We can solve for the terminal currents and voltages starting from Eq.(2.5.8), which yields a system of six equations. As explained earlier (see discussion preceding Eq.(2.4.6)) these equations are not independent and we can choose the voltage at one of the terminals to be zero and omit the row and column corresponding to that terminal. Setting $V_4 = 0$,

$$
\begin{Bmatrix} I_1 \\ I_2 \\ I_3 \\ I_5 \\ I_6 \end{Bmatrix} =
\begin{bmatrix} G_C & 0 & 0 & 0 & -G_C \\ -G_C & G_C & 0 & 0 & 0 \\ 0 & -G_C & G_C & 0 & 0 \\ 0 & 0 & 0 & G_C & 0 \\ 0 & 0 & 0 & -G_C & G_C \end{bmatrix}
\begin{Bmatrix} V_1 \\ V_2 \\ V_3 \\ V_5 \\ V_6 \end{Bmatrix}
$$

We could invert this matrix as we have done in the past, but it is unnecessary. We can easily write down the solution to the above set of equations noting that the currents at the voltage terminals are all zero $(I_2 = I_3 = I_5 = I_6 = 0)$:

$$V_2 = V_3 = V_1, \quad V_5 = V_6 = 0$$

This is of course precisely what we had assumed, namely, that any voltage probe on one side floats to a potential equal to the right contact while any probe on the other side floats to a potential equal to the left contact. Also the current is given by

$$I_1 = G_C V_1$$

so that the longitudinal resistance R_L measured between probes 2 and 3 or between 5 and 6 is zero

$$R_L = \frac{V_2 - V_3}{I_1} = \frac{V_6 - V_5}{I_1} = 0$$

while the Hall resistance R_H measured between probes 2 and 6 or between 3 and 5 has the quantized value stated earlier (see Eq.(4.1.5)).

$$R_H = \frac{V_2 - V_6}{I_1} = \frac{V_3 - V_5}{I_1} = G_C$$

Does the current flow only at the edges?

We stated above that if $\mu_1 > \mu_2$ (as shown in Fig. 4.1.1) then the states below μ_2 are all filled and do not carry any net current. Any net current can be calculated from the filled states on the left between μ_1 and μ_2. However, this does not mean that current flows only near the edge having the potential μ_1. There are currents everywhere in the sample. We could choose to do our bookkeeping in a different way so that the net current appears at a different spatial location. We have identified all the states below μ_2 (note that $\mu_1 > \mu_2$) as our Fermi sea which does not carry any net current. Consequently the net current is carried by electrons in the edge states on one side of the sample with energies lying in the range $\mu_1 > E > \mu_2$. But we could just as well identify all the states below μ_1 as our Fermi sea. The net current would then be carried by the *holes* occupying the edge states on the *other side* of the sample!

We mentioned in Section 1.7 that while the *conductance* at low temperatures is a 'Fermi surface property' the *conductivity* may not be. A quantum Hall conductor provides a very good example of this. At low temperatures the current is carried by the states near the Fermi energy. States lying deep inside the Fermi sea have no effect on the conductance. But the local current density due to these states is not zero. These states give rise to circulating currents in the sample even at equilibrium. An applied electric field can induce a change in this circulating current flow pattern thus contributing to the conductivity tensor defined by the relation $\delta \mathbf{J} = \sigma \delta \mathbf{E}$. Thus electrons deep inside the Fermi sea can contribute to the conductivity, even though they do not contribute to the conductance.

Why should the Fermi energy ever lie between Landau levels?

The above discussion shows that the longitudinal resistance can be extremely small if the electrochemical potentials μ_1 and μ_2 were located between bulk Landau levels as shown in Fig. 4.1.1. This requires that the equilibrium Fermi energy E_f must be located between the Landau levels since at low bias $\mu_1 \sim \mu_2 \sim E_f$. How is the location of E_f determined?

At low temperatures we can write

$$n_s = \int_{-\infty}^{E_f} N_s(E, B) \mathrm{d}E$$

where n_s is the electron density and N_s is the density of states. The electron density increases with the Fermi energy as shown in Fig. 4.1.4. The important point to note is that the electron density increases rapidly whenever the density of states is high. This is because a change in the electron density is related to the change in the Fermi energy by the relation

$$\delta n_s = N_s(E_f, B) \delta E_f$$

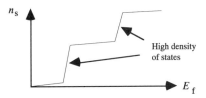

Fig. 4.1.4. Electron density vs. Fermi energy. Note that the electron density changes rapidly whenever the density of states is high.

As a result it is extremely unlikely for the Fermi energy to be located in a region where the density of states N_s is very small. A slight change in the electron density would cause a large shift in the Fermi energy. The Fermi energy thus tends to be *pinned* to energies where the density of states is high.

From this point of view we would expect the the Fermi energy to be pinned to one Landau level or another, where the density of states is high. If this were true then the low resistance condition discussed earlier would never be observed since the Fermi energy would never be located between two Landau levels. This is not a problem in narrow conductors where the edge states provide a significant density of states between the Landau levels. But in wide conductors the edge states represent a negligible fraction of the total density of states. How can the Fermi energy in a wide conductor ever lie between the bulk Landau levels, leading to the low resistance condition that we have been discussing?

It is believed that in practice the density of states between two Landau levels is quite significant because real samples have potential fluctuations leading to the formation of localized states (see for example the introductory article in Ref.[4.3] by Prange). This can be understood by noting that potential fluctuations in the interior of the sample lead to the formation of local equipotential contours that close on themselves as shown in Fig. 4.1.5a. Since cyclotron orbits drift along equipotential contours they get stuck at these spots forming localized states. These states do not contribute to the current flow but they help stabilize the Fermi energy between Landau levels by providing a respectable density of states between Landau levels as sketched in Fig. 4.1.5b.

Fractional quantum Hall effect

We have seen that in the quantum Hall regime the Hall resistance takes on quantized values given by

$$\rho_{yx} = \frac{h}{2e^2 M} = \frac{25.8128\,\text{k}\Omega}{2M}$$

where M is an integer. Actually at high fields the energy levels for the two spins split apart due to the Zeeman effect and quantized plateaus are obtained with

$$\rho_{yx} = \frac{h}{e^2 M} = \frac{25.8128\,\text{k}\Omega}{M}$$

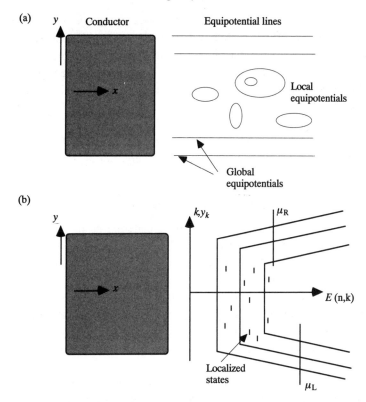

Fig. 4.1.5. (*a*) Potential fluctuations in the interior of the sample lead to local equipotentials where cyclotron orbits get stuck forming localized states. (*b*) These localized states help stabilize the Fermi energy between Landau levels.

When the magnetic field reaches a value such that the electron density $n_s = eB/h$, we will have all the electrons in a single Landau level with one spin. For a carrier density of $n_s = 2 \times 10^{11}/\text{cm}^2$, this requires a field of about 8 T. What happens if we increase the field further?

From our earlier discussion we might expect that there will be no further plateaus with increasing magnetic field since the Fermi energy now lies in the middle of a Landau level (the last!). Experimentally, however, in very pure samples one continues to observe plateaus in the Hall resistivity given by

$$\rho_{yx} = \frac{h}{e^2 p} = \frac{25.8128\,\text{k}\Omega}{p}$$

where p is a rational fraction like $\frac{1}{3}, \frac{2}{5}, \frac{4}{7}$ etc. This is referred to as the fractional quantum Hall effect (or FQHE) to distinguish it from the integral quantum Hall effect (or IQHE) that we have been discussing. The FQHE arises from the formation of a novel many-body ground state (see R. B. Laughlin (1983), *Phys. Rev. Lett.* **50**, 1395) whose quasiparticle excitations are very different from what we expect from the simple one-particle picture that we have been using to describe the IQHE. We refer the reader to the references cited at the end of this chapter and also to the book by T. Chakraborty and P. Pietilainen (1988), *The Fractional Quantum Hall Effect*, (New York, Berlin, Heidelberg, Springer-Verlag).

4.2 Effect of backscattering

So far we have assumed that there is no backscattering so that each edge state has a transmission probability of 100%. Once we make this assumption, zero longitudinal resistance and the quantized Hall resistance follow naturally from the Landauer–Büttiker formalism. However, the real power of this formalism lies in providing a clear description of the many experiments in the quantum Hall regime reported in the late 1980s where controlled amounts of backscattering are deliberately introduced. Indeed this is one of the most elegant applications of the Landauer–Büttiker formalism.

Suppose a split gate is used to pinch off the Hall bar (between probes 2 and 3, see Fig. 4.2.1) so that only N ($N < M$) edge channels can propagate through the constriction, then the remaining $(M - N)$ channels will be completely backscattered. The net current from left to right is given by

$$I_1 = \frac{2e}{h} N(\mu_L - \mu_R) = \frac{2e^2}{h} N V_1 \quad \text{(setting } \mu_L = eV_1 \text{ and } \mu_R = 0)$$

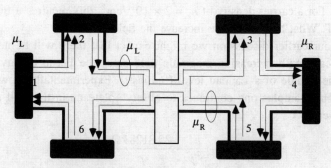

Fig. 4.2.1. Hall bridge with split gate structure used to backscatter one edge channel while the other can transmit.

We can write the current as

$$I_1 = \frac{2e^2}{h} MV_1(1-p)$$

where $\qquad p \equiv \dfrac{M-N}{M} = \dfrac{\text{No. of backscattered channels}}{\text{Total no. of channels}}$

The contact 2 'sees' only the channels originating from the left having a potential μ_L, while the contact 5 'sees' only the channels originating from the right having a potential μ_R.

$$\mu_2 = eV_1 \quad \text{and} \quad \mu_5 = 0$$

The contact 6 'sees' $(M-N)$ channels that originate from the left and have a potential μ_L and N channels that originate from the right and have a potential μ_R. Consequently it floats to a potential of

$$\mu_6 = \frac{(M-N)\mu_L + N\mu_R}{M} = eV_1 p$$

Similarly the potential at contact 3 is given by

$$\mu_3 = \frac{N\mu_L + (M-N)\mu_R}{M} = eV_1(1-p)$$

Hence the longitudinal resistance R_L measured between probes 2 and 3 or between 5 and 6 is given by

$$R_L = \frac{eV_1 p}{I_1} = \frac{h}{2e^2 M}\left[\frac{p}{1-p}\right] = \frac{h}{2e^2}\left[\frac{1}{N} - \frac{1}{M}\right] \qquad (4.2.1)$$

This 'fractional quantization' of the longitudinal resistance has been observed experimentally. The Hall resistance R_H measured between probes 2 and 6 or between 3 and 5 is unchanged from its usual quantized value:

$$R_H = \frac{eV_1(1-p)}{I_1} = \frac{h}{2e^2 M} \qquad (4.2.2)$$

Application of the Büttiker formula

The above results follow quite readily from the Büttiker formula which takes the voltage probes explicitly into account. The conductance matrix can be written as

G_{pq}:	$q=1$	$q=2$	$q=3$	$q=4$	$q=5$	$q=6$
$p=1$	0	0	0	0	0	G_C
$p=2$	G_C	0	0	0	0	0
$p=3$	0	$(1-p)G_C$	0	0	pG_C	0
$p=4$	0	0	G_C	0	0	0
$p=5$	0	0	0	G_C	0	0
$p=6$	0	pG_C	0	0	$(1-p)G_C$	0

Hence from Eq.(2.5.8) (setting $V_4 = 0$ and leaving out the rows and columns corresponding to terminal 4 as we did before)

$$\begin{Bmatrix} I_1 \\ I_2 \\ I_3 \\ I_5 \\ I_6 \end{Bmatrix} = \begin{bmatrix} G_C & 0 & 0 & 0 & -G_C \\ -G_C & G_C & 0 & 0 & 0 \\ 0 & -(1-p)G_C & G_C & -pG_C & 0 \\ 0 & 0 & 0 & G_C & 0 \\ 0 & -pG_C & 0 & -(1-p)G_C & G_C \end{bmatrix} \begin{Bmatrix} V_1 \\ V_2 \\ V_3 \\ V_5 \\ V_6 \end{Bmatrix}$$

As before it is straightforward to write down the solution, noting that the currents at the voltage terminals are all zero ($I_2 = I_3 = I_5 = I_6 = 0$):

$$V_2 = V_1, \quad V_5 = 0, \quad V_3 = (1-p)V_1, \quad V_6 = pV_1$$

and
$$I_1 = G_C(1-p)V_1$$

Eqs.(4.2.1) and (4.2.2) follow readily, noting that

$$R_L = \frac{V_2 - V_3}{I_1} = \frac{V_6 - V_5}{I_1}$$

$$R_H = \frac{V_2 - V_6}{I_1} = \frac{V_3 - V_5}{I_1}$$

Disordered contacts

The fact that the Hall resistance is unaffected by the backscattering (see Eq.(4.2.2)) may seem obvious. After all, contacts 2 and 6 are located hundreds of microns away from the split-gate scatterers. Surely the effect of the scatterers cannot be felt so far away! However, experimentally it has been observed that often the Hall resistance too is affected by the split gates. This can be understood if we postulate that there is no communica-

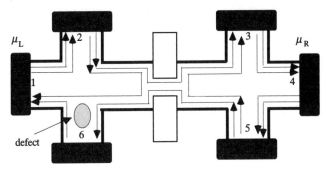

Fig. 4.2.2. Same as Fig. 4.2.1 but with contact 6 'disordered'.

tion among the edge states on the *same side* of the sample and the contacts do not communicate equally with all the edge states. For example, suppose there is a defect or an impurity near contact 6 such that it only 'sees' the outer edge states and the remaining edge states bypass it and go directly to contact 1 (see Fig. 4.2.2).

We would then expect contact 6 to float to a potential equal to $\mu_R = 0$, instead of $eV_1 p$ as we had reasoned earlier. Consequently the measured Hall resistance is given by

$$\frac{V_2 - V_6}{I_1} = \frac{V_1}{I_1} = \frac{h}{2e^2 M} \frac{1}{1-p} \quad \text{instead of} \quad \frac{h}{2e^2 M}$$

and is affected by the presence of the split-gate structure through the factor p. Note that this is only true if the edge states on the same side of the sample do not communicate with each other. If they do communicate, then they will tend to equilibrate and thereby acquire a common average potential equal to $eV_1(1 - p)$. Even if contact 6 'sees' only one of the edge states it will register this average potential, so that the measured Hall resistance will be independent of p. The fact that the measured Hall resistance is affected by the split gate shows that there is lack of equilibration between edge states on the same side of the sample. These results (as well as those for other types of disordered contacts) can be obtained readily from the Büttiker formula, as shown in Exercises E.4.1 and E.4.2 at the end of this chapter (see also Refs.[4.1], [4.2]).

Non-ohmic behavior of R_L

We know that when the Fermi energy lies on a bulk Landau level, the edge states are backscattered through the bulk level to a state on the

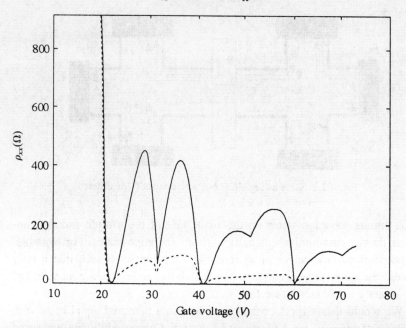

Fig. 4.2.3. Measured resistivity in two silicon field-effect transistors at a magnetic field of $B = 12$ T as a function of the gate voltage (which changes the carrier concentration). The two structures are identical and have the same width $W = 40$ μm. The only difference is the distance between the voltage probes: it is 80 μm for one sample and 2880 μm for the other. The resistivity ρ_{xx} is deduced from the measured resistance assuming the ohmic scaling law. The strong discrepancy between the resistivities in the two samples shows the breakdown of ohmic scaling. The discrepancy goes away above 4 K. Reproduced with permission from Fig. 2 of R. J. Haug and K. von Klitzing (1989), *Europhys. Lett.*, **10**, 489–92.

other side giving rise to a longitudinal resistance. These are the peaks in the SdH oscillations as discussed in Section 1.5. We would expect this peak resistance to scale linearly with the spacing between the voltage probes in accordance with Ohm's law. Experimentally it has been shown that the resistance does not increase linearly (see Fig. 4.2.3).

This non-ohmic behavior can be understood if we postulate that at high fields (when the Fermi energy lies on a bulk Landau level), it is only the innermost edge state that is backscattered through the bulk level to a state on the other side. The remaining edge states can still propagate hundreds of microns without backscattering. As a result when we make the distance between two voltage probes longer and longer, the net backscattering in the region between them does not increase asymptotically to one. Thus the longitudinal resistance measured between two voltage probes does not increase linearly with the distance between them as expected from Ohm's law. Instead it saturates to a maximum value of (see Eq.(4.2.1) with $N = M - 1$)

$$R_{\mathrm{L}}(\text{maximum}) = \frac{h}{2e^2}\left(\frac{1}{M-1} - \frac{1}{M}\right)$$

If we assume ohmic scaling and divide the measured resistance by the probe spacing to obtain the resistivity, then samples with larger probe spacing will yield smaller resistivity values as observed experimentally (see Fig. 4.2.3).

It is really quite surprising that states on the same side of the sample can travel such huge distances (1000 μm is actually 1 mm and is clearly visible to the naked eye) without equilibration. This means that even a sample 1 mm long may exhibit mesoscopic behavior. A number of experiments have been reported by different groups that support this observation (see, for example, P. L. McEuen *et al.* (1991), *Phys. Rev. Lett*, **64**, 2062).

Summary

In a two-dimensional conductor at high magnetic fields, the states carrying current in opposite directions are located on opposite sides of the sample. If the Fermi energy lies between two bulk Landau levels then the states (at the Fermi energy) are completely decoupled from each other (see Fig. 4.1.1). This leads to a complete suppression of backscattering processes resulting in a perfectly ballistic conductor. The longitudinal resistance measured with two probes placed along an edge is zero while the Hall resistance measured with two probes on opposite sides of the sample is quantized in units of $(h/2e^2)$ with an impressive accuracy that is specified in parts per million (Section 4.1). In this quantum Hall regime even conductors with dimensions of the order of millimeters exhibit 'mesoscopic' phenomena that cannot be described in terms of a conductivity tensor. For example, the longitudinal resistance does not scale linearly with length according to Ohm's law (see Fig. 4.2.3); measurements can be affected by the mere presence of a floating probe even if it is not used (see Exercise E.4.2); etc. The Landauer–Büttiker formalism provides a simple framework for the description of such phenomena.

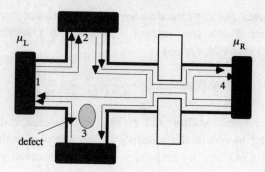

Fig. E.4.1. Same as in Fig. 4.2.2 but with terminals '3' and '5' omitted and terminal '6' renumbered as '3'.

Exercises

E.4.1 Consider a slightly simplified form of the structure shown in Fig. 4.2.2, as shown in Fig. E.4.1.

(a) Write down the conductance matrix for this structure assuming that there is no communication between edge states as they propagate from the constriction to terminal 3.

(b) Use the Büttiker formula (Eq.(2.5.8)) to show that the Hall resistance is given by

$$R_H = \frac{V_2 - V_3}{I} = \frac{h}{2e^2 M} \frac{1}{1-p}$$

as reasoned in the text.

E.4.2 Consider the same structure as in E.4.1 but with an extra terminal '5' inserted, as shown in Fig. E.4.2.

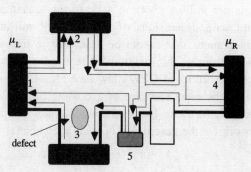

Fig. E.4.2. Same as in Fig. E.4.1 but with an extra terminal '5' inserted.

Write down the conductance matrix for this structure and show that the Hall resistance is now given by

$$R_{\mathrm{H}} = \frac{V_2 - V_3}{I} = \frac{h}{2e^2 M}$$

The extra terminal establishes equilibrium between the edge states and changes the Hall resistance. This is a rather surprising result which has been observed experimentally. In macroscopic conductors, we do not expect an extra floating probe ('5') to affect the measurement.

Further reading

A detailed review of the work in the late 1980s applying the Landauer–Büttiker formalism to the quantum Hall regime can be found in
[4.1] Beenakker, C. W. J. and van Houten, H. (1991), 'Quantum transport in semiconductor nanostructures' in *Solid State Physics*, vol.44, eds. H. Ehrenreich and D. Turnbull (New York, Academic Press) (see part IV).
[4.2] Büttiker, M. (1991). Chapter in *Nanostructured Systems*, ed. M. Reed, *Semiconductor and Semimetals*, vol.35, p.191.

A discussion of the earlier work on the quantum Hall effect (integer and fractional) can be found in
[4.3] Prange, R. E. and Girvin, S. M. (1987), eds. *The Quantum Hall Effect*, (New York, Springer).
[4.4] Chakraborty, T. (1992), 'The quantum Hall effect', in *Handbook on Semiconductors*, Chapter 19, ed. P. T. Landsberg (Amsterdam, New York, Oxford, North-Holland).

5

Localization and fluctuations

5.1 Localization length
5.2 Weak localization
5.3 Effect of magnetic field
5.4 Conductance fluctuations
5.5 Diagrammatic perturbation theory

According to Ohm's law, the resistance of an array of scatterers increases linearly with the length of the array. This describes real conductors fairly well if the phase-relaxation length is shorter than the distance between successive scatterers. But at low temperatures in low-mobility samples the phase-relaxation length can be much larger than the mean free path. The conductor can then be viewed as a series of phase-coherent units each of which contains many elastic scatterers. Electronic transport within such a phase-coherent unit belongs to the regime of quantum diffusion which has been studied by many authors since the pioneering work of Anderson (P. W. Anderson (1958), *Phys. Rev.* **109**, 1492). In this regime, interference between different scatterers leads to a decrease in the conductance. For a coherent conductor having a overall conductance much greater than $\sim (e^2/h)$ or 40 $\mu\Omega$, the decrease in the conductance is approximately (e^2/h). Such a conductor is said to be in the regime of weak localization (Section 5.2). This effect is easily destroyed by a small magnetic field (typically less than 100 G), so that it can be identified experimentally by its characteristic magnetoresistance (Section 5.3). This is a very important effect, because unlike most other transport phenomena it is sensitive to phase relaxation and not just to momentum relaxation. Indeed the weak localization effect is often used to measure the phase-relaxation length.

196

The weak localization effect was discovered in the early 1980s in two-dimensional conductors having dimensions much greater than the phase-relaxation length. In the late 1980s, it became possible to study conductors having dimensions less than the phase-relaxation length. In such conductors, fluctuations ~ (e^2/h) are observed in the conductance as a function of magnetic field or electron density, which are believed to be due to the modification of the random interference pattern between the scatterers. These fluctuations cannot be observed in large conductors because they are smoothed out by the averaging over many phase-coherent units (Section 5.4).

Throughout this chapter we have tried to convey the basic concepts using simple heuristic arguments. We end in Section 5.5 with a detailed quantitative theory of weak localization. This section is more mathematical than the rest and could be skipped on first reading.

5.1 Localization length

In Section 2.2 we showed that, if we neglect all interference between successive scatterers, the transmission probability $T(L)$ through an array of length L goes down inversely with length:

$$T(L) = \frac{L_0}{L + L_0}$$

where L_0 is of the order of a mean free path. This ensures that the resistance of an array of scatterers increases linearly with the length of the array in accordance with Ohm's law:

$$\rho(L) = \frac{1}{M} \frac{1 - T(L)}{T(L)} = \frac{L}{ML_0} \tag{5.1.1}$$

We are using ρ to denote the resistance *normalized* to $(h/2e^2)$.

What happens if we include quantum interference effects? The derivation of a quantum version of the ohmic scaling law (Eq.(5.1.1)) is conceptually more complicated. This is because the quantum mechanical resistance of an array of scatterers is not determined simply by the number of scatterers (that is, the length L of the array). Because of interference effects it also depends on the electron wavelength and the scatterer configuration. For this reason the quantum resistance $\rho(L)$ can be defined meaningfully only if we ensemble average over many conductors having the same number of scatterers, but arranged differently. This can be done analytically for a single-moded conductor ($M = 1$) as discussed below.

Quantum scaling law for a single-moded conductor

We know that if we have two scatterers (with transmission probabilities T_1 and T_2) in series, the transmission probability T through the combination is given by (see Eq.(3.2.4))

$$T = |t|^2 = \frac{T_1 T_2}{1 - 2\sqrt{R_1 R_2} \cos\theta + R_1 R_2} \tag{5.1.2}$$

where θ is the phase shift acquired in one round-trip between the scatterers. Using Eq.(5.1.2) we can write the ensemble-averaged resistance of two scatterers in series as ($\langle...\rangle$ denotes ensemble-averaging):

$$\rho_{12} = \left\langle \frac{1-T}{T} \right\rangle = \int \frac{1}{2\pi} \frac{1 + R_1 R_2 - T_1 T_2 - 2\sqrt{R_1 R_2} \cos\theta}{T_1 T_2} d\theta$$

$$= \frac{1 + R_1 R_2 - T_1 T_2}{T_1 T_2}$$

Defining the resistances of the individual scatterers as

$$\rho_1 \equiv \frac{1 - T_1}{T_1} \quad \text{and} \quad \rho_2 \equiv \frac{1 - T_2}{T_2}$$

we can write

$$\rho_{12} = \rho_1 + \rho_2 + 2\rho_1\rho_2 \tag{5.1.3}$$

To obtain an expression for the resistance as a function of the length, consider what happens if we add a short section of length ΔL to a section of length L. We then have

$$\rho_1 = \rho(L) \quad \text{and} \quad \rho_2 = (\Delta L/L_0)$$

so that from Eq.(5.1.3)

$$\rho(L + \Delta L) = \rho(L) + \left[1 + 2\rho(L)\right]\frac{\Delta L}{L_0}$$

which yields the differential equation

$$\frac{d\rho}{dL} \approx \frac{\rho(L + \Delta L) - \rho(L)}{\Delta L} = \frac{1 + 2\rho}{L_0}$$

The solution to this equation is easy to write down (R. Landauer (1970), *Philos. Mag.*, **21**, 863)

$$\rho(L) = \frac{1}{2}\left[e^{2L/L_0} - 1\right] \tag{5.1.4}$$

This derivation looks quite straightforward but there are subtle issues that we have glossed over. For example, if we ensemble-average the transmission probability directly we obtain from Eq.(5.1.2)

$$\langle T \rangle = \int \frac{T_1 T_2}{1 - 2\sqrt{R_1 R_2}\,\cos\theta + R_1 R_2} \frac{d\theta}{2\pi} = \frac{T_1 T_2}{1 - R_1 R_2}$$

so that

$$\frac{1 - \langle T \rangle}{\langle T \rangle} = \frac{1 - T_1}{T_1} + \frac{1 - T_2}{T_2}$$

If we were to define the resistance as

$$\rho = \frac{1 - \langle T \rangle}{\langle T \rangle} \quad \text{instead of} \quad \left\langle \frac{1 - T}{T} \right\rangle$$

we would obtain an ohmic scaling law:

$$\rho_{12} = \rho_1 + \rho_2 \quad \Rightarrow \quad \rho(L) = L/L_0$$

Thus a proper derivation of a quantum scaling law must also address the question of what quantity one should find the average value of. We refer the reader to Ref.[5.3] for a thorough discussion of these issues.

Quantum resistance of a multi-moded wire

Eq.(5.1.4) states that if a single-moded wire is long enough, its resistance will increase exponentially with length, rather than linearly as we would expect from Ohm's law. But is this a special result for single-moded wires or can it be observed in multi-moded wires as well? It was shown by Thouless (see *Phys. Rev. Lett.*, **39**, 1167 (1977)) that even a multi-moded wire would exhibit this behavior, if it is long enough that its resistance is $\sim (h/2e^2)$. From Eq.(5.1.1) we can see that this requires the length to exceed ML_0. This is known as the localization length, L_c.

$$L_c = ML_0 \tag{5.1.5}$$

Once the length of a phase-coherent wire becomes comparable to L_c the resistance will increase exponentially with length.

This is really rather surprising. Does it mean we could not transmit power via copper wires beyond a certain length? Not really. The key word here is phase-coherent. A real wire at a given temperature can be viewed as an ensemble of little phase-coherent segments each of length L_φ (phase-relaxation length). Significant deviations from Ohm's law are not expected unless the length (L_φ) of a phase-coherent segment becomes comparable to the localization length. This is practically impossible in metallic conductors as we will see below.

Strong and weak localization

A phase-coherent conductor is said to be in the regime of strong localization if its length is comparable to the localization length, that is if its resistance is greater than $\sim 12.5 \text{ k}\Omega$ ($= h/2e^2$) or if its conductance is less than $\sim 80 \text{ }\mu\Omega^{-1}$ ($= 2e^2/h$). Its resistance then no longer scales linearly with length and shows large fluctuations if the scatterer configuration or the electron wavelength is changed.

On the other hand, if the length of a phase-coherent conductor is much less than the localization length, the conductor is said to be in the *weakly localized* regime. In this regime we can write the resistance as (expanding Eq.(5.1.4) in a Taylor series, with L_0 replaced by L_c)

$$\rho(L) \sim \left[\frac{L}{L_c} + \left(\frac{L}{L_c} \right)^2 \right] \quad \Rightarrow \quad \rho_{CL} + \Delta\rho \qquad (5.1.6)$$

where the first term is the classical resistance and the second term represents the deviation from Ohm's law due to quantum interference:

$$\rho_{CL} = \frac{L}{L_c}, \quad \Delta\rho = \left(\frac{L}{L_c} \right)^2 = \left(\rho_{CL} \right)^2$$

Thus the quantum correction in the resistance is proportional to the square of the resistance. This means that the quantum correction to the conductance is a constant:

$$\frac{\Delta G}{2e^2/h} = \Delta\left(\frac{1}{\rho} \right) = -\frac{\Delta\rho}{\rho^2} \sim -1 \quad \text{(weakly localized regime)}$$

Strong localization requires that the length (L_φ) of a phase-coherent segment be longer than the localization length, $L_c = ML_0$. For a metallic conductor the number of modes is of the order of the number of atoms in a cross-section of the wire. A wire with cross-section ~ 2000 Å × 2000 Å has nearly $M = 10^6$ modes, so that even if the mean free path is only 10 Å the localization length is 1mm. The phase-relaxation length is usually significantly smaller than this, so that there is little danger of a metallic conductor entering the strongly localized regime. Weak localization, however, has been observed in very thin metallic wires at very low temperatures (see for example, N. Giordano (1980), *Phys. Rev. B*, **22**, 5635) and also in thin films (see for example, Ref.[5.6]). In semiconductors on the other hand the number of modes (M) is much less than that in metals, so that both weak and strong localization have been observed in 2-D and 1-D conductors.

5.2 Weak localization

Figure 5.2.1 shows the results of a 'numerical experiment' where both the classical and the quantum mechanical conductance are calculated for a conductor with 30 modes having 600 impurities. The procedure is basically the same as that described in Section 3.2. The quantum

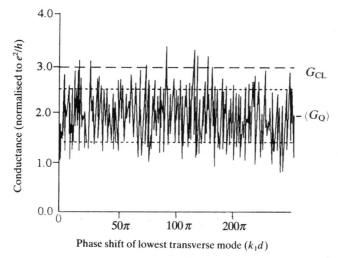

Fig. 5.2.1. Results of a numerical experiment illustrating localization and fluctuations. Calculated conductance of a random array of scatterers as the position of one impurity (in the middle) is changed. This has no effect on the semiclassical calculation which does not take phase into account but the results of the quantum calculation fluctuate about a mean value. Note that the average quantum conductance $\langle G_Q \rangle$ is smaller than the classical result G_{CL} by approximately (e^2/h). Also there are fluctuations of the order of ~ (e^2/h) in the quantum conductance. Reproduced with permission from M. Cahay, M. McLennan and S. Datta (1988), *Phys. Rev. B*, **37**, 10125.

conductance is calculated by combining the scattering matrices for successive sections, each containing one impurity, assuming complete coherence. On the other hand, the classical conductance is calculated by combining their probability matrices, assuming complete incoherence. In both cases (quantum and classical) the Landauer formula is used to obtain the conductance from the transmission:

$$G = \frac{2e^2}{h} \bar{T}(E_f)$$

The calculations are repeated many times as the middle impurity is moved around. The classical conductance is unchanged since it depends only on the number of impurities and not on their relative arrangement. But the quantum conductance fluctuates as the impurity is moved due to quantum interference effects.

It is apparent from Fig. 5.2.1 that the average quantum conductance is smaller than the classical result G_{CL} by approximately (e^2/h):

$$\Delta G = \langle G_Q \rangle - G_{CL} \sim -\frac{e^2}{h}$$

This is the 'weak localization' effect that we discussed. Note also that the fluctuations in the conductance are of order $\sim (e^2/h)$. Surprisingly, this result turns out to be independent of the background conductance and has been observed in a wide variety of conductors in the 'weak localization' regime. We will discuss this further in Section 5.4.

Why is the average quantum conductance lower?

To understand why the average quantum conductance is smaller than the classical value, we start from the Landauer formula:

$$G = \frac{2e^2}{h} MT \equiv \frac{2e^2}{h} M(1-R) \qquad (5.2.1)$$

where we have written R for the reflection probability (not to be confused with resistance). We will be discussing conductors that are many mean free paths long so that we do not need to worry about the contact resistance. Classically for a resistor of length L, the transmission and reflection probabilities are given by

$$T = \frac{L_0}{L + L_0} \quad \text{and} \quad R = \frac{L}{L + L_0}$$

where L_0 is a characteristic dimension of the order of a mean free path. Assuming that the scatterers are isotropic we would expect an incident electron in mode m to be reflected into all modes n ($n = 1, 2,..., M$) with equal probability; that is,

$$R(m \to n) = \frac{1}{M} \frac{L}{L + L_0}$$

When we take quantum interference into account this result still holds on the average for $m \neq n$. But the average probability of reflection back into the incident mode is doubled from its classical value (we will discuss the reason shortly):

$$\langle R(m \to n) \rangle = \frac{1}{M} \frac{L}{L + L_0} \quad \text{for} \quad n \neq m$$

$$= \frac{2}{M} \frac{L}{L + L_0} \quad \text{for} \quad n = m$$

This means that the average reflection probability is a little larger than the classical value:

$$\langle R \rangle = \sum_n \langle R(m \to n) \rangle = \frac{L}{L + L_0} + \frac{1}{M} \frac{L}{L + L_0}$$

and hence the transmission probability must be smaller than its classical value by the same amount:

$$\langle T_Q \rangle = T_{CL} - \frac{1}{M} \frac{L}{L + L_0} \approx T_{CL} - \frac{1}{M} \quad \text{since } L \gg L_0$$

Hence from Eq.(5.2.1) we obtain

$$\langle G_Q \rangle \approx G_{CL} - \frac{2e^2}{h} \tag{5.2.2}$$

The numerical experiment (see Fig. 5.2.1) shows a reduction in conductance that is only half as large as this. We believe that this is because the conductor is close to strong localization ($G_{CL} = 3e^2/h$). Once G_{CL} drops below $2e^2/h$, the reduction in the conductance must be less than that predicted by Eq.(5.2.2). Otherwise $\langle G_Q \rangle$ would be negative!

Enhanced backscattering

We will now explain why the probability of reflection back into the incident mode is doubled over its classical value. This enhanced backscatter-

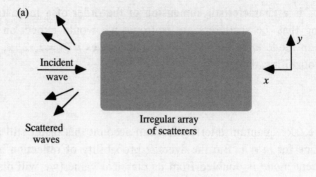

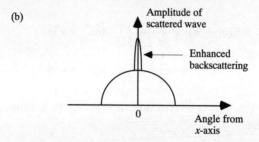

Fig. 5.2.2. Scattering from an irregular array of scatterers.

ing is a rather general phenemenon associated with electromagnetic waves as well. When an incident wave is scattered by a random array of scatterers (see Fig. 5.2.2) we expect the scattered wave to be isotropic. However, it has been observed that the backscattering is peaked along the x-axis, that is, in the direction exactly opposite the incident beam. This enhanced backscattering has been observed directly using laser light to scatter off a concentrated aqueous suspension of latex microspheres (see E. Akkermans, P. E. Wolf and R. Maynard (1986), *Phys. Rev. Lett.* **56**, 1471). It was discussed theoretically as early as 1969 in connection with radar scattering from the clouds (see K. Ishimaru (1978), *Wave Propagation and Scattering in Random Media.*, Vol.II, p.311. (New York, Academic Press)).

The enhanced backscattering is usually explained as follows. We have seen in Section 3.2 that the probability $R(m \to n)$ that an incident electron in mode m will be reflected into mode n is obtained by squaring the sum of the amplitudes A_P for all possible Feynman paths connecting the initial and final states (see Fig. (3.2.3)):

$$R(m \to n) = \left| A_1(m \to n) + A_2(m \to n) + \dots \right|^2$$

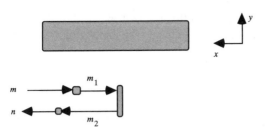

Fig. 5.2.3. The reflection $R(m{\rightarrow}n)$ from an array of scatterers can be obtained by squaring the sum of the amplitudes of all possible Feynman paths starting in mode m and ending in mode n. An example of a Feynman path is shown.

An individual path is visualized as a particular succession of scattering events at the various scatterers in the array. Figure 5.2.3 shows an example of a 'path' connecting mode m to mode n. The amplitude for this path is given by (see Section 3.2)

$$A(m \rightarrow n) = t(m \rightarrow m_1)\, r(m_1 \rightarrow m_2)\, t(m_2 \rightarrow n)$$
$$\exp\left[i\big(k_m L_m + k_{m1} L_{m1} + k_{m2} L_{m2} + k_n L_n\big)\right]$$

where L_m is the length of the section that the electron traverses while in mode m having a wavevector k_m. Since there are numerous possible paths this is not a very practical way to compute the reflection. A more efficient way is simply to combine the scattering matrices of different sections as was done in the numerical experiment. But this description in terms of paths (which is equivalent to the scattering matrix approach) gives us insight into the reason for the enhanced backscattering as we will now describe.

Usually the phases of the various paths are all random and any interference effects cancel out on the average. The square of the sum is then equal to the sum of the squares and we could just neglect the phases and add up probabilities instead of probability amplitudes. But something rather special happens when the initial and final states are the same. Consider any path starting and ending in the same mode m:

$$m \rightarrow m_1 \rightarrow m_2 \ldots m_{N-1} \rightarrow m_N \rightarrow m$$

To every such path there is a time-reversed path obtained simply by reversing all the arrows:

$$m \rightarrow m_N \rightarrow m_{N-1} \ldots m_2 \rightarrow m_1 \rightarrow m$$

We could thus group our paths into two sets such that a path '1' belonging to one set has a time-reversed path '1R' belonging to the other set:

$$R(m \rightarrow m) = \left| (A_1 + A_2 + \ldots) + (A_{1R} + A_{2R} + \ldots) \right|^2$$
$$\equiv \left| A + A_R \right|^2$$

where A is the sum of the amplitudes of all the paths in one set and A_R is the sum of the amplitudes of all the paths in the time-reversed set. It can be shown that as long as there is no magnetic field, the amplitudes of the time-reversed paths are equal ($A_1 = A_{1R}, A_2 = A_{2R}$ etc.), so that their sums are also equal: $A = A_R$. Hence

$$R(m \rightarrow m) = 4\left| A \right|^2 \qquad \text{(coherent backscattering)}$$

If there were no phase-coherence between A and A_R then we should sum the squares instead of squaring the sum. The reflection would then be only half as large:

$$R(m \rightarrow m) = \left| A \right|^2 + \left| A_R \right|^2 = 2\left| A \right|^2 \quad \text{(incoherent backscattering)}$$

Thus the perfect coherence between the pairs of time-reversed paths leads to a doubling of the probability for reflection into the incident mode $R(m \rightarrow m)$. Note that this argument could not be used for the probability of reflection $R(m \rightarrow n)$ into a different mode $n \neq m$. Time reversing a path contributing to $R(m \rightarrow n)$ does not give us another path that contributes to $R(m \rightarrow n)$; The time-reversed path contributes to $R(n \rightarrow m)$ which cannot interfere with the original path since the initial and final states are different.

Decrease in conductivity for large samples

We have seen above that due to coherent backscattering the average conductance of a phase-coherent sample is reduced by $(2e^2/h)$ from its classical value. This effect is observable even if we make conductivity measurements on large samples much greater than the phase-relaxation length. Consider, for example, a long 1-D sample having a width W that is much smaller than the phase-relaxation length (L_φ). We could view it as a series combination of many phase-coherent units, each of length L_φ (see Fig. 5.2.4). Within each unit we have enhanced backscattering and the consequent reduction in the conductivity. But different units act simply like independent classical resistors and can be stacked up according

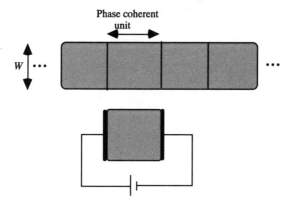

Fig. 5.2.4. A long 1-D sample ($W \ll L_\varphi$) can be conceptually divided up into individual phase-coherent units that behave like independent classical resistors. The quantum conductance of an individual phase-coherent unit is smaller than the classical conductance by $(2e^2/h)$.

to Ohm's law. The conductivity is the same for the entire sample as it is for a single unit. Since the conductance of each of these units is $(2e^2/h)$ lower than the classical conductance (see Eq.(5.2.2)), the correct conductivity σ_Q must be related to the classical conductivity σ_{CL} as follows:

$$\langle G_Q \rangle = \frac{\sigma_Q W}{L_\varphi} = \frac{\sigma_{CL} W}{L_\varphi} - \left(2e^2/h\right)$$

so that

$$\sigma_Q = \sigma_{CL} - \frac{2e^2}{h} \frac{L_\varphi}{W} \quad (1\text{-}D) \qquad (5.2.3)$$

To calculate the reduction in the conductivity of a 2-D sample having a width W that is much larger than a phase-relaxation length, it might seem that we could simply consider a rectangular phase-coherent unit, whose sides are each of length L_φ. Noting that the conductance is smaller than the classical value by $(2e^2/h)$ we would obtain for the conductivity

$$\sigma_Q = \sigma_{CL} - \frac{2e^2}{h} \qquad \text{(WRONG)}$$

This is wrong. An electron injected in the middle of a 2-D sample does not diffuse in a rectangular geometry but in a circular geometry (see Fig. 5.2.5). Consequently the correct phase-coherent unit in two dimensions is a circular conductor as pointed out in Ref.[5.3]. It is straightforward to show that the conductance G of such a circular conductor is related to its

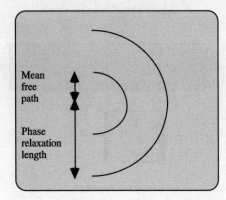

Fig. 5.2.5. Phase-coherent unit for 2-D localization.

conductivity by the relation (see Exercise E.5.1 at the end of this chapter)

$$G = \frac{\pi \sigma}{\ln(L_{max}/L_{min})}$$

where L_{max} and L_{min} are the outer and inner radii of the circular conductor. It is natural to identify the outer radius with the phase-relaxation length: $L_{max} = L_\varphi$. The inner radius is not so clear-cut. We would expect it to be of the order of a mean free path, but the precise value does not lead to any observable consequences (as long as it is unaffected by temperature or magnetic field). As we will see, a similar ambiguity arises even with the more quantitative theory discussed in Section 5.5 and it is common to set $L_{min} = L_m$ (mean free path).

For this circular phase-coherent conductor (with $L_{max} = L_\varphi$ and $L_{min} = L_m$) we could argue as before that the conductance should be reduced by $(2e^2/h)$ due to coherent backscattering, so that

$$\langle G_Q \rangle = G_{CL} - \left(2e^2/h\right)$$

that is,

$$\frac{\pi \sigma_Q}{\ln\left(L_\varphi/L_m\right)} = \frac{\pi \sigma_{CL}}{\ln\left(L_\varphi/L_m\right)} - \left(2e^2/h\right)$$

so that

$$\sigma_Q = \sigma_{CL} - \frac{2e^2}{\pi h}\ln\left(L_\varphi/L_m\right)$$

Noting that $D\tau_m = L_m^2$ and $D\tau_\varphi = L_\varphi^2$ we can write

$$\sigma_Q = \sigma_{CL} - \frac{e^2}{\pi h} \ln\left(\tau_\varphi/\tau_m\right) \quad (2\text{-}D) \tag{5.2.4}$$

Electron–electron interactions

It is apparent from Eqs.(5.2.3) and (5.2.4) that the conductivity correction due to weak localization depends on the phase-coherence length both in 1-D and in 2-D samples. As the temperature is raised the phase-coherence length decreases and the effect is suppressed. This has been observed experimentally. However, there is a separate unrelated effect which also gives rise to a logarithmic temperature dependent correction to the conductivity in two dimensions. Let us briefly outline what this effect is about.

An electron is scattered not only by the impurities but also by the other electrons. Within a one-particle picture we can represent the effect of the other electrons by including an effective potential $U_e(r)$ in the Schrödinger equation. In the Hartree approximation, this potential is calculated by inserting the electron density into the Poisson equation (in the spirit of what we did in Section 2.3 when discussing resistivity dipoles). This potential is temperature dependent and is correlated in a complicated way with the impurity scattering potential. It gives rise to a conductivity correction that looks just like the weak localization correction (see Eq.(5.2.4)) but with the phase-relaxation time replaced by an effective time related to the temperature:

$$\delta\sigma = -\frac{e^2 g}{\pi h} \ln \frac{\hbar/k_B T}{\tau_m} \tag{5.2.5}$$

The factor g depends on the Fermi wavelength and the screening length and is usually positive and of order unity. For a more detailed discussion of this effect we refer the reader to advanced review articles (such as P. A. Lee and T. V. Ramakrishnan (1985), *Rev. Mod. Phys.* **57**, 287; B. L. Al'tshuler and A. G. Aronov (1985), in *Electron–electron Interaction in Disordered Systems*, eds. M. Pollak, A. L. Efros (Amsterdam, North-Holland)).

The point we wish to make here is that due to the close similarity between the weak localization effect and the interaction effect it is difficult to separate them, simply from their temperature dependence. Luckily the first effect disappears rapidly when a small magnetic field is applied while the latter effect is relatively insensitive to low magnetic fields. This allows us to distinguish between them.

5.3 Effect of magnetic field

The unique signature of weak localization is the fact that it can be destroyed by a small magnetic field. We have seen that pairs of paths interfere constructively to produce the enhanced backscattering responsible for weak localization (Fig. 5.2.3). How is this constructive interference affected by a magnetic field? This is difficult to see in terms of the mode representation that we used in the last section. The reason is that the effect of a magnetic field on the transverse modes (see Section 1.6) is relatively complicated. It is much easier to understand the effect of a magnetic field using a real space representation. The Feynman paths in real space were derived in Section 3.8 using a series expansion for the Green's function (see Fig. 3.8.1). We could construct similar Feynman paths for the S-matrix elements since they are related to the Green's function by the Fisher–Lee relation (see Eq.(3.4.6)).

A typical reflection path is shown in Fig. 5.3.1. It starts in mode m in

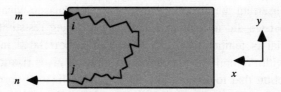

Fig. 5.3.1. Feynman paths in real space: the reflection $R(m{\to}n)$ from an array of scatterers can be obtained by squaring the sum of the amplitudes of all possible Feynman paths in real space connecting two sites i and j adjacent to the lead.

lead p, couples onto a point i in the conductor adjacent to the lead, follows some path in real space and returns to another point j adjacent to the same lead and then couples out back to mode n. These paths are somewhat different from the paths shown in Fig. 5.2.3 which go from one transverse mode to another. For wide conductors it is easier to see the effect of a magnetic field in terms of these paths (which could have been used for the discussion in Section 5.2 as well).

The effect of a magnetic field is stated very simply in a real space representation. The amplitude connecting two nearby points $\mathbf{r}_i$ and $\mathbf{r}_j$ in real space is modified by a simple phase factor (see Eq.(3.5.9b)):

$$t \to t \exp\left[ie\mathbf{A}.(\mathbf{r}_i - \mathbf{r}_j)/\hbar\right]$$

Thus the amplitude associated with any path P acquires an additional phase proportional to the line integral of the vector potential along that path.

$$A_P \rightarrow A_P \exp\left[\frac{ie}{\hbar}\int_P \mathbf{A}.\mathbf{dl}\right]$$

For a closed path the phase factor is proportional to the magnetic flux passing through the area S_P enclosed by the path:

$$A_P \rightarrow A_P \exp\left[\frac{ie}{\hbar}\oint_P \mathbf{A}.\mathbf{dl}\right] = A_P \exp\left[\frac{ie}{\hbar}BS_P\right]$$

Hence we can write

$$A_P(B) = A_P(0)\exp\left(-i2\pi B/B_P\right)$$

where B_P is a characteristic magnetic field that depends on the area enclosed by the path P:

$$\frac{|e|}{\hbar}S_P B = \frac{2\pi B}{B_P} \quad \Rightarrow \quad B_P = \frac{h}{|e|S_P} \qquad (5.3.1)$$

The phase acquired by the time-reversed path $A_P^R(B)$ is just the negative of that acquired by $A_P(B)$

$$A_P^R(B) = A_P(0)\exp\left(+i2\pi B/B_P\right)$$

so that the total amplitude becomes

$$A_P(B) + A_P^R(B) = 2A_P(0)\cos\left(2\pi B/B_P\right) \qquad (5.3.2)$$

Thus the net amplitude of a pair of paths changes in an oscillatory manner with magnetic field, the period of the oscillation being different for each path. The period of the oscillation B_P is continuously distributed from some minimum value all the way to infinity, so that the net amplitude

$$A = \sum_P 2A_P(0)\cos\left(2\pi B/B_P\right)$$

decreases monotonically with magnetic field. The critical magnetic field B_c needed to destroy the coherent backscattering is approximately equal to the smallest period B_P

$$B_c \sim B_P(\text{minimum})$$

and is determined by the coherent paths enclosing the largest area S_P. For a sample having dimensions much greater than the phase-relaxation length we have

$$B_c \sim \frac{h}{|e|S_P(\text{maximum})} \rightarrow \frac{h}{|e|L_\varphi^2} \sim 40\,\text{G if } L_\varphi \sim 1\,\mu\text{m}$$

A detailed theory of the precise functional form of the magnetoresistance requires advanced concepts that we will discuss in Section 5.5.

One small point. The reader may be bothered by the fact that the Feynman path depicted in Fig. 5.3.1 starts and ends at different lattice sites i and j and yet we have been treating it as a closed path. This can be justified by noting that the line integral of the vector potential along the straight line joining i and j is zero. As discussed in Section 2.6, we always choose a gauge such that the vector potential points along the x-direction (that is the direction perpendicular to the lead–conductor interface).

Experimental results

Figure 5.3.2 shows the measured fractional change in the resistance of a GaAs sample ($n_s = 1.6 \times 10^{11}/\text{cm}^2$, $\mu = 27\,000\ \text{cm}^2/\text{V s}$) as the magnetic field increases from zero to 140 G. The magnetic field is so low that we

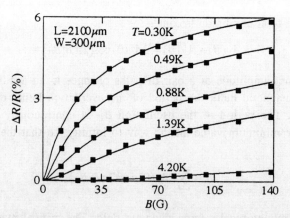

Fig. 5.3.2. Measured fractional change in the longitudinal resistance of a GaAs sample having $n_s = 1.6 \times 10^{11}/\text{cm}^2$, $\mu = 27\,000\ \text{cm}^2/\text{V s}$. The solid curves are theoretical fits with one adjustable parameter, namely, the phase-relaxation length. Adapted with permission from Fig. 1 of K.K. Choi, D. C. Tsui and K. Alavi (1987), *Phys. Rev. B*, **36**, 7751.

do not expect any quantum effects due to the formation of Landau levels. The 6% decrease in the resistance is due to the weak localization phenomenon described above. The sample used for the experimental result shown in Fig. 5.3.2 had a conductivity of about $7 \times 10^{-4}\ \Omega^{-1}$ so that a 6% change corresponds to about $40\ \mu\Omega^{-1}$ which is approximately equal to e^2/h.

In the range of conductivity we are usually interested in, the effect is a weak one as the name 'weak localization' implies. Nevertheless it is a very important effect because in contrast to most other transport phenomena, it is sensitive to phase relaxation and not just to momentum relaxation. There is a detailed theory (to be discussed in Section 5.5) describing the precise variation of the resistance with magnetic field which can be used to fit the measured magnetoresistance curves with just one adjustable parameter, namely, the phase-relaxation length. Thus weak localization measurements can be used to measure quite accurately the distance over which wavefunctions retain their coherence (see Exercise E.5.2 at the end of this chapter).

It should be mentioned that under certain conditions one can observe a positive, instead of a negative magnetoresistance, that is, one can observe *anti-localization* instead of localization. This effect arises from spin–orbit scattering but the mechanism is different in metals (see Ref.[5.6]) and in semiconductors (see G. L. Chen *et al.* (1993), *Phys. Rev. B*, **47**, 4084 and references therein).

In quantum wires (or electron waveguides) having a width W that is much less than the phase-relaxation length, the critical field B_c needed to suppress weak localization is larger:

$$B_c \sim \frac{h}{|e|WL_\varphi} \sim 1\,\text{kG} \quad \text{if} \quad W \sim 400\ \text{Å} \ \text{and}\ L_\varphi \sim 1\,\mu\text{m}$$

This increase in the critical field has been observed experimentally in narrow waveguides. Actually in 'clean' waveguides, having a width much less than the mean free path (as well as the phase-relaxation length), the critical field gets even larger. In general, the phenomena of weak localization and interactions in narrow conductors are less clearly understood than in wide conductors. One reason is that in quantum wires localization is intertwined with classical size effects related to boundary scattering, with quantum size effects, and also with Shubnikov–deHaas oscillations (see Section 1.5) due to Landau level formation. In wide conductors weak localization occurs at low magnetic fields (~ 100 G) where other effects are absent. Moreover, as we will see in the next section quantum wires exhibit large conductance fluctuations in a magnetic field which are absent in wide conductors. We refer the reader to Ref.[5.1] for a detailed review of our current understanding.

h/2e Aharanov–Bohm effect

From the above discussion it is apparent that if we were to use a cylindrical sample (see Fig. 5.3.3) then all paths would enclose approximately

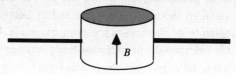

Fig. 5.3.3. The resistance of a hollow cylinder whose circumference is smaller than the phase-relaxation length oscillates as a function of the magnetic field.

the same area and would thus have approximately the same period. We should then see an oscillatory change in the resistance R of the form (r: radius of the cylinder)

$$R(B) \sim \cos^2\left(\frac{2\pi B}{B_0}\right) = \frac{1}{2}\left(1 + \cos\left(\frac{4\pi B}{B_0}\right)\right)$$

instead of the ususal monotonic decrease resulting from the superposition of many different periods. The period ΔB of the oscillations is equal to half of B_0:

$$\Delta B = \frac{B_0}{2} = \frac{h}{2|e|(\pi r^2)} \to 7G \quad \text{if} \quad r = 1\mu m$$

This oscillation has been observed experimentally using metal cylinders (see for example, A. G. Aronov and Yu. V. Sharvin (1987), *Rev. Mod. Phys.*, **59**, 755). Since the flux enclosed by the cylinder ($B.\pi r^2$) changes by $h/2e$ over one oscillation cycle this effect is referred to as the $h/2e$ Aharanov–Bohm effect to distinguish it from the h/e Aharanov–Bohm effect (to be discussed in the next section). This latter effect can be observed in individual small rings but it disappears if we average over many rings. Thus it is not observed in cylinders which can be viewed as a parallel combination of many rings.

5.4 Conductance fluctuations

We have seen how the average quantum conductance is always a little lower than the classical value leading to localization. The quantum conductance is also dependent on the relative positions of the scatterers leading to fluctuations from one sample to another as shown in the numerical experiment in Fig. 5.2.1. Experimentally these fluctuations have been observed by studying the conductance of the same sample as a function of the magnetic field or the electron density. In either case the Fermi wavelength of the electrons is changed and this changes the phase relationships among the different paths randomly. In a statistical sense this is equivalent to changing the configuration of the scatterers as was done in the numerical experiment. Interestingly enough, the size of the conductance fluctuations is about the same ($\sim 2e^2/h$) in different mesoscopic samples in the weak localization regime, even though the background conductance varies over several orders of magnitude.

Size of the fluctuations

To understand why the conductance fluctuations are of order $\sim 2e^2/h$ we start from the Landauer formula and rewrite it in the form

$$G = \frac{2e^2}{h}\overline{T} = \frac{2e^2}{h}M - \frac{2e^2}{h}\overline{R}$$

where $\overline{T}$ is the total transmission and $\overline{R}$ is the total reflection obtained by summing over all the input and output modes:

$$\overline{T} \equiv \sum_m \sum_n T(m \to n) \qquad (5.4.1a)$$

$$\overline{R} \equiv \sum_m \sum_n R(m \to n) \qquad (5.4.1b)$$

It is convenient to define a normalized conductance g as follows:

$$g \equiv \frac{G}{2e^2/h} = \overline{T} = M - \overline{R}$$

The conductance fluctuates from one sample to another because the total reflection $\overline{R}$ fluctuates due to random interference. From Eq.(5.4.1) we can write

$$\langle g \rangle = \langle \overline{T} \rangle = M - \langle \overline{R} \rangle \qquad (5.4.2a)$$

$$\langle g^2 \rangle = \langle \overline{T}^2 \rangle = M^2 - 2M\langle \overline{R} \rangle + \langle \overline{R}^2 \rangle \tag{5.4.2b}$$

where the angle brackets denote an average over the ensemble of impurity configurations. The fluctuation in any quantity X ($= g$, $\overline{R}$ or $\overline{T}$) is defined as

$$\langle \delta X^2 \rangle \equiv \langle X^2 \rangle - \langle X \rangle^2$$

From Eqs.(5.4.2a,b) we can write the conductance fluctuation as

$$\langle \delta g^2 \rangle = \langle \delta \overline{T}^2 \rangle = \langle \delta \overline{R}^2 \rangle \tag{5.4.3}$$

Next let us determine the fluctuation in $\overline{R}$. Eq.(5.4.1b) shows that $\overline{R}$ is given by the sum of M^2 quantities $R(m \to n)$ (both the indices m and n run from 1 to M). Assuming that these M^2 quantities are uncorrelated we can write

$$\langle \delta \overline{R}^2 \rangle = M^2 \langle \delta R(m \to n)^2 \rangle$$

Each of the quantities $R(m \to n)$ can be written in the form

$$R(m \to n) = \left| \sum_P A_P \right|^2$$

where the index P runs over all paths starting in mode m and ending in mode n (see Fig. 5.2.3 or 5.3.1). If we assume that the phases of the amplitudes A_P of all the paths are completely random then

$$\langle R(m \to n) \rangle = \sum_P \sum_{P'} \langle A_P A_{P'}^* \rangle = \sum_P |A_P|^2$$

while

$$\langle R(m \to n)^2 \rangle = \sum_{P,P',P'',P'''} \langle A_P A_{P'} A_{P''}^* A_{P'''}^* \rangle$$

$$= \sum_{P,P',P'',P'''} |A_P|^2 |A_{P'}|^2 \left[\delta_{P,P''} \delta_{P',P'''} + \delta_{P,P'''} \delta_{P',P''} \right]$$

$$= 2\langle R(m \to n) \rangle^2$$

Hence $\qquad \langle \delta R(m \to n)^2 \rangle = \langle R(m \to n) \rangle^2$

For a sample that is many mean free paths long,

$$\langle R(m \to n) \rangle \approx 1/M \Rightarrow \langle \delta R(m \to n)^2 \rangle \approx 1/M^2 \Rightarrow \langle \delta \overline{R}^2 \rangle \approx 1$$

and consequently the normalized conductance fluctuation is about one:

$$\langle \delta g^2 \rangle \approx 1 \tag{5.4.4}$$

The above argument (which is adapted from P. A. Lee (1986), *Physica*, **140A**, 169) shows in a fairly simple way why the conductance fluctuations are approximately $(2e^2/h)$ in magnitude. However, as pointed out there, several assumptions regarding the lack of correlations among different paths have been slipped in rather casually. We would have arrived at a very different (and wrong!) answer if we had applied a similar reasoning to the transmission instead of the reflection. The difference arises because

$$\langle T(m \rightarrow n) \rangle \approx T/M$$

T being the average transmission probability per mode which is much less than one for a sample that is many mean free paths long. Hence

$$\langle \delta T(m \rightarrow n)^2 \rangle \approx T^2/M^2 \Rightarrow \langle \delta g^2 \rangle = \langle \delta \overline{T}^2 \rangle \approx T^2$$

and consequently the fractional conductance fluctuation is about $1/M$:

$$\frac{\sqrt{\langle \delta g^2 \rangle}}{\langle g \rangle} \approx \frac{1}{M} \qquad \text{(WRONG)}$$

This is basically the result that most people would have guessed, namely that the percentage fluctuations should be of order $1/M$, M being the number of modes that carry the current. But the correct answer is that obtained earlier (see Eq.(5.4.4)). As pointed out by Lee, transmission paths have to run across the entire sample while the reflection paths with the largest contributions are only a few mean free paths long. Consequently our cavalier dismissal of correlations among the paths is more suspect for transmission than it is for reflection. Both localization and fluctuations are easier to understand in terms of the reflection paths than in terms of transmission paths. A discussion of the types of correlations that arise among different transmission paths can be found in Ref.[5.5].

Smoothing of fluctuations in large samples

We have calculated the fluctuations assuming that the entire sample is phase-coherent. Large samples can be viewed as ensembles of a large number of phase-coherent units and the fluctuations tend to be reduced or smoothed out when measurements are made on macroscopic samples. A conductor with dimensions ~ 10 μm will show hardly any fluctuations but a conductor with dimensions ~ 0.1 μm will show large reproducible fluctuations at cryogenic temperatures.

If we have a long sample such that $L \gg L_\varphi$ (we are assuming that the width W is much smaller than the phase-relaxation length) then we could view it as $N \, (= L/L_\varphi)$ resistors in series each having a (normalized) mean conductance of g_0 and a conductance fluctuation of ~ 1. The fluctuation in the resistance of each resistor is given by

$$\delta(g_0^{-1}) = \frac{\delta g_0}{g_0^2}$$

When we add N such resistors in series, the mean resistance g^{-1} is N times larger while the fluctuation in the resistance is $N^{1/2}$ times larger:

$$g^{-1} = \frac{N}{g_0} \quad \text{and} \quad \delta(g^{-1}) = \sqrt{N}\,\frac{\delta g_0}{g_0^2}$$

Hence the conductance fluctuation of the entire sample is given by

$$\delta g = g^2 \delta(g^{-1}) = \frac{\delta g_0}{N^{3/2}} \tag{5.4.5}$$

Thus the conductance fluctuations are reduced by the number, N, of independent coherent units raised to the power 1.5. As the temperature is raised, the dimensions of a coherent unit (L_φ) get smaller, N increases and the conductance fluctuations get smaller.

Another reason for the smoothing of fluctuations at higher temperatures can be energy-averaging. As we discussed earlier (see Section 1.7) current flow takes place over an energy range of a few $k_B T$ around the Fermi energy. If $k_B T > \hbar/\tau_\varphi$, then this energy range can be viewed as N uncorrelated channels in parallel, each channel having a width (in energy) of $\hbar/\tau_\varphi$:

$$N = \frac{k_B T}{\hbar/\tau_\varphi} = \left(\frac{L_\varphi}{L_T}\right)^2 \quad \text{where} \quad L_\varphi^2 = D\tau_\varphi \quad \text{and} \quad L_T^2 = \frac{D\hbar}{k_B T}$$

This reduces the conductance fluctuations by a factor of $N^{-1/2} = L_T/L_\varphi$.

h/e Aharonov–Bohm effect

Earlier in Section 3.8 we used this effect to illustrate the use of the concept of Feynman paths. This effect is observed in small ring-shaped

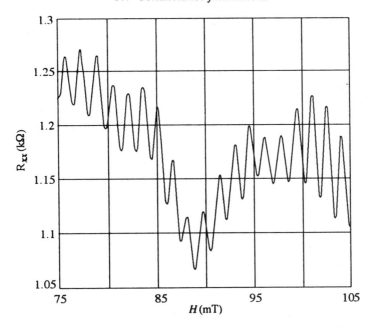

Fig. 5.4.1. A ring-shaped conductor (fabricated in high mobility GaAs–AlGaAs heterostructure) exhibits periodic oscillations in its resistance as a function of the magnetic field. The diameter of the ring is nominally 2 μm. (T = 270 mK). Adapted with permission from Fig. 2b of G. Timp *et al.* (1988), *Surf. Sci.* **196**, 68.

conductors of the type shown in Fig. 0.2. We have seen that in such structures the interference between the two arms of the ring causes the transmission from one mode in the left lead to one in the right lead to oscillate as a function of the magnetic field (see Eq.(3.8.3))

$$T(n \leftarrow m) = 2T_0 + 2T_0 \cos\left(\frac{|e|BS}{\hbar} + \varphi\right)$$

The conductance is determined by the total transmission summed over all input and output modes. That means we have to add up a total of M^2 terms of the form shown above. If we assume that the phases φ associated with every pair of modes m and n is totally random then

$$\bar{T} = \sum_m \sum_n T(m \rightarrow n) \Rightarrow 2T_0 M^2 + 2T_0 M \cos\left(\frac{|e|BS}{\hbar} + \bar{\varphi}\right)$$

where $\bar{\varphi}$ is a random phase that will vary from sample to sample. The conductance is proportional to the total transmission so that:

$$G = G_0 + \hat{G} \cos\left(\frac{|e|BS}{\hbar} + \bar{\varphi}\right) \quad \text{where} \quad \frac{\hat{G}}{G_0} \approx \frac{1}{M} \qquad (5.4.6)$$

Thus the conductance changes periodically as the magnetic field is changed. This is known as the (h/e) A–B effect, since one cycle of oscillation corresponds to a change in the enclosed magnetic flux $(= BS)$ by (h/e). From the above argument one would expect a fractional change in the conductance of the order of $(1/M)$, M being the number of modes in the leads. However, this is wrong because our assumption of complete lack of correlation among different $T(m \rightarrow n)$ is not correct (see discussion following Eq.(5.4.4)). Like conductance fluctuations, the conductance modulation $\hat{G}$ due to the A–B effect is also approximately $\sim (e^2/h)$ independent of G_0. Consequently the fractional change in the conductance is larger in semiconductors (smaller G_0) than in metals. Indeed oscillations as large as 20% have been observed in GaAs–AlGaAs conductors (see, for example, C. J. B. Ford *et al.* (1989), *Appl. Phys. Lett.* **54**, 21).

One subtle point. It would seem that the phase angle $\bar{\varphi}$ in Eq.(5.4.6) could have any value from 0 to 360 degrees depending on the sample. However, for a *two-terminal measurement* the phase angle $\bar{\varphi}$ can only have two values: 0 and 180 degrees. This is because reciprocity requires that if we interchange the voltage and current terminals and reverse the magnetic field, the measured conductance should remain unchanged (see Eq.(2.4.12)). In a two-terminal measurement the current and voltage terminals are identical, so that the conductance must remain the same when the magnetic field is reversed; that is, $G(+B) = G(-B)$. This requires the phase angle $\bar{\varphi}$ in Eq.(5.4.6) to be either 0 or 180 degrees. Actual measurements are usually made with separate voltage and current terminals, so that $G(+B)$ need not be equal to $G(-B)$. However, such measurements should be analyzed using the multiterminal formula (Eq.(2.5.7)) rather than the two-terminal formula $G = (2e^2/h)\bar{T}$.

Non-locality of mesoscopic resistance

An impressive demonstration of the non-local nature of resistance is shown in Fig. 5.4.2. Shown side-by-side are the measured magnetoresistance in two identical structures whose only difference is that one has a ring dangling on one end. In a macroscopic conductor this would have made no difference since the ring is outside the path of the current. But in a mesoscopic conductor it clearly makes a difference. Both structures exhibit similar conductance fluctuations but the one with the ring shows

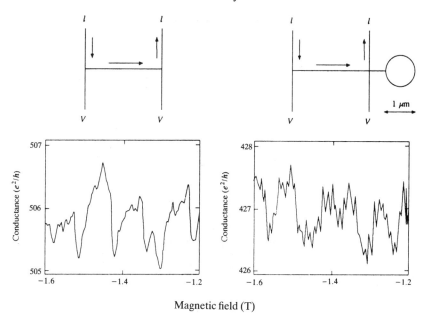

Fig. 5.4.2. Magnetoresistance of two similar wires (schematics on top), one of which has a ring dangling outside the classical path where it 'should' have no effect. Both show aperiodic fluctuations, but the wire with the ring shows additional h/e Aharonov–Bohm oscillations corresponding to the area enclosed by the ring. Reproduced with permission from R. A. Webb and S. Washburn (1988), *Physics Today*, **41**, 46.

additional oscillations superposed on the fluctuations whose period corresponds to the period expected for (h/e) oscillations in the ring. Since the ring has a diameter of about 1 μm. the expected period of the (h/e) oscillations is

$$B_p = \frac{h}{|e|S} \approx 53\,\mathrm{G} \quad \text{if } S = \frac{\pi}{4}(1\,\mu\mathrm{m})^2$$

This corresponds to the high frequency oscillations in the plot on the right, superposed on the low frequency background which is common to both plots.

This appears surprising since the ring is outside what we normally view as the current path. But the point is that every Feynman path within a phase-relaxation length contributes to the transmission. This is not at all surprising once we recognize that a phase-coherent conductor is much like an electromagnetic waveguide. As every microwave engineer knows, measurements are strongly affected if we leave unnecessary 'stuff' hanging around.

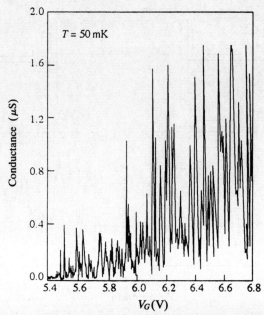

Fig. 5.4.3. Conductance vs. gate voltage (which controls the electron density) in an MOS field effect transistor at $T = 50$ mK. Reproduced with permission from Fig. 4 of A. B. Fowler, J. J. Wainer and R. A. Webb (1988), *IBM J. Res. Dev.* **32**, 372. Copyright 1988 by International Business Machines Corporation.

Strong localization

We have mentioned earlier that a phase-coherent conductor having a resistance greater than 12.5 kΩ is in the strongly localized regime. In this regime the conductance can fluctuate by several orders of magnitude as a function of the magnetic field or the electron density as shown in Fig. 5.4.3. It should be noted that in the strongly localized regime electron–electron interactions are expected to play an important role in determining the fluctuation spectra. Indeed, recent work has shown that in some cases, instead of irregular fluctuations, one can obtain regular periodic oscillations due to electron–electron interactions (see, for example, M. A. Kastner (1992), *Rev. Mod. Phys.* **64**, 849), similar to the single-electron charging effect to be discussed in Section 6.3. This is an area of mesoscopic physics where our understanding is still in the process of evolution.

5.5 Diagrammatic perturbation theory

As we have seen, experimental observations often represent ensemble averages of quantities like the conductance, where the members of the ensemble consist of individual phase-coherent units, each having a different distribution of impurities and hence a different scattering potential $U(\mathbf{r})$. The diagrammatic perturbation theory provides an elegant analytical method for computing ensemble averages. Our purpose in this section is to illustrate this formalism by deriving the weak localization correction to the conductivity with and without a magnetic field. This section is mathematically challenging and can be skipped on first reading.

The starting point for our calculations is the relation discussed earlier in Chapter 3 (see Eq.(3.7.1))

$$\sigma_{xx} = \frac{2e^2}{h}\frac{\hbar^2}{L^d}\sum_{\mathbf{k},\mathbf{k'}} v_x(\mathbf{k})v_x(\mathbf{k'})\left\langle \left| G^R(\mathbf{k},\mathbf{k'}) \right|^2 \right\rangle \qquad (5.5.1)$$

for the conductivity (at $T = 0$ K) of a phase-coherent conductor with side L in d dimensions. We will restrict our discussion to two-dimensional conductors ($d = 2$). Note that all the quantities in this expression are evaluated at the Fermi energy $E = E_f$.

Equation for the Green's function

Before we can discuss ensemble-averaging, we need to discuss how we calculate the Green's function $G^R(\mathbf{k},\mathbf{k'})$ for an individual (phase-coherent) member of the ensemble having a particular scattering potential. In Chapter 3 we showed that the Green's function is given by (see Eq.(3.5.17))

$$G^R = \left[EI - H_C - \Sigma^R \right]^{-1}$$

where H_C is the Hamiltonian operator for the conductor, I is the identity operator and the self-energy term Σ^R arises from the coupling of the conductor to the leads. In the following discussion we will not worry about the coupling to the leads explicitly since we are considering an individual phase-coherent unit in the middle of a large macroscopic sample. Instead we will just assume a small imaginary term $i\eta_\varphi$

$$G^R = \left[(E + i\eta_\varphi)I - H_C \right]^{-1} \qquad (5.5.2)$$

to represent the loss of coherence by phase-breaking processes or by coupling to the surroundings. As we will see, the weak localization correction depends on η_φ, although the classical Drude conductivity is insensitive to the precise value of η_φ.

To calculate the Green's function $G^R(\mathbf{k},\mathbf{k'})$ we need to calculate the inverse indicated in Eq.(5.5.2) using the momentum representation. In the position representation the Hamiltonian operator is given by (we set the vector potential to zero assuming there are no magnetic fields)

$$H_C = -\frac{\hbar^2}{2m}\nabla^2 + U(\mathbf{r})$$

In the momentum representation

$$H_C(\mathbf{k}',\mathbf{k}) = \frac{\hbar^2 k^2}{2m}\delta_{\mathbf{k}',\mathbf{k}} + \underline{U}(\mathbf{k}',\mathbf{k}) \tag{5.5.3}$$

where $\underline{U}$ is the matrix representation of the scattering potential:

$$\underline{U}(\mathbf{k}',\mathbf{k}) = \langle \mathbf{k}'|U(\mathbf{r})|\mathbf{k}\rangle = \int U(\mathbf{r})\exp\left[-i(\mathbf{k}'-\mathbf{k}).\mathbf{r}\right]\frac{d\mathbf{r}}{L^2} \tag{5.5.4}$$

Note that we are considering square-shaped two-dimensional conductors of side L; the wavevector takes on discrete values given by

$$k_x = n_x(2\pi/L) \quad \text{and} \quad k_y = n_y(2\pi/L) \quad (n_x, n_y \text{ are integers})$$

Later we will convert summations into integrals following the usual prescription

$$\sum_{\mathbf{k}} \;\Rightarrow\; L^2\int\frac{d\mathbf{k}}{4\pi^2}$$

Perturbation expansion for the Green's function

It is easy to see from Eqs.(5.5.2) and (5.5.3) that if the scattering potential were absent $\underline{U} = 0$ and the Green's function G_0 would be purely diagonal:

$$G_0(\mathbf{k}',\mathbf{k}) = \delta_{\mathbf{k}',\mathbf{k}}\frac{1}{E - (\hbar^2 k^2/2m) + i\eta_\varphi} \tag{5.5.5}$$

In the presence of a scattering potential we can write in matrix notation

$$\left[G^R\right] = \left[G_0^{-1} - \underline{U}\right]^{-1} \tag{5.5.6}$$

Next we expand G^R in a perturbation series using straightforward matrix algebra:

$$\begin{aligned}
G^R &= \left[G_0^{-1}[I - G_0\underline{U}]\right]^{-1} = [I - G_0\underline{U}]^{-1}G_0 \\
&= [I + G_0\underline{U} + G_0\underline{U}G_0\underline{U} + \ldots]G_0 \\
&= G_0 + G_0\underline{U}G_0 + G_0\underline{U}G_0\underline{U}G_0 + \ldots
\end{aligned}$$

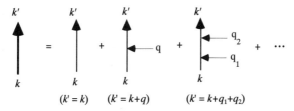

Fig. 5.5.1. Diagrammatic representation of the perturbation series (see Eq.(5.5.8)) for the Green's function.

Noting that the matrix elements $\underline{U}(\mathbf{k}',\mathbf{k})$ of the scattering potential depend only on the difference between the wavevectors $(\mathbf{k}' - \mathbf{k})$ we can simplify the notation a little by defining

$$U_{\mathbf{q}} \equiv \underline{U}(\mathbf{k}',\mathbf{k}) \quad \text{where} \quad \mathbf{q} = \mathbf{k}' - \mathbf{k} \tag{5.5.7}$$

and writing the Green's function as

$$G^{\mathrm{R}}(\mathbf{k}',\mathbf{k}) = \delta_{\mathbf{k}',\mathbf{k}}G_0(\mathbf{k}) + G_0(\mathbf{k}')U_{\mathbf{q}}G_0(\mathbf{k})$$
$$+ G_0(\mathbf{k}')U_{\mathbf{q}_2}G_0(\mathbf{k}+\mathbf{q}_1)U_{\mathbf{q}_1}G_0(\mathbf{k}) + \dots \tag{5.5.8}$$

We can depict Eq.(5.5.8) pictorially as shown in Fig. 5.5.1, noting that the wavevectors must add up so as to satisfy the relations

(1) $\mathbf{k} = \mathbf{k}'$ or   (2) $\mathbf{k} + \mathbf{q} = \mathbf{k}'$ or   (3) $\mathbf{k} + \mathbf{q}_1 + \mathbf{q}_2 = \mathbf{k}'$

just like the currents in an electrical circuit.

Statistical properties of the scattering potential

The Green's function is different for each individual phase-coherent unit since the scattering potential is different. The ensemble-average is calculated by averaging over all the phase-coherent units assuming appropriate statistical properties for the random scattering potential. In particular we

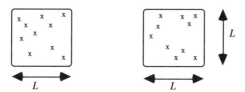

Fig. 5.5.2. Ensemble-averages are calculated by averaging over an ensemble of conductors with random impurity potentials. Two members of the ensemble are shown.

will assume that the scattering potential has zero mean and is delta-correlated:

$$\langle U(\mathbf{r}) \rangle = 0 \quad \langle U(\mathbf{r})U(\mathbf{r}') \rangle = \overline{U}^2 \delta(\mathbf{r} - \mathbf{r}') \qquad (5.5.9)$$

From Eqs.(5.5.4) and (5.5.9) it is straightforward to show that

$$\langle U_{\mathbf{q}} \rangle = 0 \quad \langle U_{\mathbf{q}} U_{\mathbf{q}'} \rangle = \frac{\overline{U}^2}{L^2} \delta_{\mathbf{q}, -\mathbf{q}'} \qquad (5.5.10)$$

The assumption of delta correlation is equivalent to assuming that the individual scatterers are point-like and hence isotropic. Although impurity scattering is often anisotropic, we make this assumption in order to simplify the details. For anisotropic scatterers, the quantity $\langle U_{\mathbf{q}} U_{\mathbf{q}'} \rangle$ is still non-zero only for $\mathbf{q} = -\mathbf{q}'$, but its magnitude is dependent on $\mathbf{q}$.

Ensemble-averaging

Making use of this property (see Eq.(5.5.10)) of the scattering potential we can calculate the ensemble-averaged Green's function by ensemble-averaging the perturbation series given in Eq.(5.5.8) term by term. For example, the second term averages to zero

$$\langle G_0(\mathbf{k}')U_{\mathbf{q}}G_0(\mathbf{k}) \rangle = 0$$

while the third term averages to a non-zero value only if $\mathbf{q}_2 = -\mathbf{q}_1$.

$$\langle G_0(\mathbf{k}')U_{\mathbf{q}_1}G_0(\mathbf{k} + \mathbf{q}_2)U_{\mathbf{q}_2}G_0(\mathbf{k}) \rangle = \frac{\overline{U}^2}{L^2} G_0(\mathbf{k})G_0(\mathbf{k} - \mathbf{q}_1)G_0(\mathbf{k})$$

Diagrammatically we can represent the effect of averaging by connecting together the scattering lines in pairs as shown in Fig. 5.5.3. Each such line contributes a factor of

$$\overline{U}^2/L^2$$

to the diagram. Any diagram with one or more free scattering lines (that is not connected to another) vanishes on ensemble averaging.

Self-energy

It is useful to define a self-energy function Σ as the sum of all diagrams of the type shown in Fig. 5.5.4a, such that any diagram contributing to the ensemble-averaged Green's function can be represented by sandwiching

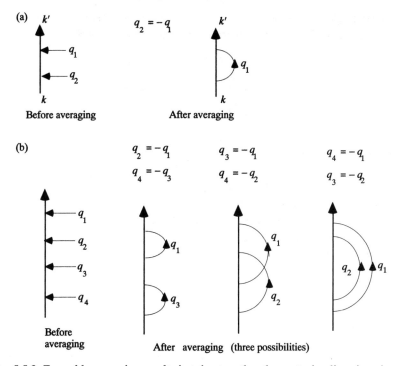

Fig. 5.5.3. Ensemble-averaging results in tying together the scattering lines in pairs. (*a*) Diagrams with two scatterings, (*b*) diagrams with four scattering lines.

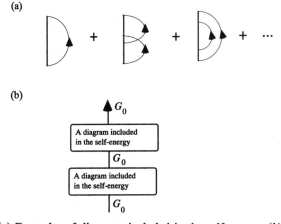

Fig. 5.5.4 (*a*) Examples of diagrams included in the self-energy. (*b*) Any diagram contributing to the ensemble-averaged Green's function can be represented by sandwiching self-energy diagrams between free propagators (G_0).

self-energy diagrams between free propagators (G_0) as shown in Fig. 5.5.4b. This allows us to express the Green's function in the form

$$\langle G^R \rangle = G_0 + G_0 \Sigma G_0 + G_0 \Sigma G_0 \Sigma G_0 + \dots$$

$$= G_0 \left[I + \Sigma G_0 + \Sigma G_0 \Sigma G_0 + \dots \right]$$

$$= G_0 \left[I - \Sigma G_0 \right]^{-1}$$

so that $$\langle G^R \rangle^{-1} = G_0^{-1} - \Sigma \qquad (5.5.11)$$

This is known as the *Dyson equation*.

Note that we must not include diagrams of the type shown in Fig. 5.5.5 when calculating the self-energy. Such diagrams can be expressed in the form $G_0 \Sigma G_0$ and must be excluded from Σ in order to avoid double-counting.

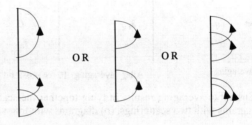

Fig. 5.5.5. Examples of diagrams that should not be included in the self-energy.

Ensemble-averaged Green's function

We are now ready to evaluate the ensemble-averaged Green's function. The first point to note is that as a result of ensemble-averaging all the off-diagonal elements of the Green's function vanish, so that we can express it in the form

$$\langle G^R(\mathbf{k}, \mathbf{k}') \rangle = \delta_{\mathbf{k}, \mathbf{k}'} G(\mathbf{k}) \qquad (5.5.12)$$

This is easy to see from the diagrams shown in Fig. 5.5.3. Ensemble-averaging requires that all scattering lines be paired, that is, every increase in momentum $+\mathbf{q}$ be matched by a subsequent decrease in momentum $-\mathbf{q}$. Hence the final momentum $\mathbf{k}'$ must be equal to the initial momentum $\mathbf{k}$. Physically we can understand this result by noting that the off-diagonal elements $G^R(\mathbf{k}', \mathbf{k})$ arise from the scattering of an electron from its initial state $\mathbf{k}$ to another state $\mathbf{k}'$. Since the phases of the scattering matrix elements $U(\mathbf{k}',\mathbf{k})$ vary randomly from one member of the ensemble to another, the ensemble-averaged value of $G^R(\mathbf{k}',\mathbf{k})$ is zero.

Making use of Eqs.(5.5.12) and (5.5.5) we can rewrite Eq.(5.5.11) as

$$\frac{1}{G(\mathbf{k})} = E - \varepsilon(\mathbf{k}) + i\eta_\varphi - \Sigma(\mathbf{k},\mathbf{k})$$

where $$\varepsilon(\mathbf{k}) \equiv (\hbar^2 k^2/2m) \qquad (5.5.13)$$

In general, the self-energy function has both real and imaginary parts and can be expressed as (the subscript 'el' is added as a reminder that it arises from elastic processes)

$$\Sigma(\mathbf{k},\mathbf{k}) = \sigma_{el}(\mathbf{k}) - i\eta_{el}(\mathbf{k}) \qquad (5.5.14)$$

so that $$\langle G^R(\mathbf{k},\mathbf{k}')\rangle = \delta_{\mathbf{k},\mathbf{k}'}\left(\frac{1}{E - (\varepsilon + \sigma_{el}) + i(\eta_\varphi + \eta_{el})}\right) \qquad (5.5.15)$$

The physical meaning of the self-energy function is easily appreciated if we Fourier transform from energy (E) to time (t):

$$\langle G^R(\mathbf{k},\mathbf{k}')\rangle \Rightarrow \delta_{\mathbf{k},\mathbf{k}'} \exp\left[i(\varepsilon + \sigma_{el})t/\hbar\right]\exp\left[-(\eta_\varphi + \eta_{el})t/\hbar\right]$$

Thus the scattering shifts the energy eigenvalues through the real part of the self-energy (σ_{el}), while the imaginary part of the self-energy causes the Green's function to decay with time. Physically this is due to the outscattering of electrons from the initial state $\mathbf{k}$. Some of these electrons could return later to the initial state through inscattering but they cancel out on ensemble-averaging because of the random phases. The square of the ensemble-averaged value of $G^R(\mathbf{k},\mathbf{k})$

$$\left|\langle G^R(\mathbf{k},\mathbf{k}')\rangle\right|^2 \Rightarrow \delta_{\mathbf{k},\mathbf{k}'} \exp\left[-2(\eta_\varphi + \eta_{el})t/\hbar\right]$$

represents the fraction of the electrons that have not yet outscattered. This fraction decays exponentially with time and we could relate the decay time constant to the lifetime (τ) of the state:

$$\frac{1}{\tau} = \frac{2}{\hbar}(\eta_\varphi + \eta_{el}) \equiv \frac{1}{\tau_\varphi} + \frac{1}{\tau_{el}} \qquad (5.5.16a)$$

As we would expect, the net lifetime is given by a parallel combination of the inelastic and the elastic lifetimes. Of course for localization phenomena to be observed the phase-relaxation time must be much longer than the elastic lifetime so that

$$\eta \equiv \eta_\varphi + \eta_{el} \approx \eta_{el} \quad \Rightarrow \quad \tau \approx \tau_{el} \ll \tau_\varphi \tag{5.5.16b}$$

To calculate the elastic lifetime

To calculate τ_{el} we need to evaluate the self-energy function by summing up the contributions from diagrams of the type shown in Fig. 5.5.6.

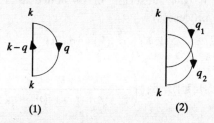

Fig. 5.5.6. Two self-energy diagrams that are evaluated to yield $\Sigma^{(1)}$ and $\Sigma^{(2)}$.

Consider the contribution $\Sigma^{(1)}$ from the first diagram:

$$\Sigma^{(1)}(\mathbf{k},\mathbf{k}) = \sum_q \frac{\overline{U}^2}{L^2} G(\mathbf{k}-\mathbf{q}) = \sum_{k'} \frac{\overline{U}^2}{L^2} G(\mathbf{k'})$$

$$= \frac{\overline{U}^2}{L^2} \sum_{k'} \frac{1}{E - (\varepsilon(k') + \sigma_{el}(k')) + i(\eta_\varphi + \eta_{el}(k'))}$$

Note that although we are evaluating the first diagram in Fig. 5.5.6, by using $G(\mathbf{k})$ instead of $G_0(\mathbf{k})$ for the propagator lines we are effectively including diagrams of the type

as well. Assuming weak scattering, we can make use of the relation (P: principal value)

$$\frac{1}{x+i\Delta} \approx P\left(\frac{1}{x}\right) - i\pi\delta(x)$$

to obtain $\sigma_{\text{el}}(\mathbf{k}) = \text{Re}\left[\Sigma^{(1)}(\mathbf{k}, \mathbf{k})\right] = \overline{U}^2 \int \frac{1}{4\pi^2} P\left(\frac{1}{E - \varepsilon(k') - \sigma_{\text{el}}(k')}\right) d\mathbf{k}$

$$\eta_{\text{el}}(\mathbf{k}) = -\text{Im}\left[\Sigma^{(1)}(\mathbf{k}, \mathbf{k})\right] = \pi \overline{U}^2 \int \frac{1}{4\pi^2} \delta\left[E - \varepsilon(k') - \sigma_{\text{el}}(k')\right] d\mathbf{k}$$

In general the above expressions have to be evaluated iteratively since the real part of the self-energy appears on the right hand side. But for weak scattering we can neglect σ_{el} and noting that

$$\int \frac{1}{4\pi^2} \delta\left[E - \varepsilon(k')\right] d\mathbf{k}' = \frac{m}{2\pi\hbar^2} \Rightarrow N_s \text{ (2-D density of states excluding spin)}$$

we obtain the standard 'golden rule' result:

$$\eta_{\text{el}} = \frac{\hbar}{2\tau_{\text{el}}} = \pi \overline{U}^2 N_s \tag{5.5.17}$$

However, the present formalism allows us to go beyond the golden rule and calculate higher-order corrections to the lifetime. For example the second term in Fig. 5.5.6 yields

$$\Sigma^{(2)}(\mathbf{k}, \mathbf{k}) = \left(\frac{\overline{U}^2}{L^2}\right)^2 \sum_{\mathbf{q}_1, \mathbf{q}_2} G_0(\mathbf{k} - \mathbf{q}_1) G_0(\mathbf{k} - \mathbf{q}_1 - \mathbf{q}_2) G_0(\mathbf{k} - \mathbf{q}_2)$$

Note that this result is of higher order than $\Sigma^{(1)}$ in the scattering potential, so that if the density of impurities is small, the golden rule result is quite accurate and we will use it for the subsequent discussions.

Drude conductivity

The conductivity formula given above (see Eq.(5.5.1)) requires us to evaluate the ensemble average of the square of the Green's function. In general this is not equal to the square of the ensemble-averaged Green's function:

$$\left\langle \left| G^R(\mathbf{k}, \mathbf{k}') \right|^2 \right\rangle \neq \left| \left\langle G^R(\mathbf{k}, \mathbf{k}') \right\rangle \right|^2$$

However, let us first calculate the conductivity assuming the two to be equal:

$$\sigma_0 = \frac{2e^2}{h} \frac{\hbar^2}{L^2} \sum_{\mathbf{k}, \mathbf{k}'} v_x(\mathbf{k}) v_x(\mathbf{k}') \left| \left\langle G^R(\mathbf{k}, \mathbf{k}') \right\rangle \right|^2 \tag{5.5.18}$$

Substituting for the ensemble-averaged Green's function from Eq.(5.5.15) we obtain

$$\sigma_0 = \frac{2e^2}{h}\frac{\hbar^2}{L^2}\sum_{\mathbf{k}}\left(\frac{\hbar k_x}{m}\right)^2\frac{1}{(E - \varepsilon(k) - \sigma(k))^2 + \eta^2(k)}$$

Converting the summation into an integral we obtain

$$\sigma_0 = \frac{e^2\hbar^3}{\pi m^2}\int\frac{k}{4\pi^2\eta}k^2\cos^2\theta\frac{\eta}{(E - \varepsilon(k) - \sigma(k))^2 + \eta^2(k)}\,d\theta dk$$

Neglecting σ and using the approximation

$$\frac{\eta}{(E - \varepsilon(k))^2 + \eta^2} \approx \pi\delta(E - \varepsilon(k))$$

it is straightforward to evaluate the integral to obtain $\sigma_0 = e^2 E/2\pi\hbar\eta$. Noting that

$$E = E_f = \frac{n_s}{m/\pi\hbar^2} \quad \text{and} \quad \eta = \frac{\hbar}{2\tau}$$

we obtain the standard Drude result (see Eq.(1.4.2))

$$\sigma_0 = \frac{n_s e^2\tau}{m} \tag{5.5.19}$$

except for one small point. The Drude conductivity involves the momentum relaxation time while the time τ appearing above represents the lifetime of a momentum state. For anisotropic scatterers, the two can be significantly different and we can obtain the correct Drude result only if we go beyond the approximation involved in replacing Eq.(5.5.1) with Eq.(5.5.18) (see Exercise E.5.4 at the end of this chapter). However, for isotropic scatterers (which is what we are considering) the two are identical and Eq.(5.5.18) yields the Drude conductivity. The difference between Eqs.(5.5.1) and (5.5.18) then leads to corrections to the Drude conductivity such as the correction due to weak localization. This is what we will discuss next.

Conductivity corrections

The difference between Eq.(5.5.1) and Eq.(5.5.18) arises from replacing the ensemble average of the square of the Green's function with the

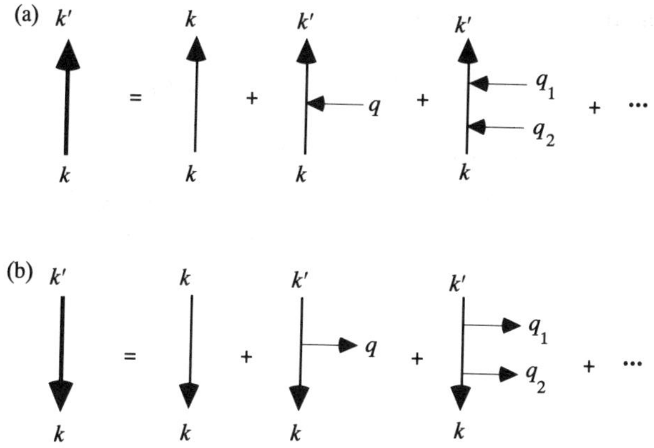

Fig. 5.5.7. Diagrammatic representation of the perturbation series (see Eq.(5.5.8)) for (*a*) the Green's function and (*b*) the complex conjugate of the Green's function.

square of the ensemble-averaged Green's function, so that we can write the correction to the conductivity as

$$\frac{\Delta\sigma}{2e^2/h} = \frac{\hbar^2}{L^2} \sum_{\mathbf{k},\mathbf{k'}} v_x(\mathbf{k}) v_x(\mathbf{k'}) \left[\left\langle \left| G^R(\mathbf{k'},\mathbf{k}) \right|^2 \right\rangle - \left| \left\langle G^R(\mathbf{k'},\mathbf{k}) \right\rangle \right|^2 \right] \quad (5.5.20)$$

Let us first try to express the quantity in parenthesis in diagrammatic terms.

Earlier we derived the perturbation expansion for the Green's function (see Eq.(5.5.8)):

$$G^R(\mathbf{k'},\mathbf{k}) = \delta_{\mathbf{k'},\mathbf{k}} G_0(\mathbf{k}) + G_0(\mathbf{k'}) U_q G_0(\mathbf{k})$$
$$+ G_0(\mathbf{k'}) U_{q_1} G_0(\mathbf{k}+\mathbf{q}_2) U_{q_2} G_0(\mathbf{k}) + \dots$$

with $\qquad \mathbf{k}+\mathbf{q} = \mathbf{k'} \quad$ and $\quad \mathbf{k}+\mathbf{q}_1+\mathbf{q}_2 = \mathbf{k'}$

and discussed the diagrammatic representation for the series (reproduced in Fig. 5.5.7a). The complex conjugate of this series is given by (note that $U_q = U_{-q}^*$)

$$G^R(\mathbf{k'},\mathbf{k})^* = \delta_{\mathbf{k'},\mathbf{k}} G_0(\mathbf{k})^* + G_0(\mathbf{k'})^* U_{-q} G_0(\mathbf{k})^*$$
$$+ G_0(\mathbf{k'})^* U_{-q_1} G_0(\mathbf{k}+\mathbf{q}_2)^* U_{-q_2} G_0(\mathbf{k})^* + \dots$$

Diagrammatically we could represent this series as shown in Fig. 5.5.7b, simply by reversing all the arrows in Fig. 5.5.7a. Note that all the wavevectors still add up like the currents in an electrical circuit.

The squared magnitude of the Green's function

$$\left| G^R(\mathbf{k},\mathbf{k}') \right|^2 = G^R(\mathbf{k},\mathbf{k}') G^R(\mathbf{k},\mathbf{k}')^*$$

consists of diagrams obtained by multiplying a diagram from Fig. 5.5.7a with one from Fig. 5.5.7b. Consider for example the diagram obtained by pairing the third diagram from each (Fig. 5.5.8a). Ensemble-averaging leads to tying together the scattering lines as discussed earlier. This can

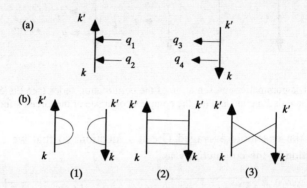

Fig. 5.5.8. (*a*) Example of a diagram that contributes to $\left| G^R(\mathbf{k},\mathbf{k}') \right|^2$ obtained by multiplying one diagram from Fig. 5.5.7a with one from Fig. 5.5.7b, (*b*) ensemble-averaging leads to tying together scattering lines, giving rise to three possibilities.

be done in one of three ways as shown in Fig. 5.5.8b. If we were to ensemble average $G^R(\mathbf{k},\mathbf{k}')$ and $G^R(\mathbf{k},\mathbf{k}')^*$ separately and multiply them, we would obtain only the first of these diagrams. Thus the second and the third terms in Fig. 5.5.8b represent the difference between

$$\left\langle \left| G^R(\mathbf{k},\mathbf{k}') \right|^2 \right\rangle \quad \text{and} \quad \left| \left\langle G^R(\mathbf{k},\mathbf{k}') \right\rangle \right|^2$$

In general the difference between the two terms can be evaluated by summing diagrams that have one or more scattering lines connecting $G^R(\mathbf{k},\mathbf{k}')$ to $G^R(\mathbf{k},\mathbf{k}')^*$, such as the one shown in Fig. 5.5.9a. Noting that each such diagram has free propagators G at the two ends, we can write

$$\frac{\Delta\sigma}{2e^2/h} = \frac{\hbar^2}{L^2} \sum_{\mathbf{k},\mathbf{k}'} v_x(\mathbf{k}) v_x(\mathbf{k}') \left| G(\mathbf{k}) \right|^2 \left| G(\mathbf{k}') \right|^2 \Gamma(\mathbf{k}',\mathbf{k}) \qquad (5.5.21)$$

where the diagrams for $\Gamma(\mathbf{k},\mathbf{k}')$ are obtained by leaving out the propagators at the ends as shown in Fig. 5.5.9b.

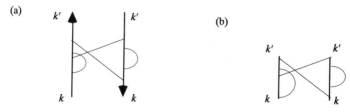

Fig. 5.5.9. (*a*) Example of a diagram that contributes to the conductivity correction due to the difference between Eq.(5.5.1) and Eq.(5.5.18). (*b*) If we leave out the propagators at the two ends we obtain the corresponding diagram for $\Gamma(\mathbf{k},\mathbf{k}')$ in Eq.(5.5.21).

To make further progress we need to evaluate the function $\Gamma(\mathbf{k},\mathbf{k}')$ by summing selected diagrams of the type shown in Fig. 5.5.9b. One group of diagrams known as the *ladder diagrams* is shown in Fig. 5.5.10. For anisotropic scatterers this set of diagrams added to our earlier result (Eq.(5.5.19)) leads to the correct Drude conductivity (see, Exercise E.5.4 at the end of this chapter). For isotropic scatterers (which is what we are considering) the ladder diagrams make no net contribution and we will not discuss them further.

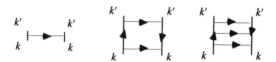

Fig. 5.5.10. Ladder diagrams contributing to $\Gamma(\mathbf{k},\mathbf{k}')$.

Maximally crossed diagrams

The weak localization correction arises from the diagrams shown in Fig. 5.5.11a, which are often referred to as the maximally crossed diagrams. These can be redrawn as shown in Fig. 5.5.11b, by reversing the line on the right. Let us write down the contributions from the diagrams one by one.

The first diagram consists of two scattering lines and two free propagator lines, so that its contribution is given by

$$\Gamma^{(1)}(\mathbf{k},\mathbf{k}') = \left(\overline{U}^2/L^2\right)^2 \sum_{\mathbf{q}} G(\mathbf{k}-\mathbf{q})\,G(\mathbf{k}'+\mathbf{q})^*$$

which can be written as

$$\Gamma^{(1)}(\mathbf{k},\mathbf{k}') = \left(\overline{U}^2/L^2\right)\Lambda(\mathbf{k},\mathbf{k}')$$

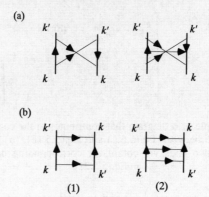

Fig. 5.5.11. (*a*) Maximally crossed diagrams contributing to $\Gamma(\mathbf{k},\mathbf{k}')$, (*b*) same diagrams redrawn so as to look like a ladder diagram. Note that the line on the right has been turned around.

where
$$\Lambda(\mathbf{k},\mathbf{k}') = \left(\overline{U}^2/L^2\right) \sum_{\mathbf{q}} G(\mathbf{k}-\mathbf{q})G(\mathbf{k}'+\mathbf{q})^* \tag{5.5.22}$$

It is straightforward to show that the contributions from the subsequent diagrams form a geometric series of the form:

$$\Gamma^{(n)}(\mathbf{k},\mathbf{k}') = \left(\overline{U}^2/L^2\right)\left[\Lambda(\mathbf{k},\mathbf{k}')\right]^n$$

This allows us to sum up all the diagrams easily to obtain

$$\Gamma(\mathbf{k},\mathbf{k}') = \sum_{n=1}^{\infty} \Gamma^{(n)}(\mathbf{k},\mathbf{k}') = \frac{\overline{U}^2}{L^2}\frac{\Lambda(\mathbf{k},\mathbf{k}')}{1-\Lambda(\mathbf{k},\mathbf{k}')} \tag{5.5.23}$$

To evaluate $\Lambda(\mathbf{k},\mathbf{k}')$ we define $\boldsymbol{\beta} \equiv \mathbf{k}+\mathbf{k}'$ and rewrite Eq.(5.5.22) in the form

$$\Lambda(\mathbf{k},\mathbf{k}') = \left(\overline{U}^2/L^2\right) \sum_{\mathbf{k}_1} G(\mathbf{k}_1)G(\boldsymbol{\beta}-\mathbf{k}_1)^*$$

$$= \overline{U}^2 \int \frac{1}{E-\varepsilon(\mathbf{k}_1)+i\eta}\frac{1}{E-\varepsilon(\boldsymbol{\beta}-\mathbf{k}_1)-i\eta}\frac{k_1 dk_1 d\theta}{4\pi^2}$$

$$= \overline{U}^2 \int \frac{1}{\varepsilon(k_1)-E-i\eta}\frac{1}{\varepsilon(k_1)-E-(\hbar^2 k_1 \beta\cos\theta/m)+(\hbar^2\beta^2/2m)+i\eta}\frac{md\varepsilon(k_1)d\theta}{4\pi^2\hbar^2}$$

The integration over $\varepsilon(k_1)$ is performed readily using the residue theorem (see G. Arfken (1970), *Mathematical Methods for Physicists*, Second Edition, Section 7.2, Academic Press) assuming that the limits extend from negative infinity to positive infinity. This makes no difference since most of the contribution comes from around the poles.

$$\Lambda(\mathbf{k},\mathbf{k}') = \frac{m\bar{U}^2}{4\pi^2\hbar^2}\int_0^{2\pi}(2\pi i)\left[\frac{1}{\varepsilon(k_1)-E-(\hbar^2k_1\beta\cos\theta/m)+(\hbar^2\beta^2/2m)+i\eta}\right]_{\varepsilon=E+i\eta}d\theta$$

$$= \frac{\eta_{el}}{\eta}\int_0^{2\pi}\frac{1}{1+i(\hbar v\beta\cos\theta/2\eta)}\frac{d\theta}{2\pi}$$

where $E = (mv^2/2)$ and we have made use of Eq.(5.5.17). The importance of the maximally crossed diagrams arises from the fact that

$$\Lambda(\mathbf{k},\mathbf{k}') = \frac{\eta_{el}}{\eta}\approx 1 \quad\text{if}\quad \boldsymbol{\beta}\equiv\mathbf{k}+\mathbf{k}'=0$$

leading to a large contribution to $\Gamma(\mathbf{k},\mathbf{k}')$ for $\mathbf{k}' = -\mathbf{k}$ (see Eq.(5.5.23)). This is a manifestation of the enhanced backscattering that leads to the phenomenon of weak localization. The conductivity correction due to weak localization arises primarily from a small angle of wavevectors around the backscattering direction defined by $\beta = 0$, so that it is sufficient to evaluate $\Lambda(\mathbf{k},\mathbf{k}')$ for small β:

$$\Lambda(\mathbf{k},\mathbf{k}')\approx\frac{\eta_{el}}{\eta}\int\left\{1-\frac{i\hbar v\beta\cos\theta}{2\eta}-\frac{\hbar^2v^2\beta^2\cos^2\theta}{4\eta^2}\right\}\frac{d\theta}{2\pi}$$

$$= \frac{\eta_{el}}{\eta}\left\{1-\frac{\hbar^2v^2\beta^2}{8\eta^2}\right\}$$

Noting that $\eta\equiv\hbar/2\tau$ and defining $D\equiv v^2\tau/2$

$$\Lambda(\mathbf{k},\mathbf{k}')\approx\frac{\eta_{el}}{\eta}(1-D\tau\beta^2)\quad\text{where}\quad\boldsymbol{\beta}=\mathbf{k}+\mathbf{k}' \tag{5.5.24}$$

Substituting into Eq.(5.5.23) we obtain

$$\Gamma(\mathbf{k},\mathbf{k}') = (\bar{U}^2/L^2)C(\beta)\quad(\boldsymbol{\beta}\equiv\mathbf{k}+\mathbf{k}') \tag{5.5.25}$$

where

$$C(\beta)\approx\frac{1/D\tau}{(1/D\tau_\varphi)+\beta^2} \tag{5.5.26}$$

noting that $\tau_{el}\ll\tau_\varphi$, and neglecting the term involving β^2 in the numerator.

Weak localization correction to the conductivity

Now we can evaluate the correction to the conductivity from Eq.(5.5.21) by substituting for $\Gamma(\mathbf{k},\mathbf{k}')$ from Eq.(5.5.24):

$$\frac{\Delta\sigma}{2e^2/h} = \frac{\hbar^2\overline{U}^2}{L^4} \sum_{\mathbf{k},\beta} v_x(\mathbf{k})v_x(\boldsymbol{\beta} - \mathbf{k})|G(\mathbf{k})|^2|G(\boldsymbol{\beta} - \mathbf{k})|^2 C(\beta)$$

As we have discussed, the primary contribution comes from small values of β, so that we can write approximately

$$\frac{\Delta\sigma}{2e^2/h} \approx \left[\frac{\hbar^2\overline{U}^2}{L^2} \sum_{\mathbf{k}} v_x(\mathbf{k})v_x(-\mathbf{k})|G(\mathbf{k})|^2|G(-\mathbf{k})|^2\right] \frac{1}{L^2}\sum_{\beta} C(\beta)$$

The term within parentheses can be written as

$$-\hbar^2\overline{U}^2 \int \frac{\hbar^2 k^2 \cos^2\theta}{m^2} \left(\frac{1}{E - \varepsilon(k) + i\eta}\right)^2 \left(\frac{1}{E - \varepsilon(k) - i\eta}\right)^2 \frac{k\,dk\,d\theta}{4\pi^2}$$

$$= \frac{-\overline{U}^2}{2\pi} \int \frac{\varepsilon(k)}{\left[\varepsilon(k) - E - i\eta\right]^2} \frac{1}{\left[\varepsilon(k) - E + i\eta\right]^2} d\varepsilon(k)$$

This integral can be performed using contour integration to yield (again assuming the limits to extend from negative to positive infinity)

$$-\frac{\overline{U}^2}{2\pi}(2\pi i)\left[\frac{d}{d\varepsilon} \frac{\varepsilon}{\left[\varepsilon - E + i\eta\right]^2}\right]_{\varepsilon = E + i\eta}$$

$$= -\overline{U}^2 E/4\eta^3 = -2D\tau$$

For the last step we have made use of Eqs.(5.5.16), (5.5.17) and the relations $E = mv^2/2$ and $D = v^2\tau/2$. We thus obtain

$$\frac{\Delta\sigma}{2e^2/h} \approx -\frac{2D\tau}{L^2}\sum_{\beta} C(\beta) \tag{5.5.27}$$

Substituting from Eq.(5.5.26) and converting the summation into an integral

$$\frac{\Delta\sigma}{2e^2/h} = -\frac{1}{2\pi}\int_0^{(D\tau)^{-1}} \frac{d(\beta^2)}{\beta^2 + (1/D\tau_\varphi)} \tag{5.5.28}$$

An upper cutoff on the integral is necessary since the expression derived above for $\Lambda(\mathbf{k},\mathbf{k}')$ (see Eq.(5.5.24)) is only valid for small values of β. A reasonable cutoff seems to be

$$\beta^2 = (D\tau)^{-1}$$

as indicated, since the expression for $\Lambda(\mathbf{k},\mathbf{k}')$ goes to zero and changes sign at this point. The precise value of this cutoff does not lead to any observable consequences, as long as it is unaffected by temperature and magnetic field. Performing the integral we obtain

$$\Delta\sigma = -\frac{e^2}{\pi h}\ln\frac{\tau_\varphi}{\tau}$$

in agreement with the result obtained earlier from heuristic arguments (see Eq.(5.2.4)).

Effect of magnetic field

So far we have neglected the magnetic field and set the vector potential to zero. A simple heuristic argument can be used to arrive at the correct results quite simply. As we have discussed earlier the effect of a magnetic field can be understood in terms of a quantization of the allowed wavevectors so that an integer number of wavelengths can fit into the cyclotron orbit (see Fig. 1.5.2). The allowed wavevectors are obtained by equating the kinetic energy to the energy of the Landau levels:

$$\frac{\hbar^2 k_n^2}{2m} = \left(n+\frac{1}{2}\right)\hbar\omega_c \quad\Rightarrow\quad k_n^2 = \left(n+\frac{1}{2}\right)\frac{2|e|B}{\hbar}$$

In a magnetic field the quantized values of β are given by a similar expression but with the magnetic field B replaced by $2B$:

$$\beta_n^2 = \left(n+\frac{1}{2}\right)\frac{4|e|B}{\hbar} \tag{5.5.29}$$

This can be understood by noting that a vector potential has the effect of replacing all wavevectors

$$\mathbf{k} \quad\text{with}\quad \mathbf{k}+(e\mathbf{A}/\hbar)$$

Consequently we would expect the vector $\boldsymbol{\beta} = (\mathbf{k}+\mathbf{k}')$ to be replaced by $\boldsymbol{\beta}+(2e\mathbf{A}/\hbar)$ as if the vector potential (and hence the magnetic field) were twice as large.

The conductivity correction is obtained by modifying Eq.(5.5.28) to take into account this quantization of β:

$$\frac{\Delta\sigma}{2e^2/h} = -\frac{1}{2\pi}\frac{4|e|B}{\hbar}\sum_{n=0}^{a}\frac{1}{\beta_n^2+(1/D\tau_\varphi)} \quad\text{where}\quad a \equiv \frac{\hbar}{4|e|BD\tau}$$

Substituting from Eq.(5.5.29) we obtain

$$\Delta\sigma = -\frac{e^2}{\pi h} \sum_{n=0}^{a} \frac{1}{(n+\frac{1}{2})+b} \quad \text{where} \quad b \equiv \frac{\hbar}{4|e|BD\tau_\varphi}$$

The sum over n can be expressed in terms of the digamma functions Ψ (see for example Section 10.2 of the book by Arfken cited after Eq.(5.5.23)) to yield

$$\Delta\sigma = -\frac{e^2}{\pi h}\left[\Psi\left(\frac{1}{2}+a+b\right) - \Psi\left(\frac{1}{2}+b\right)\right]$$

$$= -\frac{e^2}{\pi h}\left[\Psi\left(\frac{1}{2}+\frac{\hbar}{4|e|BD\tau}\right) - \Psi\left(\frac{1}{2}+\frac{\hbar}{4|e|BD\tau_\varphi}\right)\right] \tag{5.5.30}$$

This is the expression that was used to fit the experimental data in Fig. 5.3.2 with one adjustable parameter, namely τ_φ (see Exercise E.5.2 at the end of this chapter). The excellent quality of the fit suggests that the essential physics is captured quite well by the theoretical treatment described above.

Real space interpretation

Our entire discussion so far has been based on the momentum representation. However, it is easy to transform to real space and interpret the results physically. For example if we Fourier transform Eq.(5.5.26)

$$\left[D\beta^2 + \frac{1}{\tau_\varphi}\right]C(\boldsymbol{\beta}) \approx \frac{1}{\tau} \quad (\boldsymbol{\beta} = \mathbf{k}+\mathbf{k}')$$

we obtain $\quad \left[-D\nabla^2 + \frac{1}{\tau_\varphi}\right]C(\boldsymbol{\rho}) \approx \frac{1}{\tau}\delta(\boldsymbol{\rho}) \quad (\boldsymbol{\rho} = \mathbf{r}-\mathbf{r}') \tag{5.5.31}$

This is basically the diffusion equation for particles (having a lifetime of τ_φ), so that the function $C(\mathbf{r},\mathbf{r}')$ represents the probability of finding an electron at $\mathbf{r}$ if it is introduced at $\mathbf{r}'$. Ensemble-averaging makes the medium appear homogeneous so that $C(\mathbf{r},\mathbf{r}')$ depends only on the difference coordinate $\mathbf{r}-\mathbf{r}'$. In the momentum representation this is reflected in the fact that $C(\mathbf{k},\mathbf{k}')$ depends only on $(\mathbf{k}+\mathbf{k}')$ and not on $(\mathbf{k}-\mathbf{k}')$.

In the presence of a weak magnetic field Eq.(5.5.31) is changed to

$$\left[-D\left(\nabla - \frac{2ie\mathbf{A}(\mathbf{r})}{\hbar}\right)^2 + \frac{1}{\tau_\varphi}\right]C(\mathbf{r},\mathbf{r}') \approx \frac{1}{\tau}\delta(\mathbf{r}-\mathbf{r}') \tag{5.5.32}$$

This can be understood by noting that a vector potential has the effect of replacing all wavevectors

$$\mathbf{k} \quad \text{with} \quad \mathbf{k} + (e\mathbf{A}/\hbar)$$

Consequently we would expect the vector $\beta = (\mathbf{k} + \mathbf{k}')$ to be replaced by $\beta + (2e\mathbf{A}/\hbar)$ and hence the gradient operator ∇ by $\nabla - (2ie\mathbf{A}/\hbar)$. This argument is strictly correct only if the vector potential is spatially uniform (zero magnetic field) but is expected to be reasonably accurate for weak magnetic fields. The function $C(\mathbf{r},\mathbf{r}')$ is often referred to as the 'Cooperon' due to its similarity to a quantity appearing in the theory of Cooper pairs in superconductivity.

We can express the weak localization correction to the conductivity in terms of $C(\mathbf{r},\mathbf{r}')$ by Fourier transforming Eq.(5.5.27)

$$\frac{\Delta\sigma}{2e^2/h} \approx -D\tau\big[C(\rho)\big]_{\rho=0} = -D\tau\big[C(\mathbf{r},\mathbf{r})\big] \tag{5.5.33}$$

Thus the weak localization correction is proportional to the probability of a particle returning to its point of origin. This is quite reasonable since paths that return to the point of origin interfere constructively with their time-reversed counterparts thus slowing down the process of diffusion and reducing the conductivity. The correction to the conductivity (Eq.(5.5.30)) can be calculated from this real space formulation using Eqs.(5.5.32) and (5.5.33).

Electron–electron interactions

The diagrammatic theory has also been used to calculate the effect of electron–electron interactions on the conductivity. It has been shown that such interactions affect the self-energy in a disordered medium resulting in a temperature-dependent correction to the conductivity as stated earlier in Eq.(5.2.5). However, this is beyond the scope of this book and we refer the reader to the references cited earlier.

Conductance fluctuations

We have seen that the calculation of conductance requires us to evaluate the ensemble average of the product of two Green's functions since the transmission function T is proportional to G^R times its complex conjugate. The theory of conductance fluctuations requires us to compute the ensemble average of T^2 which is proportional to the product of four Green's functions. Consequently the details are much more complicated. We refer the reader to P. A. Lee, A. D. Stone and H. Fukuyama (1987), *Phys. Rev. B*, **35**, 1039 (see also Ref.[5.5] and references therein).

Summary

According to Ohm's law, the resistance of an array of scatterers increases linearly with the length of the array. But if the array is phase-coherent then its resistance will increase exponentially once the resistance reaches a value of $\sim (h/2e^2)$, that is, about 12.5 kΩ. In this regime of strong localization quantum interference between the scattered waves from different scatterers plays an important role leading to large fluctuations in the resistance if the scatterers are moved around or if the wavelength of the electrons is changed by changing the Fermi energy or the magnetic field. Phase-coherent conductors with conductances much larger than $(2e^2/h)$ are said to be in the regime of weak localization. In this regime quantum interference effects make the conductance $\sim (2e^2/h)$ less than what we would expect from a semiclassical theory of particle diffusion; interference also leads to fluctuations $\sim (2e^2/h)$ in the conductance. These fluctuations are averaged out in large samples consisting of many phase-coherent units and can only be observed in small samples. The decrease in conductance, however, can be observed even in large samples. This effect is easily destroyed by a small magnetic field (typically less than 100 G), so that it can be identified experimentally by its characteristic negative magnetoresistance. This effect is often used to measure the phase-relaxation length in low-mobility semiconductors.

Exercises

E.5.1 Show that the conductance G of a circular conductor (see Fig. 5.2.5) is related to its conductivity σ by the relation

$$G = \frac{\pi\sigma}{\ln\left(L_{max}/L_{min}\right)}$$

where L_{max} and L_{min} are the outer and inner radii of the circular conductor.

E.5.2 Consider the experimental data shown in Fig. 5.3.2. Use Eq.(5.5.30) to fit the data at $T = 0.3$ K, with the phase-relaxation time as an adjustable parameter. Note that the digamma function obeys the recurrence relation

$$\Psi(1 + x) = \Psi(x) + \frac{1}{x}$$

and can be approximated quite accurately by $\Psi(x) = \ln(x - 0.5)$ for $x > 4$.

E.5.3 (a) Consider the data shown in Fig. 5.4.1 for the (h/e) A–B effect. Estimate the period of the oscillations and thereby deduce the diameter of the ring.
(b) Suppose we were to use a thick ring such that the conducting channel has an inner diameter of ~ 1 μm that is only 50% of the outer diameter of ~ 2 μm. Would you expect to see A–B oscillations?

E.5.4 *Anisotropic scatterers* For our calculations in Section 5.5 we assumed isotropic scatterers. For anisotropic scatterers we can write

$$\left| U_{\mathbf{k}-\mathbf{k}'} \right|^2 \equiv U_0^2 P(\theta)$$

where θ is the angle between $\mathbf{k}'$ and $\mathbf{k}$ which have the same magnitude ($k' = k$). Instead of Eq.(5.5.17) we have for the lifetime

$$\frac{1}{\tau} = \frac{U_0^2}{\hbar} N_s \int P(\theta) d\theta$$

while the momentum relaxation time is given by

$$\frac{1}{\tau_m} = \frac{U_0^2}{\hbar} N_s \int P(\theta) \left[1 - \cos\theta \right] d\theta$$

If we simply evaluate Eq.(5.5.18) we obtain for the conductivity

$$\sigma = \frac{n_s e^2 \tau}{m}$$

while the correct Drude conductivity is given by

$$\sigma = \frac{n_s e^2 \tau_m}{m}$$

Evaluate the conductivity correction from Eq.(5.5.21) by summing the ladder diagrams shown in Fig. 5.5.10 to obtain the correct Drude conductivity for anisotropic scatterers.

Further reading

[5.1] Beenakker, C. W. J. and van Houten, H. (1991). 'Quantum transport in semiconductor nanostructures', *Solid State Physics*, vol. 44, eds. H. Ehrenreich and D. Turnbull (New York, Academic Press) (see part II).

[5.2] *Mesoscopic Phenomena in Solids*, eds. B. L. Al'tshuler, P. A. Lee and R. A. Webb, (New York, North-Holland, 1991).

Washburn, S. and Webb, R. A. (1992). 'Quantum transport in small disordered samples from the diffusive to the ballistic regime.' *Rep. Prog. Phys.*, **55**, 1311.

For a thorough discussion of the scaling theory of localization based on the Landauer approach we refer the reader to
[5.3] Anderson, P. W., Thouless, D. J., Abrahams, E. and Fisher, D. S. (1980): 'New method for a scaling theory of localization', *Phys. Rev. B*, **22**, 3519; Anderson, P. W. (1981). 'New method for a scaling theory of localization. II. Multichannel theory of a wire and possible extension to higher dimensionality', *Phys. Rev. B*, **23**, 4828.

An elementary description of weak localization and electron–electron interactions in disordered conductors can be found in
[5.4] Al'tshuler, B. L. and Lee, P. A. (1988). 'Disordered electronic systems.' *Physics Today*, **41**, 36–45. See also other articles in this issue.

A physical discussion of the correlations between different transmission paths that give rise to conductance fluctuations can be found in
[5.5] Feng, S. and Lee, P. A. (1991). 'Mesoscopic conductors and correlations in laser speckle patterns.' *Science*, **251**, 633. See also Imry, Y. (1986). 'Active transmission channels and universal conductance fluctuations'. *Europhys. Lett.*, **1**, 249.

A lucid account of the experiments (on metallic films) as well as the theory can be found in
[5.6] Bergmann, G. (1984). 'Weak localization in thin films.' *Phys. Rep.*, **107**, 1. Section 3 of this article can be used to supplement our discussion in Section 5.5. Also many citations to the original literature (which we have not provided) can be found.

The reader can also supplement our discussion in Section 5.5 with other tutorial discussions such as

[5.7] Doniach, S. and Sondheimer, E. H. (1974). Green's functions for Solid State Physicists, Chapter 5, in *Frontiers in Physics*. Benjamin/ Cummings.

Bagwell, P. F. (1988). M.S.Thesis, Chapter 7, *Massachussetts Institute of Technology*.

Finally we note that there have been many interesting developments in the field that we have not discussed at all, such as persistent currents in normal metal rings (see for example, Mailly, D., Chapelier, C. and Benoit, A. (1993), *Phys. Rev. Lett.*, **70**, 2020) or quantum chaos in microstructures (see for example, the three articles in *Chaos* (1993), **3**, 643, 655 and 664).

6

Double-barrier tunneling

6.1 Coherent resonant tunneling
6.2 Effect of scattering
6.3 Single-electron tunneling

Tunneling is perhaps the oldest example of mesoscopic transport. Single-barrier tunneling has found widespread applications in both basic and applied research. The latest example is scanning tunneling microscopy which has made it possible to image on an atomic scale. However, our purpose in this chapter is not to discuss single-barrier tunneling; the field is far too large and well-developed. Instead we will focus on what is called a double-barrier structure, consisting of two tunneling barriers in series. Since the pioneering work of Chang, Esaki and Tsu (*Appl. Phys. Lett.* **24**, (1974) 593) much research has been devoted to the study of such structures. Two important paradigms of mesoscopic transport have emerged from this study, namely, resonant tunneling and single-electron tunneling. At the same time, the current–voltage characteristics of these structures exhibit useful features at room temperature and high bias, unlike most other mesoscopic phenomena which are limited to the low temperature linear response regime.

We start in Section 6.1 with a discussion of current flow through a double-barrier structure, assuming that transport is coherent. The current can then be obtained by calculating the coherent transmission through the structure from the Schrödinger equation. In Section 6.2 we discuss how scattering processes inside the well affect the peak current and the valley current.

Finally in Section 6.3 we briefly introduce the phenomenon of single-electron tunneling which is one of the most active areas of research at

this time. Like resonant tunneling, it is observed in double-barrier structures. But the physical mechanisms underlying the two phenomena are
fundamentally different. Resonant tunneling arises from the *wave* nature
'of electrons which gives rise to energy quantization in confined structures, while single-electron tunneling arises from the *particle* nature of
electrons which gives rise to charge quantization. Resonant tunneling is
not observed if the distance between the barriers is long enough that the
spacing between the allowed energy levels is negligible compared to $k_B T$.
But single-electron tunneling can still be observed, as long as the capacitance is small enough that the electrostatic energy of a single electron
(e^2/C, C: capacitance) exceeds $k_B T$. This effect would be absent if charge
were not quantized, that is, if e were equal to zero.

6.1 Coherent resonant tunneling

Mesoscopic conductors are usually fabricated lithographically on a 2-
DEG such that the current flows laterally along the film. By contrast the
device we will now discuss is usually implemented in a 'vertical'
configuration where the current flows perpendicular to the plane of the
films (Fig. 6.1.1). It consists of two potential barriers in series, the barriers
being formed by thin layers of a wide-gap material like AlGaAs sandwiched between layers of a material like GaAs having a smaller gap.
Both barriers are thin enough for electrons to tunnel through. It might
seem that the current–voltage characteristics of two barriers in series

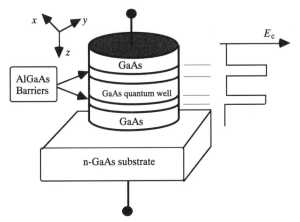

Fig. 6.1.1. Resonant tunneling device. A GaAs layer a few nanometers thick is
sandwiched between two AlGaAs barrier layers of similar thickness. Adapted with
permission from Fig. 2 of F. Capasso and S. Datta (1990), *Physics Today*, **43**, 74.

cannot be any more interesting than those of a single barrier. Ohm's law would suggest that we would simply need twice the voltage to get the same current. That is exactly what would happen if the region between the two barriers were microns in length. But if this region is only a few nanometers long (which is a fraction of the de Broglie wavelength), the current–voltage characteristics are qualitatively different from those of a single barrier, underlining once more the failure of Ohm's law on a mesoscopic scale.

Principle of operation

The current–voltage $(I-V)$ characteristics of a double-barrier structure are easily understood if we note that the region between the barriers acts like a 'quantum box' that traps electrons. It is known from elementary quantum mechanics that a particle in a box has discrete energy levels whose spacing increases as the box gets smaller. We assume that the box is small enough that there is only one allowed energy E_r in the energy range of interest (see Fig. 6.1.2). The structure then acts as a filter that only lets electrons with energy E_r transmit. An applied bias lowers the resonant energy relative to the energy of the incident electrons from the emitter. When the bias exceeds the threshold voltage V_T, the resonant energy falls below the conduction band edge in the emitter and there is a sharp drop in the current. The current–voltage characteristics thus exhibit a negative differential resistance. This was first demonstrated in 1974 by Chang, Esaki and Tsu, but it was a small effect limited to low temperatures. Since then there has been significant improvement in the material quality and fabrication techniques and large effects at room temperature are quite common today (see Fig. 6.1.3). The negative differential resistance forms the basis for practical applications as a switching device and in high frequency oscillators (see Refs.[6.1] and [6.2]).

One small point. The band diagrams drawn in Fig. 6.1.2 are greatly simplified versions where we have assumed that the applied voltage drops linearly across the device. This would be true if there were no space charge inside the device and the surrounding regions were very highly conducting. For quantitative calculations (like the theoretical curve shown in Fig. 6.1.3) it is necessary to compute the charge density everywhere and use it in the Poisson equation to obtain the actual band diagram (see for example, M. Cahay *et al.* (1987), *Appl. Phys. Lett.* **50**, 612). We will generally ignore these important 'details', in order to emphasize the basic conceptual issues.

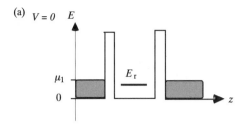

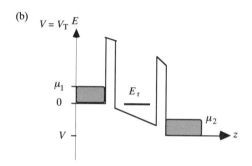

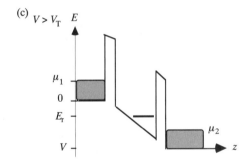

Fig. 6.1.2. Conduction band diagram for a resonant tunneling diode with (a) $V = 0$, (b) $V = V_T$ and (c) $V > V_T$.

Quantitative calculations of the I–V characteristics are usually performed using the expression for the current derived in Chapter 2 (see Eq.(2.5.1)):

$$I = \frac{2e}{h} \int \bar{T}(E)\big[f_1(E) - f_2(E)\big]\,dE \qquad (6.1.1)$$

where $f_1(E)$ and $f_2(E)$ are the Fermi functions in the two contacts and $\bar{T}(E)$ is the transmission function obtained by summing the transmission probability $T_{nm}(E)$ over all input and output modes:

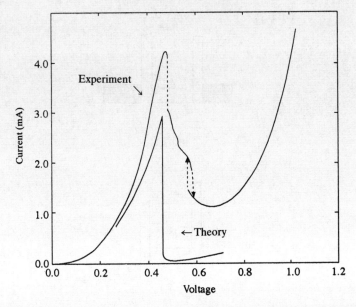

Fig. 6.1.3. *I–V* characteristics of a resonant tunneling diode 8 μm in diameter. Reprinted with permission of MIT Lincoln Laboratory, Lexington, Massachussetts, from E. R. Brown, C. D. Parker, T. C. L. G. Sollner, C. I. Huang and C. E. Stutz (1989), *Proceedings of the OSA Topical Meeting on Picosecond Electronics and Optoelectronics*, March 1989, Salt Lake City, Utah.

$$\bar{T}(E) = \sum_m \sum_n T_{n \leftarrow m}(E) \qquad (6.1.2)$$

The transmission function is usually calculated from the Schrödinger equation, neglecting all scattering processes. The first part of the current–voltage curve is described fairly well by this approach. But once the bias exceeds the threshold voltage, the theory predicts a much smaller 'valley current' than what is observed experimentally. The reason is that the valley current is determined largely by inelastic scattering processes which are completely ignored in the theoretical treatment. We will discuss the effect of scattering in Section 6.2.

In this section we will neglect all scattering processes. For simplicity we will also assume the temperature to be low enough that the Fermi functions can be approximated by step functions: $f_1(E) \approx \vartheta(\mu_1 - E)$ and $f_2(E) \approx \vartheta(\mu_2 - E)$ (see Eq.(1.2.9)). Eq.(6.1.1) then simplifies to

$$I = \frac{2e}{h} \int_{\mu_2}^{\mu_1} \bar{T}(E) \mathrm{d}E \qquad (6.1.3)$$

Next, we need the transmission function $\bar{T}(E)$.

Transmission function $\bar{T}(E)$

To evaluate the transmission function we start from the effective mass equation

$$\left[E_c + \frac{(i\hbar\nabla + e\mathbf{A})^2}{2m} + U(r)\right]\Psi(r) = E\Psi(r) \quad \text{(same as Eq.(1.2.1))}$$

We will assume that there are no magnetic fields so that the vector potential $\mathbf{A}$ can be set to zero. Also we will set the band-edge energy E_c at the left equal to zero as shown in Fig. 6.1.2.

If we assume that the potential U can be written as the sum of a transverse (cross-sectional) confining potential $U_T(x,y)$ and a longitudinal potential $U_L(z)$ then we can separate the problem into two parts. The transverse potential determines the transverse mode energies

$$\left[-\frac{\hbar^2}{2m}\left(\frac{\partial^2}{\partial x^2} + \frac{\partial^2}{\partial y^2}\right) + U_T(x,y)\right]\phi_m(x,y) = \varepsilon_m\phi_m(x,y) \quad (6.1.4)$$

while the longitudinal potential defines the scattering problem:

$$\left[E_L - \frac{\hbar^2}{2m}\frac{\partial^2}{\partial z^2} + U_L(z)\right]\psi_m(z) = 0 \quad \text{where} \quad E_L = E - \varepsilon_m \quad (6.1.5)$$

Knowing the potential $U_L(z)$ we can solve Eq.(6.1.5) to obtain the transmission probability $T_L(E_L)$ for an electron incident (in any mode m) with a longitudinal energy $E_L = E - \varepsilon_m$. Since we have assumed that the transverse confining potential $U_T(x,y)$ does not change along the z-direction, there is no scattering from one transverse mode to another. The transmission probability can thus be written as

$$T_{n\leftarrow m}(E) = T_L(E - \varepsilon_m)\delta_{nm}$$

The total transmission function $\bar{T}(E)$ is obtained by summing the transmission probabilities over all initial and final modes as indicated in Eq.(6.1.2):

$$\bar{T}(E) = \sum_m \sum_n T_{n\leftarrow m}(E) = \sum_m T_L(E - \varepsilon_m) \quad (6.1.6)$$

Thus the transmission function at a particular energy E is obtained by summing the longitudinal transmission probability $T_L(E_L)$ for each incident mode m having a longitudinal energy $E_L = E - \varepsilon_m$. Next we need $T_L(E_L)$.

We can write down the longitudinal transmission probability $T_L(E_L)$ for a double barrier structure by noting that it consists of two barriers (or scatterers) in series. From our discussion on the quantum addition of series resistors we know that the transmission probability through two scatterers is given by (see Eq.(3.2.4))

$$T_L(E_L) = \frac{T_1 T_2}{1 - 2\sqrt{R_1 R_2}\,\cos\theta(E_L) + R_1 R_2} \tag{6.1.7}$$

where T_1, T_2 are the transmission probabilities through barriers 1 and 2 individually and R_1, R_2 are the corresponding reflection probabilities. θ is the phase shift acquired in one round-trip between the scatterers.

Lorentzian approximation

The longitudinal transmission probability $T_L(E_L)$ given by Eq.(6.1.7) usually has sharp peaks at specific values of E_L. To see this we rewrite Eq.(6.1.7), assuming that R_1, $R_2 \sim 1$ (as is normally the case):

$$T_L(E_L) = \frac{T_1 T_2}{\left[1 - \sqrt{R_1 R_2}\right]^2 + 2\sqrt{R_1 R_2}\,(1 - \cos\theta(E_L))}$$

$$\approx \frac{T_1 T_2}{\left[\dfrac{T_1 + T_2}{2}\right]^2 + 2(1 - \cos\theta(E_L))}$$

The sharpness of the resonance arises from the fact that since R_1, $R_2 \sim 1$, T_1 and T_2 are very small, so that the denominator is very small every time the round-trip phase shift $\theta(E_L)$ is close to a multiple of 2π:

$$\left[\cos\theta(E_L)\right]_{E_L = E_r} = 1 \quad \rightarrow \quad \left[\theta(E_L)\right]_{E_L = E_r} = 2n\pi$$

Close to this resonance value we can expand the cosine function in a Taylor series

$$1 - \cos\theta(E_L) \approx \frac{1}{2}(\theta(E_L) - 2n\pi)^2 \approx \frac{1}{2}\left(\frac{d\theta}{dE_L}\right)^2 (E_L - E_r)^2$$

and write the transmission $T_L(E_L)$ as

$$T_L(E_L) \approx \frac{\Gamma_1 \Gamma_2}{(E_L - E_r)^2 + \left(\dfrac{\Gamma_1 + \Gamma_2}{2}\right)^2}$$

where
$$\Gamma_1 \equiv \frac{dE_L}{d\theta} T_1 \quad \text{and} \quad \Gamma_2 \equiv \frac{dE_L}{d\theta} T_2$$

This approximate result is often used (neglecting any energy dependence of Γ_1 and Γ_2) in place of the exact result (Eq.(6.1.7)) for analytical calculations. We can write it in the form

$$T_L(E_L) \approx \frac{\Gamma_1 \Gamma_2}{\Gamma_1 + \Gamma_2} A(E_L - E_r) \qquad (6.1.8)$$

where $A(\varepsilon)$ is a Lorentzian function:

$$A(\varepsilon) = \frac{\Gamma}{\varepsilon^2 + (\Gamma/2)^2} \quad (\Gamma \equiv \Gamma_1 + \Gamma_2) \qquad (6.1.9)$$

Substituting Eq.(6.1.8) into Eq.(6.1.6) we obtain the total transmission function:

$$\overline{T}(E) = \sum_m T_L(E - \varepsilon_m) = \frac{\Gamma_1 \Gamma_2}{\Gamma_1 + \Gamma_2} \sum_m A(E - E_m) \qquad (6.1.10)$$

where
$$E_m \equiv E_r + \varepsilon_m \qquad (6.1.11)$$

The final result for the transmission function is intuitively quite clear. Every transverse mode m gives rise to a peak in the transmission if the total energy equals the longitudinal resonance energy plus the transverse mode energy: $E_r + \varepsilon_m$. The magnitude of the transmission is determined by the parallel combination of Γ_1 and Γ_2 while the width of the peak depends on the sum of Γ_1 and Γ_2.

Significance of Γ_1 and Γ_2

The Lorentzian approximation for the transmission function (see Eq.(6.1.10)) is often used for analytical calculations. It is reasonably accurate close to the resonance, but should not be used far from resonance. One advantage of this approximation is that the entire physics is now characterized by just two parameters, Γ_1 and Γ_2, which are defined by

$$\Gamma_1 \equiv \frac{dE_L}{d\theta} T_1 \quad \text{and} \quad \Gamma_2 \equiv \frac{dE_L}{d\theta} T_2 \qquad (6.1.12)$$

Physically Γ_1 and Γ_2 (divided by $\hbar$) represent the rate at which an electron placed between the barriers would leak out through the barriers into lead 1 and lead 2 respectively.

To see this, we first note that if we write the round-trip phase shift as $\theta \approx 2kw$ where w is the effective width of the well (any phase shifts associated with the reflections at the barriers are included in w), then,

$$\frac{dE_L}{d\theta} = \frac{1}{2w}\left(\frac{dE}{dk}\right) = \hbar v \quad \text{where} \quad v \equiv \frac{v}{2w} \tag{6.1.13}$$

where $v \equiv dE/\hbar dk$ is the velocity with which an electron bounces back and forth between the barriers. The quantity v is called the *attempt frequency* which tells us the number of times per second that the electron impinges on one of the barriers (that is, attempts to escape). It is equal to the inverse of the time that the electron takes to travel from one barrier to another and back.

The physical significance of Γ_1 and Γ_2 is now easy to see. From Eqs.(6.1.12) and (6.1.13) we can write

$$\frac{\Gamma_1}{\hbar} \equiv vT_1 \quad \text{and} \quad \frac{\Gamma_2}{\hbar} \equiv vT_2 \tag{6.1.14}$$

The attempt frequency tells us the number of times per second that an electron attempts to escape. A fraction T_1 of the attempts on barrier 1 are successful while a fraction T_2 of the attempts on barrier 2 are successful. Hence $\Gamma_1/\hbar$ and $\Gamma_2/\hbar$ tell us the number of times per second that an electron succeeds in escaping through barriers 1 and 2 respectively.

.

Current

The current is obtained by integrating the transmission function from μ_2 to μ_1 as indicated in Eq.(6.1.3). Since there is no transmission unless the longitudinal energy E_L is greater than zero (that is, the total energy is greater than ε_m),we can integrate from ε_m to μ_1:

$$I = \sum_m I_m$$

$$I_m = \frac{2e}{h} \int\limits_{\varepsilon_m}^{\mu_1} \bar{T}(E)dE = \frac{2e}{h} \frac{\Gamma_1\Gamma_2}{\Gamma_1 + \Gamma_2} \int\limits_{\varepsilon_m}^{\mu_1} A(E - E_m)dE \tag{6.1.15}$$

Thus the current carried by a mode depends on the area under its spectral

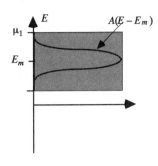 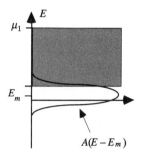

(a) Resonant condition (b) Non-resonant condition

Fig. 6.1.4. The current carried by a mode m is obtained by integrating the corresponding spectral function $A(E-E_m)$ over the energy range $\varepsilon_m < E < \mu_1$.

function $A(E - E_m)$ inside the energy window from ε_m to μ_1 (Fig. 6.1.4). If the lineshape function for a mode is completely inside the window, then the integral is equal to 2π. It then carries the maximum current, I_P that can be carried by a single mode:

$$I_P = \frac{2e}{\hbar} \frac{\Gamma_1 \Gamma_2}{\Gamma_1 + \Gamma_2} \tag{6.1.16}$$

The current carried by a mode m is approximately equal to I_P if it is 'resonant', and approximately zero if it is non-resonant (see Fig. 6.1.4). The total current can be obtained approximately by adding up the currents carried by all the resonant modes.

Peak current

In any given bias condition, the resonant modes are characterized by

$$\mu_1 > E_m > \varepsilon_m \quad \text{that is} \quad \mu_1 - \varepsilon_m > E_r > 0 \tag{6.1.17}$$

As the bias is increased E_r decreases. Different modes start conducting at different values of E_r (when $\mu_1 - E_r$ exceeds ε_m), so that the current increases gradually at first. But they all stop conducting simultaneously when E_r becomes less than zero, making the current drop sharply at a certain value of bias (see Fig. 6.1.3).

For a large cross-section device we can convert the the summation over modes in Eq.(6.1.15) into an integral (using the two-dimensional density of states $mS/\pi\hbar^2$) to obtain

$$I = \frac{mS}{\pi \hbar^2} \int_0^{\mu_1 - E_r} I_m \mathrm{d}\varepsilon_m = I_P \frac{mS}{\pi \hbar^2} \left[\mu_1 - E_r \right] \qquad (6.1.18)$$

From Eq.(6.1.18) we can see that the current increases linearly as the bias is increased. The current, I_P, carried per resonant mode remains the same, but the number of resonant modes increases linearly with bias. The peak current is reached when $E_r = 0$ at $V = V_T$:

$$I_{\text{peak}} = I_P \frac{mS\mu_1}{\pi \hbar^2} \qquad (6.1.19)$$

Note that for simplicity we assumed 'zero' temperature throughout this discussion. The peak current is modified at non-zero temperatures due to the spread in the energy of the incident electrons.

6.2 Effect of scattering

We have so far assumed that transport is fully coherent; that is, the electron transmits from the left to the right in a single quantum mechanical process whose probability can be calculated from the Schrödinger equation. This is a reasonably accurate picture if the average time an electron spends in the resonant state (called the eigenstate lifetime) is much less than the scattering time τ_φ. Otherwise, a significant fraction of the current is due to 'sequential tunneling' where an electron first tunnels into the well and then, after losing memory of its phase, tunnels out of the well. The difference between coherent resonant tunneling and sequential resonant tunneling is somewhat like the difference between a two-photon process and two one-photon processes in optics. Coherent resonant tunneling is like a two-photon process (with a photon energy of zero) where the resonant level E_r in the device acts as a virtual state while sequential resonant tunneling is like two one-photon processes where an electron makes a real transition into the resonant level and another real transition out of it.

   In Section 6.1 we showed that the current carried by a resonant mode is given by

$$I_P = \frac{2e}{\hbar} \frac{\Gamma_1 \Gamma_2}{\Gamma_1 + \Gamma_2} \qquad \text{(same as Eq.(6.1.16))}$$

How is this result modified by scattering processes? This is the question we will first address in this section. We will then discuss the effect of scattering on the off-resonant current.

The two parameters Γ_1 and Γ_2 represent the rate (times $\hbar$) at which an electron placed between the barriers would leak out through the barriers into lead 1 and lead 2 respectively (see Eq.(6.1.14)). Scattering introduces a new parameter into the picture, namely the scattering rate, which is related to the inverse of the scattering time:

$$\Gamma_\varphi \equiv \hbar/\tau_\varphi$$

The coherent tunneling picture is appropriate for devices with thin barriers having

$$\Gamma_1 + \Gamma_2 \gg \Gamma_\varphi \quad \text{that is} \quad (T_1 + T_2) \gg \frac{1}{v\tau_\varphi}$$

while for devices having

$$\Gamma_1 + \Gamma_2 \le \Gamma_\varphi \quad \text{that is} \quad (T_1 + T_2) \le \frac{1}{v\tau_\varphi}$$

a significant fraction of the current is due to sequential tunneling.

Sequential model for the resonant current

Consider a device biased such that the resonant energy lies within the energy range of the incident electrons (see Fig. 6.2.1). In Section 6.1 we assumed a coherent model with no scattering processes inside the well. At the other extreme, with lots of scattering inside the well, we could ignore coherent transmission altogether and calculate the current by writing a simple *rate equation* for the rate at which electrons enter and leave the

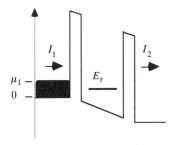

Fig. 6.2.1. A double-barrier diode biased such that the resonant level E_r lies within the energy range of the incident electrons. We consider a single transverse mode with its transverse energy ε_m set equal to zero.

resonant level through the two leads. Noting that the rate constants for the two barriers are given by $\Gamma_1/\hbar$ and $\Gamma_2/\hbar$, we can write

$$I_1 = 2e\frac{\Gamma_1}{\hbar}\left[f_1(1-f_r) - f_r(1-f_1)\right]$$

$$I_2 = 2e\frac{\Gamma_2}{\hbar}\left[f_2(1-f_r) - f_r(1-f_2)\right]$$

(6.2.1)

where f_r is the probability that the resonant level is occupied and the factor of two accounts for the two spins. Assuming low temperature and high bias, we can set the Fermi factor in lead 1 equal to one and that in lead 2 equal to zero in the energy range of interest ($f_1 = 1$, $f_2 = 0$) to obtain

$$I_1 \approx 2e\frac{\Gamma_1}{\hbar}(1-f_r) \quad \text{and} \quad I_2 \approx -2e\frac{\Gamma_2}{\hbar}f_r$$

(6.2.2)

For current conservation we must have $I_1 + I_2 = 0$. Hence

$$\Gamma_1(1-f_r) = \Gamma_2 f_r \quad \Rightarrow \quad f_r = \frac{\Gamma_1}{\Gamma_1 + \Gamma_2}$$

so that

$$I_1 = -I_2 = \frac{2e}{\hbar}\frac{\Gamma_1\Gamma_2}{\Gamma_1 + \Gamma_2}$$

(6.2.3)

This is the same answer that we obtained for the resonant current per mode assuming coherent transmission (see Eq.(6.1.16))! It is surprising that the current calculated from a fully sequential model agrees precisely with that obtained from a fully coherent model. This suggests that phase-breaking processes should have little effect on the resonant current (T. Weil and B. Vinter (1987), *Appl. Phys. Lett.* **50**, 1281). This conclusion is approximately true though detailed calculations show small effects (see S. M. Booker *et al.* (1992), *Semicond. Sci. Technol.*, **7**, B439).

We will now justify this conclusion using a general model that includes both coherent and sequential tunneling. This problem has been addressed by many authors (see, for example, A. D. Stone and P. A. Lee (1985), *Phys. Rev. Lett.* **54**, 1196 and M. Jonson and A. Grincwajg (1987), *Appl. Phys. Lett.* **51**, 1729).

Resonant current with both coherent and sequential components

In Section 6.1 we assumed that all the electrons could transmit coherently through the structure without scattering. In the presence of scattering

(a)

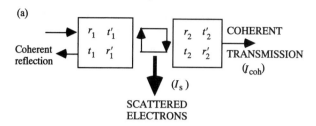

SCATTERED
ELECTRONS

(b)

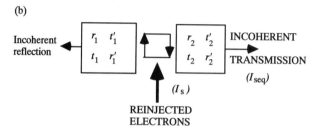

REINJECTED
ELECTRONS

Fig. 6.2.2. A resonant tunneling diode consists of two barriers with scattering matrices as shown in series. (*a*) Scattering processes cause electrons to leak out of the coherent stream as shown. (*b*) The scattered electrons are returned partially to terminal 1 and partially to terminal 2.

processes, we have the situation shown in Fig. 6.2.2a where only a fraction of the electrons transmit coherently while the remainder get scattered inside the well and effectively leak out of the coherent stream. We will assume that the scattering rate is small compared to the attempt frequency

$$\Gamma_\varphi \ll \hbar v \quad \text{that is} \quad v\tau_\varphi \gg 1$$

This ensures that an electron can bounce back and forth several times inside the well before it is scattered, so that it still makes sense to talk about a resonant level.

We will show that in the presence of scattering processes, the peak coherent current per mode (I_{coh}) is given by (cf. Eq.(6.1.16))

$$I_{coh} = \frac{2e}{\hbar} \frac{\Gamma_1\Gamma_2}{\Gamma_1 + \Gamma_2 + \Gamma_\varphi} \tag{6.2.4}$$

while the scattered current per mode is given by

$$I_S = \frac{2e}{\hbar} \frac{\Gamma_1\Gamma_\varphi}{\Gamma_1 + \Gamma_2 + \Gamma_\varphi} \tag{6.2.5}$$

Since the scattered current has no place to go it must be reinjected back into the structure (see Fig. 6.2.2b). When this current is reinjected it comes partially out of terminal '1' and partially out of '2' in the ratio $\Gamma_1:\Gamma_2$. The part coming out of '2' comprises the sequential current (I_{seq}), so that

$$I_{seq} = \frac{\Gamma_2}{\Gamma_1 + \Gamma_2} I_S = \frac{\Gamma_\varphi}{\Gamma_1 + \Gamma_2} I_{coh} \qquad (6.2.6)$$

The total current is obtained by adding Eqs.(6.2.4) and (6.2.6):

$$I_0 = I_{coh} + I_{seq} = \frac{2e}{\hbar} \frac{\Gamma_1 \Gamma_2}{\Gamma_1 + \Gamma_2} \qquad (6.2.7)$$

This result is *independent of* Γ_φ, showing that scattering processes have no significant effect on the total resonant current through a level.

Derivation of Equations (6.2.4) and (6.2.5)

Equations (6.2.4) and (6.2.5) are the key results that we used in the above argument. These are very similar to the Breit–Wigner formulae for nuclear scattering (see for example, Section 145 of L. D. Landau and E. M. Lifshitz (1977), *Quantum Mechanics*, Third Edition (Oxford, Pergamon Press)). We will briefly outline the derivation here; the details are spelt out in Exercise E.6.3 at the end of this chapter. We model the 'leaking out' of electrons due to scattering by introducing an attenuation constant per unit length α. The amplitude for coherent transmission can be calculated by summing the different paths shown in Fig. 3.2.2, taking care to insert a factor of $\exp[-2\alpha W]$ every time we traverse the well. We then obtain

$$\frac{t_1 t_2}{1 - r_1' \, r_2 \exp\left[-2\alpha W\right]} \quad \text{instead of} \quad \frac{t_1 t_2}{1 - r_1' \, r_2}$$

W being the width of the well. The transmission probability (obtained by squaring the amplitude) is modified from the earlier expression (Eq.(6.1.7)):

$$T_L(E_L) = \frac{T_1 T_2}{1 - 2\sqrt{R_1 R_2} \exp\left[-2\alpha W\right]\cos\theta(E_L) + R_1 R_2 \exp\left[-4\alpha W\right]}$$

If we approximate the cosine function by a quadratic function as we did before and assume that $\alpha W \ll 1$ (this is equivalent to the condition stated earlier, namely, $\Gamma_\varphi \ll \hbar v$, as evident from Eq.(6.2.9) below), then we can reduce this expression to a Lorentzian but with an increased linewidth:

$$T_L(E_L) \approx \frac{\Gamma_1\Gamma_2}{\left(E_L - E_r\right)^2 + \left(\dfrac{\Gamma_1 + \Gamma_2 + \Gamma_\varphi}{2}\right)^2} \tag{6.2.8}$$

where Γ_1 and Γ_2 are defined as before (see Eq.(6.1.12)) and

$$\Gamma_\varphi \equiv \hbar v(4\alpha W) \tag{6.2.9}$$

We can write the transmission probability in the form (cf. Eq.(6.1.8))

$$T_L(E_L) \approx \frac{\Gamma_1\Gamma_2}{\Gamma_1 + \Gamma_2 + \Gamma_\varphi} A_\varphi(E_L - E_r) \tag{6.2.10}$$

where $A_\varphi(\varepsilon)$ is a Lorentzian function but with a larger linewidth (cf. Eq.(6.1.9)):

$$A_\varphi(\varepsilon) \approx \frac{\Gamma}{\varepsilon^2 + \left(\Gamma/2\right)^2} \quad (\Gamma \equiv \Gamma_1 + \Gamma_2 + \Gamma_\varphi) \tag{6.2.11}$$

Hence the coherent transmission function is given by (cf. Eq.(6.1.10))

$$\left[\bar{T}(E)\right]_{\text{coh}} = \frac{\Gamma_1\Gamma_2}{\Gamma_1 + \Gamma_2 + \Gamma_\varphi} \sum_m A_\varphi(E - E_m) \tag{6.2.12}$$

The coherent current is obtained by integrating the coherent transmission function as before. Instead of Eq.(6.1.15) we now obtain

$$\left[I_m\right]_{\text{coh}} = \frac{2e}{h} \frac{\Gamma_1\Gamma_2}{\Gamma_1 + \Gamma_2 + \Gamma_\varphi} \int_{\varepsilon_m}^{\mu_1} A_\varphi(E - E_m)\mathrm{d}E$$

It is easy to see that the peak current per mode is given by Eq.(6.2.4).

The derivation of Eq.(6.2.5) for the scattered current I_S follows along the same lines. We first show that the transmission probability T_S corresponding to the scattered current is given by (see Exercise E.6.3 at the end of this chapter)

$$T_S(E_L) \approx \frac{\Gamma_1\Gamma_\varphi}{\Gamma_1 + \Gamma_2 + \Gamma_\varphi} A_\varphi(E_L - E_r) \tag{6.2.13}$$

This is basically the same as the result for the coherent current (Eq.(6.2.4)), but with Γ_2 replaced by Γ_φ. This replacement seems reasonable if we think of the scattered current as transmitting into a third terminal with coupling Γ_φ, just as the coherent current transmits into a terminal with coupling Γ_2. Using the same replacement we obtain the scattered current (Eq.(6.2.5)) from the coherent current (Eq.(6.2.4)).

Linear response

It will be noted that the spectral function A_φ does get broader due to the scattering processes. But because the energy range of the incident electrons is typically much larger than Γ_φ, the measured current reflects the area under the spectral function and remains unchanged. Instead, if we were to measure the linear response conductance, the result may not be independent of Γ_φ. Using the Landauer formula (see Eq.(2.5.3)), we can write the linear response conductance corresponding to the coherent current as

$$
\begin{aligned}
\left[G(E_\mathrm{f}) \right]_\mathrm{coh} &= \frac{2e^2}{h} \left[\overline{T}(E_\mathrm{f}) \right]_\mathrm{coh} \\
&= \sum_m \frac{2e^2}{h} \frac{\Gamma_1 \Gamma_2}{\Gamma_1 + \Gamma_2 + \Gamma_\varphi} A_\varphi (E_\mathrm{f} - E_m)
\end{aligned}
$$

Making use of Eq.(6.2.6) we obtain the total (coherent + sequential) conductance:

$$
\begin{aligned}
G(E_\mathrm{f}) &= \left[G(E_\mathrm{f}) \right]_\mathrm{coh} \left[1 + \frac{\Gamma_\varphi}{\Gamma_1 + \Gamma_2} \right] \\
&= \sum_m \frac{2e^2}{h} \frac{\Gamma_1 \Gamma_2}{\Gamma_1 + \Gamma_2} A_\varphi (E_\mathrm{f} - E_m)
\end{aligned}
$$

By measuring the linear response conductance as a function of the Fermi energy we could measure the spectral function A_φ, and not just the integrated area. Since the spectral function is significantly broadened by the scattering if $\Gamma_\varphi > \Gamma_1 + \Gamma_2$ (see Eq.(6.2.11)), we would expect the measurement to be sensitive to scattering processes.

However, it is also essential that the temperature be sufficiently low ($k_\mathrm{B}T < \Gamma_\varphi$). Otherwise the spread in the energy of the incident electrons wipes out any effect due to the broadening of the spectral function. To see this, we note that at non-zero temperatures, neglecting inelastic processes, the conductance can be written as (see Eqs.(2.5.4)–(2.5.6))

$$
G(E_\mathrm{f}) = \sum_m \frac{2e^2}{h} \frac{\Gamma_1 \Gamma_2}{\Gamma_1 + \Gamma_2} L(E_\mathrm{f} - E_m) \tag{6.2.14}
$$

where $L(\varepsilon)$ is the composite lineshape function obtained by convolving the spectral function with the thermal broadening function $F_T(\varepsilon)$:

$$L(\varepsilon) = \int A_\varphi(\varepsilon') F_T(\varepsilon - \varepsilon') d\varepsilon'$$

At high temperatures the lineshape is dominated by F_T and becomes insensitive to any broadening of the spectral function A_φ due to scattering processes.

Off-resonant current

So far we have considered only the resonant current that flows when the resonant energy lies within the energy range of the incident electrons. Consider next the off-resonant current (see Fig. 6.2.3). The off-resonant

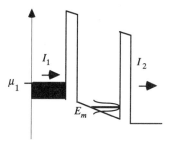

Fig. 6.2.3. A double-barrier diode biased such that the resonant level E_r lies outside the energy range of the incident electrons.

current is very small if we neglect scattering processes, as in the theoretical *I–V* characteristics shown in Fig. 6.1.3. Evidently there is a large discrepancy between theory (which neglects scattering) and experiment for the valley current, showing the importance of including scattering processes in this regime (see for example, F. Chevoir and B. Vinter (1993). *Phys. Rev. B*, **47**, 7260 and references therein). It might seem that we could include such processes using the same procedure as described above for the resonant current. However, there are a few difficulties which we will now describe.

The basic difference between the resonant (Fig. 6.2.1) and the off-resonant (Fig. 6.2.3) situation is that in one case the linewidth of the resonance lies within the energy range of the incident electrons, while in the latter case it lies largely outside the resonance. Thus if we follow the same analysis as before we find instead of Eq.(6.2.4)

$$I_{V,\text{coh}} = \frac{2e}{\hbar} K \Gamma_1 \quad \text{where} \quad K \approx \int_0^{\mu_1} \frac{\Gamma_2 dE}{(E - E_r)^2} \qquad (6.2.15)$$

assuming that the linewidth is much smaller than the energy difference, $E - E_r$, between the incident electrons and the resonant energy. The scattered current is obtained by replacing Γ_2 with Γ_φ, just as we did before:

$$I_s = \frac{2e}{\hbar} g \Gamma_1 \quad \text{where} \quad g \approx \int_0^{\mu_1} \frac{\Gamma_\varphi dE}{(E - E_r)^2} \qquad (6.2.16)$$

Actually we should not be using the Lorentzian approximation far from resonance, but that only changes the precise values of K and g and does not affect the following argument.

Earlier we obtained the sequential current by arguing that the scattered current, when reinjected, comes out of terminals 1 and 2 in the ratio $\Gamma_1 : \Gamma_2$ (see Eq.(6.2.6)). However, in the present case the scattered current is reinjected at the resonance energy E_r where it must come entirely out of terminal 2, since there are no states in terminal 1 corresponding to this energy. Hence the sequential current should simply be equal to the scattered current:

$$I_{V,\text{seq}} = I_s = \frac{2e}{\hbar} g \Gamma_1 \qquad (6.2.17)$$

We are assuming the temperature to be low enough that once the electron has emitted a phonon and reached E_r, it cannot absorb a phonon and return to the upper energy. Adding Eqs.(6.2.16) and (6.2.17) we obtain the total valley current (per mode)

$$I_V = I_{V,\text{coh}} + I_{V,\text{seq}} = \frac{2e}{\hbar} \big[K \Gamma_1 + g \Gamma_1 \big] \qquad (6.2.18)$$

A little thinking shows, however, that there is something wrong with Eq.(6.2.18). Suppose we make barrier 2 so thick that Γ_2 (and hence K) is essentially zero. Eq.(6.2.18) still predicts a non-zero valley current equal to $(2e/\hbar)g\Gamma_1$! But how can any current flow if electrons cannot transmit across barrier 2? What happens for very thick barriers is that the resonant level fills up, since it cannot empty anywhere. Once the level is full, electrons can no longer scatter down into it so that the scattering rate $\Gamma_\varphi/\hbar$ becomes zero and the valley current goes to zero as expected. To include this effect we should modify Eq.(6.2.17) to write the sequential current as

$$I_{V,\text{seq}} = \frac{2e}{\hbar} g \Gamma_1 (1 - f_r) \qquad (6.2.19)$$

To complete the story we need to calculate the occupation factor f_r for the resonant level, which requires us to go beyond the scattering theory we have been using in most of this book. This is the basic difficulty with transport involving vertical flow that we discussed in Section 2.7.

It has been shown (see R. Lake *et al.* (1993), *Phys. Rev. B*, **48**, 15 132) that the correct occupation factor f_r for the resonant level can be obtained quite simply from the sequential model by modifying the rate equation as follows (cf. Eq.(6.2.2)):

$$I_1 \approx 2e\frac{g\Gamma_1}{\hbar}(1 - f_r) \quad \text{and} \quad I_2 \approx -2e\frac{\Gamma_2}{\hbar}f_r \qquad (6.2.20)$$

All that we have done is to reduce the rate constant for barrier 1 from Γ_1 to $g\Gamma_1$. This seems reasonable since an electron in order to get into the resonant level from terminal 1 must tunnel through barrier 1 *and* emit a phonon. Setting $I_1 + I_2 = 0$ as before we obtain

$$f_r = \frac{g\Gamma_1}{g\Gamma_1 + \Gamma_2} \qquad (6.2.21)$$

As long as $\Gamma_2 \gg g\Gamma_1$, the resonant level is essentially empty ($f_r \sim 0$) and the valley current is independent of Γ_2 as we had argued earlier. But once Γ_2 becomes comparable to $g\Gamma_1$ the level starts to fill up and the current decreases. Since the factor g is usually quite small, this requires very thick barriers.

Substituting from Eq.(6.2.21) into Eq.(6.2.19) we obtain a simple expression for the sequential component of the valley current (per mode):

$$I_{V,seq} = \frac{2e}{\hbar}\frac{g\Gamma_1\Gamma_2}{g\Gamma_1 + \Gamma_2} \qquad (6.2.22)$$

Generally we assume that the valley current will be a small fraction of the peak current, independent of Γ_1 and Γ_2. But it is easy to see from Eqs.(6.2.7) and (6.2.22) that this is not true in general:

$$I_{peak} \sim \frac{1}{\Gamma_1} + \frac{1}{\Gamma_2} \quad \text{and} \quad I_{valley} \sim \frac{1}{g\Gamma_1} + \frac{1}{\Gamma_2}$$

For very asymmetric devices having $\Gamma_2 \ll g\Gamma_1$, the peak and valley currents can even become equal! There is some experimental evidence for this (see P. J. Turley *et al.* (1993), *Phys. Rev. B*, **47**, 12 640).

Note that the simple result given in Eq.(6.2.22) could be derived only in the low temperature regime assuming that $f_1 = 1$ and $f_2 = 0$ in the energy range of interest. In general the sequential current depends on f_1 and f_2 in a complicated manner, as we had argued in Section 2.6, following Eq.(2.6.12).

6.3 Single-electron tunneling

Double-barrier devices are normally fabricated in a vertical configuration as shown in Fig. 6.1.1. However, similar devices can also be fabricated in a lateral configuration using metallic electrodes on a 2-DEG to implement potential barriers (Fig. 6.3.1). One advantage of these lateral structures is that a back-gate voltage can be used to change the electron density in the leads, so that we can study the change in the current flow through the structure as a function of the equilibrium Fermi energy.

Figure 6.3.2 shows the experimentally measured conductance as a function of the back-gate voltage. The conductance shows a series of evenly spaced sharp spikes. The question is whether we can understand this observation in terms of the resonant tunneling model that we have been

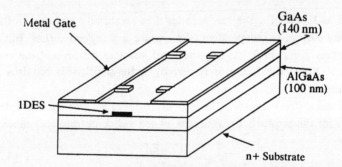

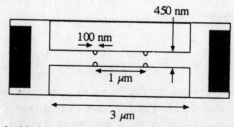

Fig. 6.3.1. Lateral double-barrier structure fabricated lithographically on a 2-DEG in a GaAs–AlGaAs heterostructure. Reproduced with permission from Fig. 7 of M. A. Kastner (1992), *Rev. Mod. Phys.* **64**, 849.

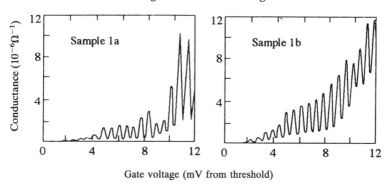

Fig. 6.3.2. Measured conductance vs. back gate voltage at $T \sim 100$ mK for two structures of the type shown in Fig. 6.3.1. Although the amplitudes of the peaks are different, the period of the oscillations is the same for both. Reproduced with permission from Fig. 8 of M. A. Kastner (1992), *Rev. Mod. Phys.* **64**, 849.

discussing. From this model we can write the linear response conductance as

$$G(E_f) = \sum_m \frac{e^2}{h} \frac{\Gamma_1^m \Gamma_2^m}{\Gamma_1^m + \Gamma_2^m} L(E_f - E_m) \qquad (6.3.1)$$

This is basically the same as the result we obtained for the vertical structures (see Eq.(6.2.14)). There are a few differences. Firstly we have left out the factor of two arising from the degeneracy of the spin levels; instead we let the index m denote a summation over spins as well. Secondly, the resonant energies for the vertical structures are given by $E_m = E_r + \varepsilon_m$ where E_r is the longitudinal (z-directed) resonance energy and ε_m is the subband energy arising from the x–y confinement. Usually the well width is small enough (a few nanometers) that the longitudinal resonances are hundreds of meV apart and we only need to consider the lowest of these (E_r). The lateral structures, on the other hand, are quite long (hundreds of nanometers) so that the longitudinal energies are quite close together. Thus in calculating the resonant energies E_m we have to consider both the longitudinal (x-directed) and the transverse (y-directed) confining potentials.

The third difference arises from the fact that since different longitudinal energies are involved there can be significant differences in the Γ_1 and Γ_2 for the different resonances m. This is because energy levels with higher longitudinal energies can penetrate the barriers much better. For this reason we have added a superscript m to the Γ_1 and Γ_2.

At first sight it might seem that we can understand the experimental data in Fig. 6.3.2 in terms of Eq.(6.3.1). As the Fermi energy is changed we expect a series of spikes corresponding to the resonance energies E_m. The difficulty with this interpretation is that for a 1 μm long well, the spacing ΔE between one-particle levels is about 20 μeV = $k_B(0.3\text{ K})$. Since this is comparable to the temperature k_BT, we do not expect to see such clear, well separated peaks. Moreover, we do not expect the peaks to be evenly spaced, since the spacing between one-particle energy levels is usually non-uniform. It is now believed that the experimental observations cannot be explained satisfactorily within the non-interacting resonant tunneling model described above. We need to include electron–electron interactions.

Electron–electron interactions

The usual procedure for including electron–electron interactions is to calculate the electron density in the structure and insert it into the Poisson equation to obtain the potential energy U_W in the quantum well. The conductance can then be written as

$$G(E_f) = \frac{e^2}{h} \sum_m \frac{\Gamma_1^m \Gamma_2^m}{\Gamma_1^m + \Gamma_2^m} L(E_f - E_m - U_W(E_f))$$

assuming that the resonance energy E_m simply floats up by an amount equal to the potential energy, U_W, in the well. As the Fermi energy is raised, the number of electrons, $\overline{N}$, in the well increases, thus increasing the well potential U_W.

We can use a simple model (see Fig. 6.3.3) to write down the potential in the well in terms of the number of electrons in the well

$$U_W(E_f) = U_0 \overline{N}(E_f) \quad \text{where} \quad C = C_1 + C_2 \quad \text{and} \quad U_0 = e^2/C$$

so that

$$G(E_f) = \frac{e^2}{h} \sum_m \frac{\Gamma_1^m \Gamma_2^m}{\Gamma_1^m + \Gamma_2^m} L\left[E_f - E_m - U_0 \overline{N}(E_f)\right] \tag{6.3.2}$$

However, this will not help us explain the experimental results any better than the non-interacting model. To see this let us assume that the number of electrons in the well increases linearly with the Fermi energy: $\overline{N}(E_f) = \alpha E_f$. The conductance in Eq.(6.3.2) can then be written as

$$G(E_f) = \frac{e^2}{h} \sum_m \frac{\Gamma_1^m \Gamma_2^m}{\Gamma_1^m + \Gamma_2^m} L\left[E_f(1 - \alpha U_0) - E_m\right]$$

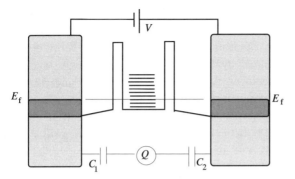

Fig. 6.3.3. Simple model for a double-barrier device at zero bias. Under bias a charge Q appears inside the well and the corresponding image charges appear in contacts 1 and 2 in the ratio C_1:C_2.

All that the interaction does is to stretch out the horizontal axis by the factor $(1 - \alpha U_0)^{-1}$ in a plot of the conductance vs. Fermi energy. This cannot lead to peaks in the conductance that were not there in the absence of interactions.

Charge quantization

The experimental results can be explained if we recognize that the number of electrons, $\overline{N}$, in the well is not a continuous variable as assumed above. It can only take on discrete values $N = 0, 1, 2$ etc. with different probabilities P_N, such that the average number $\overline{N}$ is given by

$$\overline{N}(E_f) = \sum_N NP_N(E_f)$$

From this point of view it seems that we should modify Eq.(6.3.2) for the conductance as follows:

$$G(E_f) = \frac{e^2}{h} \sum_m \frac{\Gamma_1^m \Gamma_2^m}{\Gamma_1^m + \Gamma_2^m} \sum_N P_N(E_f) L\big[E_f - E_m - NU_0\big] \qquad (6.3.3)$$

Suppose the dimension of the well is very large compared to a wavelength so that size quantization effects are negligible and the single particle levels E_m are very close together. Setting $E_m = 0$, we obtain

$$G(E_f) = \frac{e^2}{h} \sum_m \frac{\Gamma_1^m \Gamma_2^m}{\Gamma_1^m + \Gamma_2^m} \sum_N P_N L(E_f - NU_0)$$

so that there are peaks in the conductance every time the Fermi energy is an integer multiple of the single-electron charging energy $U_0 = e^2/C$. Thus we get peaks in a plot of the conductance versus Fermi energy which are uniformly separated by U_0, even though the energy levels E_m obtained from the Schrödinger equation are very close together. This single-electron tunneling effect arises from charge quantization, while the resonant tunneling phenomenon we discussed earlier arises from size quantization. It is believed that the experimental results shown in Fig. 6.3.2 are largely due to single-electron tunneling rather than resonant tunneling. These small-area (and hence low capacitance) double-barrier structures are often referred to as *quantum dots*.

It is important to recognize the profound conceptual step involved in going from Eq.(6.3.2) to Eq.(6.3.3).

$$L\big[E_f - E_m - \overline{N}U_0\big] \;\Rightarrow\; \sum_N P_N L\big[E_f - E_m - NU_0\big]$$

We usually associate the Hartree potential with the average number of electrons $\overline{N}$. But when the electron–electron interactions are strong (large U_0) we need to think of the Hartree potential as a statistical variable that can have different values NU_0 with different probabilities P_N. This is a major departure from the usual one-particle picture where it is assumed that individual particles feel some average potential due to their interaction with the surroundings.

The conductance expression given above (see Eq.(6.3.3)) is actually not quite right. The correct expression is (see Y. Meir, N. S. Wingreen and P. A. Lee (1991), *Phys. Rev. Lett.* **66**, 3048 and C. W. J. Beenakker (1991), *Phys. Rev. B*, **44**, 1646)

$$G(E_f) = \frac{e^2}{h} \sum_m \frac{\Gamma_1^m \Gamma_2^m}{\Gamma_1^m + \Gamma_2^m} \sum_N P_{N,m}(E_f) L\big[E_f - E_m - (N + 0.5)U_0\big] \quad (6.3.4)$$

where $P_{N,m}(E_f)$ is the probability that there are N electrons in the well, *not counting* any electron that may be occupying the one-particle level E_m. Comparing Eqs.(6.3.3) and (6.3.4) we notice two differences. The argument of the lineshape function L is a little different and the prefactor is a little different:

$$E_m + NU_0 \rightarrow E_m + (N + 0.5)U_0 \quad \text{and} \quad P_N \rightarrow P_{N,m}$$

To understand Eq.(6.3.4) we note that in an interacting system, peaks appear in the conductance whenever the Fermi energy coincides with a *one-particle excitation energy*, that is, whenever,

$$E_f = E(N+1,\alpha) - E(N,\beta)$$

where $E(N+1,\alpha)$ represents the energy of a particular $(N+1)$-particle state while $E(N,\beta)$ represents the energy of an N-particle state. Consider the states $(N+1,1_m)$ having the one-particle state E_m occupied and the state $(N,0_m)$ having the one-particle state E_m unoccupied:

$$E(N+1,1_m) - E(N,0_m) = \left[E_m + \frac{(N+1)^2 e^2}{2C} \right] - \frac{N^2 e^2}{2C} = E_m + (N+0.5)U_0$$

Consequently peaks appear in the conductance whenever the Fermi energy coincides with $E_m + (N+0.5)U_0$ so that the argument of the line-shape function is given by

$$E_f - E_m - (N+0.5)U_0$$

as stated in Eq.(6.3.4).

The strength of the peak is proportional to the sum of the probabilities of the two many-particle states contributing to this peak: $P(N+1,1_m)$ is the probability that there are $(N+1)$ electrons in the well with the level E_m occupied, and $P(N,0_m)$ is the probability that there are N electrons in the well with the level E_m empty. The reason we sum the two probabilities is that the device can conduct if it is in either of these states. If the device is in the state $(N,0_m)$ it conducts by having an electron hop into the level E_m and then hop out again:

$$(N,0_m) \;\; \rightarrow \;\; (N+1,1_m) \;\; \rightarrow \;\; (N,0_m)$$

If the device is in the state $(N+1,1_m)$ then it can conduct by having an electron hop out *first* and then hop in:

$$(N+1,1_m) \;\; \rightarrow \;\; (N,0_m) \;\; \rightarrow \;\; (N+1,1_m)$$

Both states thus contribute identically to the conductance which consequently is proportional to the sum of the probabilities of the two states:

$$P(N+1,1_m) + P(N,0_m) = P_{N,m}$$

This is slightly different from the factor, P_N, appearing in Eq.(6.3.3) which is given by

$$P_N = P(N,1_m) + P(N,0_m)$$

To apply Eq.(6.3.4) we need $P_{N,m}$, which are equilibrium probabilities, and can be evaluated by applying the standard methods of equilibrium statistical mechanics to the many-particle states of the quantum dot (see Exercise E.6.4 at the end of this chapter). This approach has been widely used to describe the experiments on semiconductor quantum dots. Note that we have only discussed the linear response of the system which is an equilibrium property. To go beyond linear response and calculate the full current–voltage characteristics we need to consider the detailed kinetics of the many-particle system. Even for linear response, Eq.(6.3.4) needs to be modified if we consider more complicated structures like quantum dot arrays where the charging energy does not have the same value U_0 for all states (see for example G. Klimeck *et al.* (1994), *Phys. Rev. B*, **50**, 2316, 5484 and references therein).

The differences between Eqs.(6.3.3) and (6.3.4) are insignificant if a large number of one-particle levels are involved and N is large. This is usually the case in metallic systems. It should be noted that the work on semiconductor quantum dots that we have described above was preceded by a lot of work on arrays of *metallic islands* where the one-particle levels are essentially continuous. For a discussion of the 'orthodox theory' applicable to this regime we refer the reader to Ref.[6.7].

We will not go into the phenomenon of single-electron tunneling in any more detail since this is a frontier topic in which our understanding is still in a state of rapid evolution. It is quite likely that in the coming years we will see many exciting developments in this field that will have important implications for both basic and applied physics. The above discussion is simply intended to alert the reader to this new direction in mesoscopic physics that requires a change in our way of thinking. We generally assume that even in an interacting system we can describe conduction processes in terms of individual quasi-particles that move in an average field (to be calculated self-consistently) produced by the surroundings. This viewpoint has been adequate for everything else that we have discussed in this book, but single-electron tunneling requires us to revise this conceptual framework.

Summary

Two important paradigms of mesoscopic transport have emerged from the study of artificial double-barrier structures, namely, resonant tunneling and single-electron tunneling. Resonant tunneling involves the flow of current through the discrete states formed between the barriers. The resonant tunneling current per transverse mode is given by

$$I_P = \frac{2e}{\hbar} \frac{\Gamma_1 \Gamma_2}{\Gamma_1 + \Gamma_2} \qquad \text{(same as Eq.(6.1.16))}$$

where Γ_1 and Γ_2 are the energy broadening of the level due to the coupling to the leads through barriers 1 and 2 respectively. Scattering processes inside the well have little effect on the total resonant current. However, this is not true when the device is biased past threshold. The off-resonant current flowing under these conditions depends strongly on the scattering rate, Γ_φ. The physics of the scattering must be properly taken into account in order to understand the dependence of the off-resonant current on the different parameters (Γ_1, Γ_2 and Γ_φ).

Single-electron tunneling is observed in double-barrier structures having a small cross-sectional area. If the well is long enough that the energy levels are very close together we do not expect to be able to resolve the resonant tunneling current through the individual levels. But if the capacitance (C) of the structure is small enough, we can still observe tunneling through discrete levels spaced by e^2/C. These discrete levels arise not from the wave nature of electrons (size quantization) but from their particle nature (charge quantization). This effect requires us to revise the one-particle picture that we are accustomed to.

Exercises

E.6.1 Consider an AlGaAs–GaAs–AlGaAs resonant tunneling diode with barrier widths of 50 Å and a well width of 50 Å. Assuming that the barrier height is 300 meV estimate $\Gamma_L = \Gamma_R$. Calculate the fraction of the current that is coherent, if the phase-relaxation time is ~ 1 ps.

E.6.2 Estimate the peak current density for the diode in E.6.1 neglecting the change in Γ_L and Γ_R under bias. Assume $\mu_1 = 10$ meV.

E.6.3 Show that the transmission probability corresponding to the scattered current is given by

$$T_S(E_L) \approx \frac{\Gamma_1 \Gamma_\varphi}{\Gamma_1 + \Gamma_2 + \Gamma_\varphi} A_\varphi (E_L - E_r) \qquad (6.2.13)$$

E.6.4 Single-electron tunneling Consider a double-barrier structure with two states having energies E_1 and E_2 as shown in Fig. E.6.4 (*neglect spin*). Apply the principles of equilibrium statistical mechanics to calculate the number of electrons inside the structure at $T = 0$ K, as a function of the Fermi energy E_f. Use Eq.(6.3.4) to show that there are two peaks in the

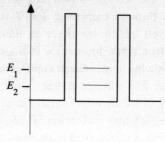

Fig. E.6.4. Double-barrier structure with two discrete states.

conductance spectrum $G(E_f)$ which occur at the energies where the number of electrons inside the structure changes from zero to one and from one to two. This is a consequence of the general principle that a structure can conduct only if the number of particles inside it can fluctuate between (at least) two possible values.

Further reading

A good collection of articles on the physics and device applications of resonant tunneling devices can be found in
[6.1] *Physics of Quantum Electron Devices*, ed. F. Capasso (1990), Heidelberg, Springer-Verlag.
Resonant Tunneling in Semiconductors: Physics and Applications, eds. L. L. Chang, E. E. Mendez and C. Tejedor (1991). New York, Plenum Press.
[6.2] *The Physics and Applications of Resonant Tunnelling Diodes*, H. Mizuta and T. Tanoue (1995), Cambridge University Press.

Some recent references on single-electron tunneling are
[6.3] *Single Charge Tunneling: Coulomb Blockade Phenomena in Nanostructures*, eds. H. Grabert and M. H. Devoret (1992), NATO ASI Series B, Physics, vol. 294, New York, Plenum Press. In the last chapter of this book Averin and Likharev explore the possibility of device applications based on single-electron tunneling.

[6.4] *Z. Phys. B* (1991). Special Issue on Single Charge Tunneling, **85**, 317.

[6.5] Staring, A. A. M. (1992). 'Coulomb blockade oscillations in quantum dots and wires' Ph.D. Thesis, Technical University at Eindhoven.

[6.6] Kastner, M. (1993). 'Artificial atoms', *Physics Today*, **46**, 24.

We have only discussed the recent work on single-electron tunneling in semiconductors. For an account of the earlier work on metals, see

[6.7] Averin, D. V. and Likharev, K. (1991) in *Mesoscopic Phenomena in Solids*, eds. B. L. Altshuler, P. A. Lee and R. A. Webb, Amsterdam Elsevier.

7

Optical analogies

The propagation of electrons has many interesting similarities with the propagation of light. Mesoscopic phenomena often have familiar optical analogies. We have generally not emphasized these analogies since they can be somewhat distracting especially if one is unfamiliar with the optical counterpart. However, the analogies are quite interesting and can provide useful insights. In this chapter we will briefly explore the similarities and differences between electron waves and electromagnetic waves. The discussion is qualitative and can be read without much reference to any other chapter in this book.

7.1 Electrons and photons: conceptual similarities

The propagation of photons is described by the Maxwell equation (in simplified form)

$$\frac{\partial^2 \mathbf{E}}{\partial t^2} = \frac{1}{\mu \varepsilon} \nabla^2 \mathbf{E} \qquad (7.1.1a)$$

just as the propagation of electrons is described by the Schrödinger equation

$$i\hbar \frac{\partial \Psi}{\partial t} = -\frac{\hbar^2}{2m} \nabla^2 \Psi + U\Psi \qquad (7.1.1b)$$

276

This analogy may not seem quite right since Ψ is the wavefunction whose square gives the probability while the electric field $\mathbf{E}$ is a measurable macroscopic field. However, we can view the electric field as the wavefunction of the photon although this is not the interpretation Maxwell had in mind when he wrote his famous equations. Suppose we have solved Maxwell's equations to find that a given medium has a transmission probability of 50%. This means that if an electromagnetic wave carrying 1 W/cm^2 is incident on the medium, the transmitted wave will be carrying only 0.5 W/cm^2. However, we could also interpret it as saying that if a photon is incident on the medium then there is a 50% probability that it will be transmitted. If one million photons are incident then on the average half a million will be transmitted. With this interpretation, $\mathbf{E}$ for photons becomes a concept analogous to the concept of Ψ for electrons and the Maxwell equation plays the same role for photons as the Schrödinger equation plays for electrons. For further discussion of this interpretation of the Maxwell equation we refer the reader to D. Marcuse, (1980), *Principles of Quantum Electronics*, Chapter 10, (New York, Academic Press) and to R. P. Feynman (1985), *QED: The Strange Theory of Light and Matter* (Princeton University Press).

Frequency vs. energy

In describing steady-state phenomena involving electromagnetic waves it is common to talk in terms of monochromatic waves having a single frequency. Similarly with electrons it is convenient to talk in terms of monoenergetic waves:

$$\mathbf{E}(r,t) \rightarrow \mathbf{E}(r)\exp(-i\omega t) \quad \Psi(r,t) \rightarrow \Psi(r)\exp(-iEt/\hbar)$$

Substituting these forms for the solution into Eqs.(7.1.1a,b) we obtain the time-independent forms of these equations:

$$\nabla^2\Psi = -\frac{2m}{\hbar^2}(E-U)\Psi \qquad (7.1.2a)$$

and
$$\nabla^2\mathbf{E} = -\omega^2\mu\varepsilon\,\mathbf{E} \qquad (7.1.2b)$$

Dispersion relation

In a homogeneous medium the solutions to Eqs.(7.1.2a,b) can be expressed in the form of plane waves $\exp[i\mathbf{k}.\mathbf{r}]$, with the magnitude of the wavevector **k** related to the frequency (or energy) by the dispersion relation:

$$k^2 = \omega^2 \mu \varepsilon \qquad \text{and} \qquad k^2 = \frac{2m}{\hbar^2}[E - U] \qquad (7.1.3)$$

The correspondence between photons and electrons is summarized in Fig. 7.1.1.

Before proceeding we should mention that the Schrödinger equation in the presence of a vector potential (representing magnetic fields)

$$\left[\frac{(\mathbf{p} - e\mathbf{A})^2}{2m} + U\right]\Psi(\mathbf{r}) = E\Psi(\mathbf{r}) \qquad (7.1.4)$$

looks somewhat different and we are not aware of any simple analogies with the propagation of light that would help us understand the behavior of electrons in a magnetic field. As we know, the magnetic field affects

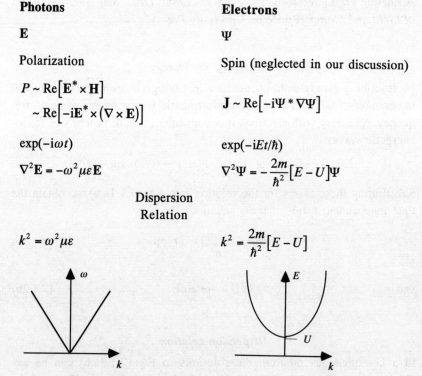

Photons   **Electrons**

E Ψ

Polarization   Spin (neglected in our discussion)

$P \sim \text{Re}\left[\mathbf{E}^* \times \mathbf{H}\right]$
$\sim \text{Re}\left[-i\mathbf{E}^* \times (\nabla \times \mathbf{E})\right]$ $J \sim \text{Re}\left[-i\Psi^* \nabla \Psi\right]$

$\exp(-i\omega t)$ $\exp(-iEt/\hbar)$

$\nabla^2 \mathbf{E} = -\omega^2 \mu \varepsilon \mathbf{E}$ $\nabla^2 \Psi = -\dfrac{2m}{\hbar^2}[E - U]\Psi$

Dispersion
Relation

$k^2 = \omega^2 \mu \varepsilon$ $k^2 = \dfrac{2m}{\hbar^2}[E - U]$

Fig. 7.1.1. Correspondence between photons and electrons.

the electronic motion differently depending on its strength. At low magnetic fields one is in the 'geometrical optics' limit and the motion of electrons is described just as well by Newton's laws. An electron with velocity v in a magnetic field B describes a circular trajectory of radius mv/eB. But at high magnetic fields, Landau levels form and there is a spectacular suppression of backscattering leading to the quantum Hall effect. None of this, we believe, has any optical analogs.

7.2 Linear optics

Most interesting structures involve wave propagation through an inhomogeneous medium of some kind. For photons this could mean a medium with a spatially varying dielectric constant $\varepsilon(\mathbf{r})$ while for electrons it could mean a medium with a spatially varying potential $U(\mathbf{r})$. Looking at Eqs.(7.1.2a,b) it is apparent that the two are analogous. Photons propagating through a medium with a non-uniform dielectric constant $\varepsilon(\mathbf{r})$ and electrons propagating through a medium with a non-uniform potential $U(\mathbf{r})$ are described by similar looking equations with the following correspondence:

$$\omega\sqrt{\mu\varepsilon(\mathbf{r})} \quad \Leftrightarrow \quad \frac{\sqrt{2m(E-U(r))}}{\hbar} \tag{7.2.1}$$

A simple example of this analogy can be found in the way *waveguides* are constructed. We know that electrons tend to be confined in layers having a lower potential energy than the surroundings. For example, quantum wells are formed by sandwiching a layer of GaAs (with a smaller potential energy) between two layers of AlGaAs. From the correspondence stated above (see Eq.(7.2.1)) we would expect light to be guided in a region with a higher dielectric constant $\varepsilon(\mathbf{r})$ (see Fig. 7.2.1). Indeed this

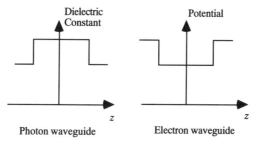

Photon waveguide   Electron waveguide

Fig. 7.2.1. Light is guided in the layer with the higher dielectric constant ($\varepsilon(\mathbf{r})$) while electrons are guided in the region with the lower potential ($U(\mathbf{r})$).

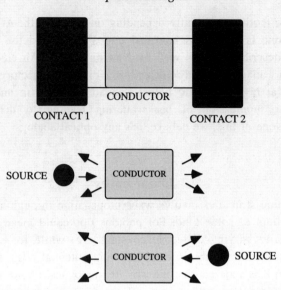

Fig. 7.2.2. Resistance measurement can be viewed as a scattering problem where each contact launches electrons toward the conductor which acts as a scatterer.

is precisely how optical waveguides are constructed and the mathematics for calculating the wavefunctions in quantum wells is very similar to the mathematics used in calculating the field profile in optical waveguides.

The Landauer approach to transport underlines the fact that any conductor is essentially a scatterer and that the measurement of conductance is essentially a measurement of the scattering properties of the conductor (Fig. 7.2.2). The Landauer formula (see Eq.(2.5.3)) states that the conductance G at low temperatures is proportional to the sum of the transmission probabilities $T(m \leftarrow n)$ from all possible input modes n to all possible output modes m at the Fermi energy E_f.

$$G \sim \sum_m \sum_n T(m \leftarrow n; E_f)$$

On the other hand, light scattering experiments allow us to probe the individual transmission probabilities $T(m \leftarrow n; \omega)$ by measuring the scattered wave amplitudes in various directions (modes) in response to incident waves in a particular direction (mode). In conductance measurements we do not get this 'mode-resolved' information: what we measure represents a sum over all input and output modes.

This drawback is partially overcome by studying conductance as a function of magnetic field. For example, the weak localization effect discussed in Chapter 5 can be detected in optical experiments by measuring the backscattering as a function of angle (see Fig. 5.2.2). But conductance measurements do not give such angle-resolved information. The conductance is a little lower due to the enhanced backscattering, but that could easily be ascribed to a variety of other causes (maybe the conductor is not as wide as we think!). What allows us to identify the effect unambiguously is its variation with magnetic field as discussed in Section 5.3.

Another point to note is that at higher temperatures the conductance is not determined just by the scattering properties of the electrons with energy E_f. All electrons having energies within a few $k_B T$ of the Fermi energy contribute. In optics this is analogous to using a non-monochromatic source with a spread in the frequency ω. The scattering viewpoint helps emphasize the many analogies between the propagation of electron waves through different types of conductors and the propagation of electromagnetic (or photon) waves through different types of media.

Let us look at a few examples.

Geometrical optics

Focusing by optical lenses is a well-known phenomenon. From the correspondence stated above (see Eq.(7.2.1)), we would expect that an electron lens would be formed by a convex metallic gate with a voltage applied to it such that the potential energy U underneath it is lower than that in the surrounding region. This requires an increase in the electron density (or in other words an accumulation of electrons) under the lens. In practice it is usually more convenient to deplete the lens area. To obtain a focusing action it is then necessary to use a concave lens as shown in Fig. 7.2.3. With light too we would obtain focusing with concave lenses if the lens were made of a material with a dielectric constant lower than that of the surroundings (which is not the usual case).

Electron focusing has recently been demonstrated in a high mobility 2-DEG ($\mu = 10^7$ cm^2/V s) in GaAs. It was shown that the current reaching the detector (marked 'D' in Fig. 7.2.3) from the emitter (marked 'E') exhibits a peak at a particular voltage on the lens when the focal length is just right. Of course, electron focusing experiments as such are not novel. Far better focusing apparatus are routinely manufactured in television sets or electron microscopes. What is novel is the fact that the

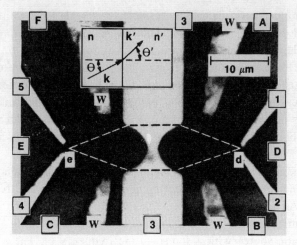

Fig. 7.2.3. Photograph of the sample used to demonstrate focusing. Reproduced with permission from J. Spector, H. L. Stormer, K. W. Baldwin, L. N. Pfeiffer and K. W. West (1990), *Appl. Phys. Lett.* **56**, 1290. Similar results were reported by U. Sivan, M. Heiblum, C. P. Umbach and H. Shtrikman (1990), *Phys. Rev. B*, **41**, 7937.

experiments are conducted in a solid and not in vacuum. Typically such experiments would not work in solids because of the strong scattering processes. However, the extremely high mobility of the samples used in these experiments leads to scattering lengths ~ 60 μm. which is longer than the length of the device. Consequently a significant fraction of the injected electrons can get from one contact to another without being scattered.

Electron focusing experiments belong to the category of 'geometrical optics', where the potential energy varies slowly on the scale of an electron wavelength. Such phenomena are not really quantum mechanical in origin. They can be understood without invoking the wave nature of electrons. In other words we can describe focusing phenomena simply in terms of Newton's laws; it is not necessary to use the Schrödinger equation. Also, such phenomena can be observed even if the phase-relaxation length is short, since phase plays no role here. The relevant length scale is the mean free path determined by the momentum relaxation length.

Wave optics

If the potential varies rapidly on the scale of a wavelength then we have phenomena analogous to 'wave optics' which cannot be described by Newton's laws. The classic wave optical experiment is the Young's

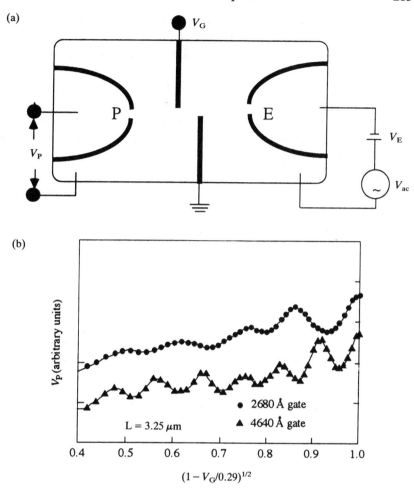

Fig. 7.2.4. A 'double-slit' experiment in high mobility GaAs. (a) Schematic picture of the structure. The dark lines are metallic gates deposited on the surface of a GaAs/AlGaAs heterostructure (see Section 1.1) that are used to deplete the underlying layer. (b) Measured V_P as a function of V_G. L is the separation between the emitter (E) and the detecting probe (P). (Adapted with permission from Figs. 1 and 2 of A. Yacoby, U. Sivan, C. P. Umbach and J. M. Hong (1991), *Phys. Rev. Lett.* **66**, 1938)

double-slit experiment (for a very interesting discussion comparing the double-slit experiment for electrons and photons, see R. P. Feynman (1965), *Lectures on Physics*, vol.III, Chapter 1 (New York, Addison–Wesley)). Similar experiments with electrons in GaAs have been reported using very high mobility samples (see Fig. 7.2.4).

Electrons are injected from the emitter (marked E) through a small orifice between two gates (which deplete the underlying layer) and are detected by the probe at the other end (marked P). The probability that an injected electron from the emitter will transmit to the probe is determined by the sum of the amplitudes associated with all the classical paths connecting the emitter and collector orifices. We could divide all such paths into two distinct groups: those that lie in the upper half and those that lie in the lower half. Since we neglect all scattering processes we do not need to worry about complicated paths that lie in both halves. Let a_1 and a_2 be the total amplitudes associated with the two groups of paths, so that the probability of an electron being detected at the collector is proportional to the squared magnitude of the sum of a_1 and a_2. We can write the transmission function as

$$\overline{T} \sim |a_1 + a_2|^2 = |a_1|^2 + |a_2|^2 + 2|a_1||a_2|\cos\theta$$

where θ is the phase difference between a_1 and a_2. Now a potential eV_G applied to the top gate changes the magnitude of the wavevector $\mathbf{k}$ for all the paths lying in the upper half (see Eq.(7.1.3)), thus changing θ.

$$\Delta k = \frac{1}{h}\left[\sqrt{2mE} - \sqrt{2m(E - eV_G)}\right] \Rightarrow \Delta\theta = \Delta k.d = \sqrt{\frac{2mE}{h}}d\left[1 - \sqrt{1 - \frac{eV_G}{E}}\right]$$

where d is the gate length. We thus expect the voltage V_P (which is proportional to the transmission) to oscillate periodically as a function of $[1 - (eV_G/E)]^{1/2}$. This is observed experimentally as shown in Fig. 7.2.4. Also the period of oscillation is smaller if the gate length d is made longer, as we would expect from the above discussion.

Diffusive optics

So far we have been talking about clean ballistic conductors where electrons can propagate from one contact to another without appreciable scattering. Ballistic transport is a relatively recent phenomenon. More commonly we are interested in electronic transport in diffusive conductors which is like the passage of photons through 'foggy' media. Propagation of electromagnetic waves through a random medium with multiple scattering is a problem of great practical importance that occurs in many different contexts. For example, as we discussed in Chapter 5, the phenomenon of weak localization arises from the enhanced backscattering from a random array of scatterers as shown in Fig. 5.2.2. Similar effects have been observed in connection with radar scattering from the clouds. It is interesting to note that a figure similar to Fig. 5.2.2 appears in the book by Ishimaru entitled *Wave Propagation and Scattering in Random Media* (New York, Academic Press, 1978) (see Fig. 15-3, p.311). However, the effect of magnetic field on weak localization (which makes it observable in solids) has no optical counterpart.

One of the significant discoveries in mesoscopic physics is the phenomenon of conductance fluctuations (see Section 5.4) which arises from correlations between the transmission probabilities $T(m,n)$ for different output and input modes m and n. In optical experiments the same basic phenomenon shows up in the form of correlations between the intensities of light transmitted through a random medium in different directions (see Ref. [5.5]).

Classical wave propagation through random media has been described theoretically from two different approaches. One is the transport theory approach (also known as radiative transfer) where the basic quantity is the intensity distribution $I(\mathbf{r},\mathbf{k})$ at different points $\mathbf{r}$ and wavevectors $\mathbf{k}$. We could describe this as a Boltzmann approach to photon transport. The other approach, based on the wave equation, is often referred to as multiple scattering theory. The basic quantity in this approach is the coherence function describing the correlation between the electric fields at different points in space and time. This is analogous to the Green's function formalism that we will discuss in Chapter 8. In the book by Ishimaru (see Section 14-7) quoted above, there is an interesting discussion of the conditions under which the equations of multiple scattering theory can be reduced to the equations of radiative transfer. Similar questions regarding the conditions under which the equations of quantum transport (the Green's function formalism) reduce to the Boltzmann equation have occupied many theorists interested in electronic transport. We will also discuss this question briefly in Chapter 8.

7.3 Non-linear optics

Earlier we pointed out the similarity between the Schrödinger equation and Maxwell's equation and noted that the potential energy for electrons played a role similar to that played by the dielectric constant for light:

$$\omega\sqrt{\mu\varepsilon(\mathbf{r})} \quad \Leftrightarrow \quad \frac{\sqrt{2m(E-U(r))}}{\hbar} \quad \text{(same as Eq.(7.2.1))}$$

The field of non-linear optics relies on materials having a dielectric

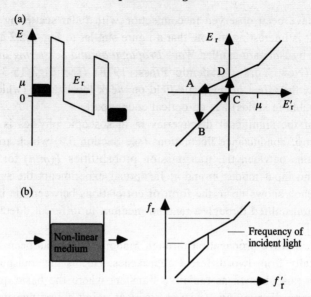

Fig. 7.3.1 (*a*) Bistable path of the resonant energy as a function of the applied bias in a resonant tunneling device. (*b*) Optical analogy.

constant that is a function of the light intensity. For electrons this corresponds to having a potential U (analogous to the refractive index) that is a function of the electron density (analogous to the intensity of light). Since electrons are charged particles this is a strong zero-order effect in contrast to optical non-linearities which are weak second-order effects.

In Chapter 6 we discussed the resonant tunneling diode where an electron is trapped in the region between two barriers (Fig. 7.3.1a). As a result only electrons having an energy equal to the resonant energy E_r can transmit through the structure. A build-up of electron density in the resonant level causes an increase in the potential U in the well thus shifting the resonant level to higher energies. We can write

$$E_r = E'_r + Kn_w \tag{7.3.1a}$$

where E'_r denotes the resonant energy if charging effects were absent, n_w denotes the electron density inside the well and K is a constant depending on the effective capacitance. As we increase the applied bias we pull down the resonant energy. Once E'_r crosses μ there is charge accumulation in the well, n_w increases and so the resonant energy E_r changes slowly as shown. Eventually the resonant energy becomes zero and discharges suddenly causing a sudden lowering of E_r from 'A' to 'B'. If we now decrease the bias, the resonant energy will follow a different path from 'B' to 'C' since there is no charge in the well. It will then charge up to 'D'. This difference in the path of the resonance for increasing and decreasing bias can lead to bistability in the current–voltage characteristics. Specially designed double-barrier devices are needed to distinguish this intrinsic bistability from the extrinsic bistability exhibited by any device having a negative differential resistance. We will not discuss these details since our main purpose here is to point out the analogy with optics. For further details see Zaslavsky *et al.* (1988), *Appl. Phys. Lett.* **53**, 1408; Alves *et al.*, (1988), *Electron. Lett.*, **24**, 1190; and G. A. Toombs and F. W. Sheard (1990), Chapter in *Electronic Properties of Multilayers and Low-dimensional Semiconductor Structures*, eds. J. M. Chamberlain *et al.*, Plenum, New York.

Optical bistability in non-linear Fabry–Perot interferometers is a very well-known phenomenon in optics. The resonant frequency f_r of a Fabry–Perot interferometer is related to the distance d between the two mirrors by the relation

$$f_r = \frac{c}{2nd} \times (\text{integer})$$

where n is the refractive index of the medium and c is the velocity of light. In a non-linear medium the index is intensity dependent so that we could express the resonant frequency in a form analogous to Eq.(7.3.1a) for the resonant energy of a resonant tunneling diode:

$$f_r = f'_r + KI_w \tag{7.3.1b}$$

where f'_r denotes the resonant frequency if non-linear effects were absent, I_w denotes the light intensity inside the interferometer and K is a constant depending on the strength of the non-linearity. Suppose we could change f'_r by some external means. We would then observe a bistable behavior of the resonant frequency analogous to that described above for the electronic case. Once f'_r comes close to the frequency of the incident light, the light intensity will build up and cause the resonant frequency to change slowly (or faster depending on the sign of K). But the intensity drops quickly to zero when the resonant frequency has changed sufficiently that the incident frequency falls outside the pass-band of the interferometer. In practice the experiment is usually done by changing the frequency of the incident light instead of changing f'_r, but the effect is the same.

Another example of the analogy between charging effects and nonlinear optics can be found in the article by Bergmann (see *Phys. Rev. B*, **35**, 4205 (1987)) where the effect of electron–electron interaction on weak localization phenomena has been shown to be similar to diffraction by holograms.

However, it should be noted that there are subtle differences between electronic and optical phenomena. With light, non-linear effects are usually accompanied by a large build-up in the intensity in a particular mode. But with electrons there can be no more than one electron in one mode. Non-linear effects usually arise from a build-up of electron density in many modes. Recent work has shown that with electrons the non-linearity is large enough that even a single electron can cause a significant shift in the resonant energy, leading to the single-electron tunneling effect discussed in Section 6.3. But this phenomenon is very different from the non-linear optical phenomena discussed above. We are not aware of *single-photon* effects analogous to the single-electron tunneling phenomenon. An interesting question is whether miniature optical cavities can be constructed with a non-linearity strong enough for this to be observed.

7.4 Coherent sources

An important ingredient in interference experiments is a 'coherent' source. Let us try to explain the meaning of the word 'coherent' which can denote different things in different contexts. Generally a light beam will contain many temporal and spatial frequencies, that is, its intensity $I(\omega,\mathbf{k})$ will have a large spread in both the frequency ω and the wavevector $\mathbf{k}$. By passing it through a filter or monochromator we can generate a beam with a very narrow frequency spectrum. Such a beam has high temporal coherence (time is the Fourier transform of ω) but it may not have much spatial coherence since there can still be wavevector components $\mathbf{k}$ pointing in all directions.

To generate a beam having spatial coherence as well, we need a collimator that selects out only the wavevectors lying within a narrow range of angles. The easiest way to do so is to use the light from a source that is located very far away. The region of observation then occupies a very small fraction of the total solid angle surrounding the source. Sunlight, for example, has excellent spatial coherence.

Similar arguments hold for electron waves as well. Electron diffraction experiments are routinely used to probe crystal structures and monoenergetic well-collimated sources (that is, sources with high spatial and temporal coherence) are used for the purpose. In mesoscopic conductors temporal coherence is achieved at low temperatures since only the electrons having the Fermi energy contribute to the conductance. Spatial coherence is achieved by going to small-area devices having a limited number of transverse modes. However, there is a basic difference with light, as we saw in Section 7.3. There is no limit to the number of photons one can put into a particular mode (ω,$\mathbf{k}$) but the exclusion principle limits the number of electrons that can occupy a particular mode. As a result, there is no limit to the amount of power per unit frequency that a single-moded fiber can transmit but a single-moded quantum wire can only transmit a maximum of 80 nA in an energy range of 1 meV (= $2e/h$).

One of the most significant developments in the history of optics is the invention of the laser which has made possible a wide variety of experiments that were inconceivable before the advent of the laser. This is largely because laser sources concentrate very large amounts of power into a single coherent beam, while coherent beams derived from thermal sources by filtering and collimating tend to be very weak since we are selecting a small fraction of the total incoherent output. But there is a more fundamental difference between laser sources and filtered thermal sources.

With thermal sources interference phenomena can only be observed if the interfering beams are derived from the same source as in the famous double-slit experiment (see Fig. 7.4.1a). Interference requires that there be multiple paths from the same source (initial state) to the same detector (final state). But with laser sources it is possible to observe interference between separate sources as shown in Fig. 7.4.1b. The reason is that all the photons from a laser are in a coherent state (this is different from the coherence discussed above) such that their electric fields add up to give a macroscopic field with a definite magnitude and phase, just like the output from low frequency oscillators. Two separate sources thus have a definite phase relationship that can be detected in an interference experiment. But the electric fields due to photons from a thermal source are all randomly phased so that there is no definite phase relationship between independent sources. For further discussion of this point see D. Marcuse (1980), *Principles of Quantum Electronics*, Chapter 5 (New York, Academic Press).

With normal electrons there is no equivalent of a laser source because of the exclusion principle. Thus all the interference phenomena observed with normal electrons are of the single source variety (Fig. 7.4.1a).

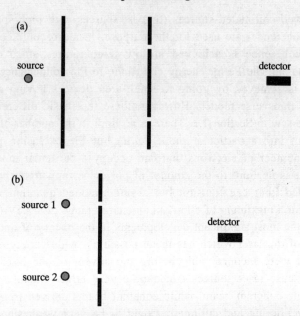

Fig. 7.4.1. Double-slit interference experiment with (*a*) single source and (*b*) separate sources. In (*b*) interference can only be observed with laser sources or superconducting electrons but not with thermal sources or normal electrons.

However, superconducting electrons are analogous to laser sources where numerous electrons occupy a single state coherently to build up a macroscopic wavefunction with a definite phase. This makes it possible to observe interference between separate sources. The Josephson effect is an example of an interference phenomenon between two separate sources, namely, the two contacts (see R. P. Feynman, *Lectures on Physics* (1965), vol.III, p.21–14, (New York, Addison–Wesley)). An interesting direction in which mesoscopic physics is currently evolving is the study of current flow in small devices with *superconducting contacts* (see, for example, P. G. N. de Vegvar *et al.* (1994), *Phys. Rev. Lett.* **73**, 1416).

Summary

Maxwell's equation for photons is analogous to the Schrödinger equation for electrons, with the electric field **E** for photons playing the role played by the wavefunction Ψ for electrons. Comparing the time-independent forms of Maxwell's equation and the Schrödinger equation it is evident that the propagation of electrons through a region with a varying potential energy is analogous to the propagation of light through a region with a varying refractive index. Experiments on mesoscopic samples have demonstrated the electronic analog of geometrical optics, wave optics, guided waves, diffusive optics etc.

Optical non-linearity is a weak second-order effect arising from the dependence of the refractive index on the light intensity. The electronic analog of this is the dependence of the potential on the electron density which is a strong effect because electrons are charged particles. Indeed the effect is so strong that even single-electron charging effects can be observed in appropriate structures. The author is not aware of analogous single photon non-linearities.

An important point to bear in mind is that normal electrons are like photons from thermal sources. Superconducting electrons, on the other hand, are like laser sources and could lead to interesting new phenomena when combined with mesoscopic structures.

Exercises

E.7.1 (a) Consider a beam of electrons with energy E incident at an angle on an interface where the potential energy steps up from 0 to U as shown in Fig. E.7.1. Derive a 'Snell's law' relating the incident and transmitted angles.

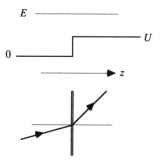

Fig. E.7.1. A beam of electrons with energy E is incident at an angle on an interface where the potential energy steps up from 0 to U.

(b) Suppose the potential energy is constant across the interface but the effective mass changes from m_1 to m_2. What is the new 'Snell's law'?

Further reading

[7.1] Henderson G. N., Gaylord T. K. and Glytsis E. N. (1991). 'Ballistic electron transport in semiconductor heterostructures and its analogies in electromagnetic propagation in dielectrics' *Proc. IEEE*, **79**, 1643.

[7.2] *Analogies in Optics and Microelectronics*, eds. W. van Haeringen and D. Lenstra (1990), Kluwer Academic Publishers, Netherlands.

8

Non-equilibrium Green's function formalism

So far in this book we have described the effect of electron–phonon or electron–electron interactions in phenomenological terms, through a phase-relaxation time. In this chapter we will describe the non-equilibrium Green's function (NEGF) formalism which provides a microscopic theory for quantum transport including interactions. We will introduce this formalism using simple kinetic arguments based on a one-particle picture that are only slightly more difficult than those used to derive semiclassical transport theories like the Boltzmann equation. This heuristic description is not intended as a substitute for the rigorous descriptions available in the literature [8.1–8.8]. Our intention is simply to make the formalism accessible to readers unfamiliar with the language of second quantization. We will restrict our discussion to steady-state transport as we have done throughout this book.

The NEGF formalism (sometimes referred to as the Keldysh formalism) requires a number of new concepts like correlation functions which we introduce in Sections 8.1 and 8.2. We then describe the formalism in

293

Sections 8.3–8.6. In Section 8.7 we relate it to the Landauer–Büttiker formalism which, as we have seen, has been very successful in describing mesoscopic phenomena. For non-interacting transport the two are equivalent, and the added conceptual complexity of the NEGF formalism is not necessary. The real power of this formalism lies in providing a general approach for describing quantum transport in the presence of interactions.

The Boltzmann formalism has found widespread use in describing a wide variety of semiclassical transport phenomena. It combines Newton's law with a probabilistic description of the dissipative interaction with random scattering forces. The NEGF formalism, on the other hand, combines quantum dynamics with a statistical description of the dissipative interactions. In Section 8.8 we discuss the conceptual similarities and differences between the two formalisms.

Much of our understanding of electronic transport is based on the one-particle picture where an individual particle is assumed to move in an effective potential due to its interaction with the surroundings. As mesoscopic conductors get smaller and begin to resemble large molecules, this concept of an effective potential is breaking down as we saw in Section 6.3 in our discussion of single-electron tunneling. In Section 8.9 we briefly discuss the applicability of the NEGF formalism to this novel transport regime involving strong interactions.

We end this chapter with a simple analytical example illustrating the application of this formalism to a concrete problem, namely, the effect of phonon scattering on the current–voltage characteristics of resonant tunneling diodes (Section 8.10).

8.1 Correlation and scattering functions

A number of different Green's functions like G^R, G^A, $G^<$ and $G^>$ appear in the NEGF formalism which all look similar to the beginner. However, the physical significance of G^R and G^A is very different from that of $G^<$ and $G^>$. Similarly the physical significance of the self-energy functions Σ^R, Σ^A is very different from that of $\Sigma^<$, $\Sigma^>$. We will only discuss the functions $G^<$, $G^>$, $\Sigma^<$ and $\Sigma^>$ in this section. In the next section we will discuss the retarded and advanced functions G^R, G^A, Σ^R and Σ^A. Let us start with a concept that is closely related to the correlation function $G^<$, namely, the density matrix.

Density matrix

Consider a homogeneous conductor whose eigenstates are plane waves labeled by their wavevector **k**:

$$|\mathbf{k}\rangle \equiv \frac{1}{\sqrt{V}} e^{i\mathbf{k}.\mathbf{r}} \quad (V = \text{normalization volume})$$

In the semiclassical picture we can describe the electrons by specifying the distribution function $f(\mathbf{k})$ which tells us the number of electrons occupying a particular state **k**. But in the quantum mechanical picture this is not enough. We also need to specify the phase-relationship among the different states. To describe the state of the electrons fully we have to define a density matrix $\rho(\mathbf{k},\mathbf{k}')$. An electron with a wavefunction

$$\sum_{\mathbf{k}} \Psi_{\mathbf{k}} |\mathbf{k}\rangle$$

has a density matrix $\rho(\mathbf{k},\mathbf{k}') = \Psi_{\mathbf{k}}\Psi_{\mathbf{k}'}^{*}$, so that an electron occupying a standing wave state

$$\frac{1}{\sqrt{2}}|+\mathbf{k}\rangle + \frac{1}{\sqrt{2}}|-\mathbf{k}\rangle = \sqrt{2}\cos(\mathbf{k}.\mathbf{r})$$

has a density matrix of the form

$$\begin{array}{c} \quad (+\mathbf{k} \quad -\mathbf{k}) \\ \begin{pmatrix} +\mathbf{k} \\ -\mathbf{k} \end{pmatrix} \begin{bmatrix} 0.5 & 0.5 \\ 0.5 & 0.5 \end{bmatrix} \end{array}$$

This, however, is very different from having a system of electrons 50% of which occupy the state $|\mathbf{k}\rangle$ and 50% occupy the state $|-\mathbf{k}\rangle$. The density matrix is then given by

$$0.5 \times \begin{bmatrix} 1 & 0 \\ 0 & 0 \end{bmatrix} + 0.5 \times \begin{bmatrix} 0 & 0 \\ 0 & 1 \end{bmatrix} = \begin{bmatrix} 0.5 & 0 \\ 0 & 0.5 \end{bmatrix}$$

In the semiclassical picture we implicitly assume a situation like this, where all off-diagonal elements are negligible. We can then describe the electrons by a distribution function $f(\mathbf{k})$ given by the diagonal elements of the density matrix $\rho(\mathbf{k},\mathbf{k})$. This is an accurate picture if the phase-relaxation length is very short. But in phase-coherent conductors the off-diagonal elements cannot be ignored.

As we mentioned in Chapter 3, there is a simple optical analogy to this. A beam of unpolarized light is a 50-50 mixture of photons that are polarized in the x-direction and photons that are polarized in the y-direction. So is a beam of light that is polarized at 45 degrees to the x-axis. But the two are physically very different and the difference is clearly expressed in terms of the density matrix:

$$
\begin{array}{cc} (x & y) \end{array} \qquad\qquad \begin{array}{cc} (x & y) \end{array}
$$

$$
\begin{pmatrix} x \\ y \end{pmatrix} \begin{bmatrix} 0.5 & 0 \\ 0 & 0.5 \end{bmatrix} \qquad \begin{pmatrix} x \\ y \end{pmatrix} \begin{bmatrix} 0.5 & 0.5 \\ 0.5 & 0.5 \end{bmatrix}
$$

(unpolarized)   (45-degree polarized)

Different representations

It will be noted that although we have used a representation in terms of **k**-states to define the correlation function, we can always transform to other representations using an appropriate unitary transformation. For example we could transform to a real space representation as follows:

$$
\rho(\mathbf{r}, \mathbf{r}') = \langle \mathbf{r} | \rho | \mathbf{r}' \rangle = \sum_{\mathbf{k},\mathbf{k}'} \langle \mathbf{r} | \mathbf{k} \rangle \langle \mathbf{k} | \rho | \mathbf{k}' \rangle \langle \mathbf{k}' | \mathbf{r} \rangle
$$

$$
= \frac{1}{V} \sum_{\mathbf{k},\mathbf{k}'} \rho(\mathbf{k}, \mathbf{k}') \exp\left[i(\mathbf{k}\mathbf{r} - \mathbf{k}' \, \mathbf{r}') \right] \quad (V \equiv \text{normalization volume})
$$

In principle it is possible to find a representation that diagonalizes the density matrix. In such a representation there are no phase-correlations to worry about and we could use semiclassical reasoning. In practice it may not always be convenient to find this special representation or to use it.

Correlation function (G^n or $-iG^<$)

We have seen that we need to generalize the concept of a distribution function $f(\mathbf{k})$ into a density matrix $\rho(\mathbf{k}, \mathbf{k}')$ in order to include additional information regarding the phase-correlations. We could include the time coordinate in our description by defining a time-varying density matrix: $\rho(\mathbf{k}, \mathbf{k}'; t)$. Although this concept is often used, especially in the treatment of systems with discrete levels, it is not fully general. In general we need a two-time correlation function of the form $G^n(\mathbf{k}, \mathbf{k}'; t, t')$

$$
f(\mathbf{k}; t) \rightarrow G^n(\mathbf{k}, \mathbf{k}'; t, t')
$$

that tells us the correlation between the amplitude in state **k** at time t and that in state **k'** at time t'. In steady-state problems, the correlation function depends only on the *difference* between the two times and can be Fourier transformed to yield

$$G^n(\mathbf{k}, \mathbf{k}'; E) \equiv \int \frac{1}{\hbar} G^n(\mathbf{k}, \mathbf{k}'; \tau) e^{-iE\tau/\hbar} d\tau \quad (\tau \equiv t - t') \qquad (8.1.1)$$

One way to understand the Fourier transform relationship between the energy E and the difference time coordinate $(t - t')$ is to note that the wavefunction of a particle with energy E evolves in time with a phase factor of $\exp[-iEt/\hbar]$. Consequently

$$\Psi(t)\Psi^*(t') \sim \exp\left[-iE(t - t')/\hbar\right]$$

This suggests that the Fourier transform of the correlation function with respect to $(t - t')$ should yield the energy spectrum. It is interesting to note that this is exactly how we find the frequency spectrum of the noise current $I(t)$ in a device. We Fourier transform the current correlation function:

$$\bar{I}(\omega) \equiv \int \langle I(t)I(t + \tau) \rangle e^{-i\omega\tau} d\tau$$

As we mentioned earlier, some treatments of quantum transport are based on the density matrix: $\rho(\mathbf{k},\mathbf{k}';t)$ which is a 'subset' of the correlation function obtained by setting $t' = t$.

$$\rho(\mathbf{k}, \mathbf{k}'; t) = \left[G^n(\mathbf{k}, \mathbf{k}'; t, t') \right]_{t' = t}$$

It is straightforward to show from Eq.(8.1.1) that this is equivalent to integrating $G^n(\mathbf{k},\mathbf{k}';E)$ over all energy:

$$\left[G^n(\mathbf{k}, \mathbf{k}'; t, t') \right]_{t = t'} = \int \frac{1}{2\pi} G^n(\mathbf{k}, \mathbf{k}'; E) dE \qquad (8.1.2)$$

As a result the energy-resolved information is lost making it difficult to describe scattering processes which transfer electrons from one energy to another.

In general to describe time-varying transport, we need to use the full two-time correlation function. Since our interest is confined to steady-state transport, the correlation function depends only on the time difference $(t - t')$ and can be Fourier transformed to obtain $G^n(\mathbf{k},\mathbf{k}';E)$

as described above. This energy-dependent correlation function is what we will use throughout this chapter.

We know that the diagonal elements of the correlation function give us the number of electrons occupying a particular state. From Eq.(8.1.2) we can write

$$f(\mathbf{k}) = \left[G^n(\mathbf{k}, \mathbf{k}'; t, t') \right]_{\mathbf{k}=\mathbf{k}', t=t'} = \int \frac{1}{2\pi} G^n(\mathbf{k}, \mathbf{k}; E) dE$$

This is true not just in the k-representation, but in any other representation as well. For example we can write the electron density in real space as

$$\underline{n}(\mathbf{r}) = 2 \text{ (for spin)} \times \int \frac{1}{2\pi} G^n(\mathbf{r}, \mathbf{r}; E) dE$$

where we have included a factor of 2 for the two spin components assuming these to be degenerate. We can write this in terms of the electron density per unit energy $n(r;E)$ as follows:

$$\underline{n}(r) = 2 \int n(\mathbf{r}; E) dE \quad \text{where} \quad 2\pi n(\mathbf{r}; E) = G^n(\mathbf{r}, \mathbf{r}; E) \tag{8.1.3}$$

Hole correlation function (G^p or $+iG^<$)

In deriving semiclassical kinetic equations we usually balance the outflow of electrons against the inflow of electrons. The inflow of electrons can alternatively be viewed as an outflow of 'holes' (whose number is given by $(1 - f)$). We use the quotes as a reminder that we are talking about holes in the conduction band itself (we are considering only one band) and not in some other valence band. To describe the outflow of holes in the quantum formalism we define a hole correlation function G^p using the same argument as we used above for electrons.

It is difficult to state the difference between the electron correlation function G^n and the hole correlation function G^p precisely without using the language of second quantization, as we are trying to do. For the benefit of readers familiar with this language, let us state the difference in terms of the creation and annihilation operators $a_\mathbf{k}$ and $a_\mathbf{k}^+$:

$$G^n(\mathbf{k}, \mathbf{k}'; t, t') = \left\langle a_{\mathbf{k}'}^+(t') a_\mathbf{k}(t) \right\rangle \quad \text{and} \quad G^p(\mathbf{k}, \mathbf{k}'; t, t') = \left\langle a_\mathbf{k}(t) a_{\mathbf{k}'}^+(t') \right\rangle$$

where $\langle ... \rangle$ denotes the expectation value. In a one-particle description, the annihilation and creation operators reduce to the one-particle wavefunction and its complex conjugate respectively. Since wavefunctions (unlike operators) commute, both G^n and G^p reduce to the same quantity

$$\Psi_{k'}^*(t)\Psi_k(t)$$

in a one-particle description.

Scattering functions

In the semiclassical picture we can define a function $S^{out}(k,t)$ that tells us the rate at which electrons are scattered out of a state $\mathbf{k}$ assuming it is initially full. In a quantum mechanical description we have to generalize this concept, too, to include phase-correlations:

$$S^{out}(\mathbf{k},t) \quad \rightarrow \quad \Sigma^{out}(\mathbf{k},\mathbf{k}';t,t')$$

Once again for steady-state problems the outscattering function depends only on the difference time coordinate and can be Fourier transformed to yield an energy-dependent outscattering function $\Sigma^{out}(\mathbf{k},\mathbf{k}';E)$. Similarly we can generalize the semiclassical concept of an inscattering function $S^{in}(\mathbf{k},t)$ to include phase-correlations, and then Fourier transform to obtain the function $\Sigma^{in}(\mathbf{k},\mathbf{k}';E)$, which can be viewed either as an electron inscattering function or as a hole outscattering function.

A word about notation

Before proceeding further we should point out that we are using a notation that is slightly different from the standard notation in the literature. We have deliberately chosen the notation to reflect the physical meaning of these functions. The correspondence, however, is quite straightforward:

Classical analog	Our notation	Standard notation	
$f, (1-f)$	G^n, G^p	$-iG^<, +iG^>$	(8.1.4)
S^{in}, S^{out}	$\Sigma^{in}, \Sigma^{out}$	$-i\Sigma^<, +i\Sigma^>$	

The matrices G^n, G^p, Σ^{in} and Σ^{out} are all Hermitian so that their diagonal elements in any representation are purely real. This set of four functions G^n, G^p, Σ^{in} and Σ^{out} provide us with the language needed to include phase-correlations into a transport theory. If we represent our device by a set of N nodes (in real space or in momentum space or in some other representation), then each of these quantities is a matrix of dimensions

($N \times N$) at a given energy E. *From hereon we will generally not write the energy coordinate E explicitly for clarity.*

In the following section we will introduce the retarded and advanced Green's functions (G^R, G^A) and self-energies (Σ^R and Σ^A), following the standard notation. In addition, the time ordered (G^T, Σ^T) and anti-time ordered functions ($G^{\bar{T}}$, $\Sigma^{\bar{T}}$) also appear in the literature. These are related to the other functions by the relations

$$G^T = G^R + iG^n = G^A - iG^p$$
$$G^{\bar{T}} = -G^R - iG^p = -G^A + iG^n$$

(8.1.5)

and similarly for the self-energy functions. We will be presenting the equations in a form that does *not* require the time ordered and anti-time ordered functions. However, these equations can be easily converted to the form commonly found in the literature (see Exercise E.8.4 at the end of this chapter).

8.2 Self-energy and the Green's function

In the last section we have introduced the functions G^n, G^p, Σ^{in} and Σ^{out} (that is, $-iG^<$, $+iG^>$, $-i\Sigma^<$ and $+i\Sigma^>$) which allow us to 'count the beans', that is to keep track of the comings and goings of individual particles. In addition to these concepts we need another set of functions, namely the retarded (and advanced) functions (G^R, G^A, Σ^R and Σ^A) which allow us to describe the dynamics of the electrons when they are inside the conductor. We have already encountered these functions in Chapter 3 (see Sections 3.3–3.6). To summarize the basic results in matrix notation (see Eq.(3.5.17))

$$\left[EI - H_C - \Sigma^R\right]G^R = I \quad \Rightarrow \quad G^R = \left[EI - H_C - \Sigma^R\right]^{-1} \quad (8.2.1)$$

The advanced Green's function G^A and the advanced self-energy are the Hermitian adjoints of the corresponding retarded functions

$$G^A = \left[G^R\right]^+ \quad \text{and} \quad \Sigma^A = \left[\Sigma^R\right]^+ \quad (8.2.2)$$

The Green's function G^R describes the coherent evolution of an electron from the moment it is injected till it loses coherence either by disappearing into a lead or by scattering into a different state (due to electron–phonon or electron–electron interactions) where a new coherent trajectory is initiated. The self-energy Σ^R describes the effect of the leads and the interactions on the electron dynamics.

We have seen in Section 3.6 that the spectral function A $(= i[G^R - G^A])$ represents a generalized density of states. Since the density of states is equal to the sum of the electron and hole densities, it seems reasonable that the spectral function should equal the sum of the electron and hole correlation functions defined in the last section:

$$G^n + G^p = i\left[G^R - G^A\right] \equiv A \qquad (8.2.3)$$

The spectral function A, tells us the *nature of the allowed electronic states*, regardless of whether they are occupied or not. The electron and hole correlation functions G^n and G^p, on the other hand, tell us how many of these states are *occupied or empty*.

Eq.(8.2.1) is written in matrix notation, which can be translated into any desired representation. In position representation we can write

$$\left[E - H_C\right]G^R(\mathbf{r}, \mathbf{r}') - \int \Sigma^R(\mathbf{r}, \mathbf{r}_1) G^R(\mathbf{r}_1, \mathbf{r}') d\mathbf{r}_1 = \delta(\mathbf{r} - \mathbf{r}') \qquad (8.2.4)$$

where H_C is the usual Hamiltonian operator describing the conductor:

$$H_C \equiv \frac{(i\hbar\nabla + e\mathbf{A}(\mathbf{r}))^2}{2m} + U(\mathbf{r}) \qquad (8.2.5)$$

$\mathbf{A}$ is the vector potential representing any magnetic fields while the potential energy U arises from impurities, boundaries, applied bias etc.

We could visualize the Green's function $G^R(\mathbf{r}, \mathbf{r}')$ as the wavefunction at the point $\mathbf{r}$ due to a unit excitation at $\mathbf{r}'$ (Fig. 8.2.1). If we leave out the source term on the right of Eq.(8.2.4) we obtain a Schrödinger-like equation describing the dynamics of an electron inside the conductor:

$$E\Psi(\mathbf{r}) = H_C\Psi(\mathbf{r}) + \int \Sigma^R(\mathbf{r}, \mathbf{r}_1)\Psi(\mathbf{r}_1)d\mathbf{r}_1 \qquad (8.2.6)$$

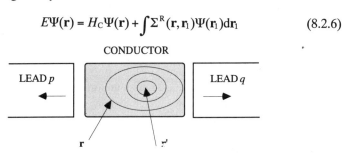

Fig. 8.2.1. A conductor and two leads p and q connected to it. An excitation at $\mathbf{r}'$ sets up 'ripples' that spread outwards. The Green's function $G^R(\mathbf{r}, \mathbf{r}')$ represents the resulting wavefunction at the point $\mathbf{r}$ due to a unit excitation at $\mathbf{r}'$.

The self-energy term Σ^R acts like an effective potential representing the effect of the interactions. In Chapter 3, the self-energy Σ^R arose solely from the coupling of the conductor with the leads (see Eq.(3.5.19)), since we were neglecting any interactions inside the conductor. Electron–electron and electron–phonon interactions give rise to an additional component in the self-energy which has to be added to the self-energy due to the leads. In Section 8.4 we will discuss how these self-energy functions are calculated. In this section we will just explore the significance of the self-energy function a little further.

The self-energy 'potential' Σ^R has two characteristics that distinguish it from the everyday potential energy terms encountered in quantum mechanics. Firstly, it is a *non-local* potential unlike the usual potential energy term $U(\mathbf{r})\Psi(\mathbf{r})$. This, however, is not a very fundamental difference. Even the usual potential energy term would appear non-local if we were to adopt a momentum representation instead of the position representation:

$$U(\mathbf{r})\Psi(\mathbf{r}) \quad \leftrightarrow \quad \int \overline{U}(\mathbf{k}, \mathbf{k}_1)\overline{\Psi}(\mathbf{k}_1)d\mathbf{k}_1$$

A more fundamental difference is that the self-energy potential is not Hermitian. One of the basic facts we learn in elementary quantum mechanics is that the Hamiltonian has to be Hermitian in order that probability be conserved. At steady-state, this means that the divergence of the probability current density (see Eq.(2.6.6))

$$\mathbf{J}(\mathbf{r}) = \frac{1}{2m}\left(\left[(\mathbf{p} - e\mathbf{A})\Psi\right]^* \Psi - \Psi^*\left[(\mathbf{p} - e\mathbf{A})\Psi\right]\right)$$

($\mathbf{p} \equiv -i\hbar\nabla$ is the momentum operator) is equal to zero as long as the wavefunction Ψ obeys the time-independent Schrödinger equation. But if the wavefunction obeys the modified Schrödinger equation given in Eq.(8.2.6) then the divergence is non-zero. To show this we first note that (see Exercise E.8.1 at the end of this chapter)

$$i\hbar\nabla.\mathbf{J}(\mathbf{r}) = \left[H_C\Psi\right]^* \Psi - \Psi^*\left[H_C\Psi\right] \tag{8.2.7}$$

Using Eq.(8.2.6) to replace $H_C\Psi$ we obtain

$$\nabla.\mathbf{J}(\mathbf{r}) = \frac{i}{\hbar}\int\left[\Sigma^R(\mathbf{r},\mathbf{r}')\Psi^*(\mathbf{r})\Psi(\mathbf{r}') - \Sigma^A(\mathbf{r}',\mathbf{r})\Psi^*(\mathbf{r}')\Psi(\mathbf{r})\right]d\mathbf{r}'$$

making use of the fact that Σ^A is the Hermitian conjugate of Σ^R. Hence

$$\int \nabla . \mathbf{J}(\mathbf{r}) d\mathbf{r} = \frac{i}{\hbar} \iint \Psi^*(\mathbf{r}) \Psi(\mathbf{r'}) [\Sigma^R(\mathbf{r}, \mathbf{r'}) - \Sigma^A(\mathbf{r}, \mathbf{r'})] d\mathbf{r} d\mathbf{r'}$$

where we have interchanged the variables of integration for the second term. The right hand side is zero if the self-energy is Hermitian ($\Sigma^A = \Sigma^R$). The quantity ($\Sigma^A - \Sigma^R$) is represented by a special symbol:

$$\Sigma^A - \Sigma^R = i\Gamma$$

so that we can write

$$\int \nabla . \mathbf{J}(\mathbf{r}) d\mathbf{r} = \frac{1}{\hbar} \iint \Psi^*(\mathbf{r}) \Psi(\mathbf{r'}) \left[\Gamma(\mathbf{r}, \mathbf{r'}) \right] d\mathbf{r} d\mathbf{r'}$$

A non-zero Γ results in the loss of electrons. Physically this loss represents the end of a coherent trajectory either through the leads or by scattering.

A useful relation

Since the function Γ determines the rate of loss of electrons by scattering (as we have just seen), it seems reasonable to expect that it should equal the outscattering function defined in the last section:

$$\Gamma = i \left[\Sigma^R - \Sigma^A \right] = \Sigma^{out} \qquad \text{(WRONG)}$$

From this point of view it would seem that if we were to describe the propagation of holes instead of electrons then we ought to use a different self-energy function such that

$$\Gamma = i \left[\Sigma^R - \Sigma^A \right] = \Sigma^{in} \qquad \text{(WRONG)}$$

since the electron inscattering function represents the hole outscattering function. The correct answer, however, is that electrons and holes are all described by the same self-energy function obeying the relation

$$\Gamma = i \left[\Sigma^R - \Sigma^A \right] = \Sigma^{in} + \Sigma^{out} \qquad (8.2.8)$$

The reason is rather subtle. The Green's function describes the coherent propagation of an injected electron. Any time the electron exits into a lead, or interacts with its surroundings, the coherent evolution is over. That is why it seems that the appropriate lifetime is determined by Σ^{out}. But suppose another electron tries to enter the conductor from the lead. It will be blocked by the electron already present inside the conductor and it might seem that nothing happened. But the point is that something does happen. A one-electron state retains coherence as long as it does not interfere with the free evolution of the rest of the world (often referred to as the reservoir). A blocked transition makes the system evolve differently from the way it would have evolved if the added electron were not there. Consequently it terminates one coherent evolution and starts another. Thus the appropriate inverse lifetime of an electron is given by the sum of the rate at which it scatters out plus the rate at which it blocks out electrons from other states trying to displace it (this is equal to the rate at which a hole, if present, would scatter out). This is the physical justification for the result stated above. The reader may find the discussion in Section 4.4 of Ref.[8.2] helpful (the author is grateful to Roger Lake for bringing this discussion to his attention).

8.3 Kinetic equation

The central result of the NEGF formalism is a kinetic equation relating the correlation functions G^n and G^p to the scattering functions Σ^{in}, Σ^{out}:

$$G^n = G^R \Sigma^{in} G^A, \quad G^p = G^R \Sigma^{out} G^A \tag{8.3.1}$$

The inscattering function Σ^{in} tells us the rate at which electrons come in, so that it seems reasonable that the electron correlation function G^n should be proportional to it. Similarly it is reasonable that the hole correlation function G^p be proportional to the outscattering function Σ^{out}.

To derive the actual relationship between Σ^{in} and G^n stated above, we consider the wavefunction $\Psi(\mathbf{r})$ due to a source term $S(\mathbf{r})$ (cf. Eq.(8.2.6)):

$$[E - H_C]\Psi(\mathbf{r}) - \int \Sigma^R(\mathbf{r}, \mathbf{r}_1)\Psi(\mathbf{r}_1)d\mathbf{r}_1 = S(\mathbf{r})$$

Since the Green's function $G^R(\mathbf{r},\mathbf{r}')$ represents the wavefunction due to a delta function source we can write

$$\Psi(\mathbf{r}) = \int G^R(\mathbf{r}, \mathbf{r}_1)S(\mathbf{r}_1)d\mathbf{r}_1$$

Multiplying by the complex conjugate we obtain

$$\Psi(\mathbf{r})\Psi(\mathbf{r}')^* = \iint G^R(\mathbf{r}, \mathbf{r}_1)G^R(\mathbf{r}', \mathbf{r}_1')^* S(\mathbf{r}_1)S(\mathbf{r}_1')^* d\mathbf{r}_1 d\mathbf{r}_1'$$

Noting that G^n represents the correlation between wavefunctions while Σ^{in} represents the correlation between sources

$$G^n(\mathbf{r},\mathbf{r}') \sim \Psi(\mathbf{r})\Psi(\mathbf{r}')^* \quad \text{and} \quad \Sigma^{in}(\mathbf{r}_1,\mathbf{r}_1') \sim S(\mathbf{r}_1)S(\mathbf{r}_1')^*$$

we can write

$$G^n(\mathbf{r},\mathbf{r}') = \iint G^R(\mathbf{r},\mathbf{r}_1)\Sigma^{in}(\mathbf{r}_1,\mathbf{r}_1')G^R(\mathbf{r}',\mathbf{r}_1')^* \, d\mathbf{r}_1 d\mathbf{r}_1'$$

$$= \iint G^R(\mathbf{r},\mathbf{r}_1)\Sigma^{in}(\mathbf{r}_1,\mathbf{r}_1')G^A(\mathbf{r}_1',\mathbf{r}') d\mathbf{r}_1 d\mathbf{r}_1'$$

In matrix notation this can be written as

$$G^n = G^R \, \Sigma^{in} \, G^A$$

as stated above. One could argue similarly in terms of injected holes to obtain the second half of Eq.(8.3.1).

$$G^p = G^R \, \Sigma^{out} \, G^A$$

It should be mentioned that in Eqs.(8.3.1) we have left out an additional 'boundary term' that often appears in the literature. This term could contribute to the initial transient in time-dependent problems, but is not relevant to the steady-state problems we are discussing (see Eq.(2.1.6) in Ref.[8.8]).

A useful identity

We can prove the following identity

$$A = G^R \, \Gamma \, G^A = G^A \, \Gamma \, G^R \tag{8.3.2}$$

following exactly the same procedure as we did in Chapter 3 (see Eq.(3.6.4)). Actually the first half of this relation can be obtained simply by adding Eqs.(8.3.1). However, we can derive both parts of this relation starting from the definition of the retarded Green's function (Eq.(8.2.1)), as we did earlier.

Equilibrium solution

At equilibrium, all states are occupied according to a single Fermi function $f_0(E)$ determined by the electrochemical potential. Since the correlation functions G^n and G^p are like the electron and hole densities, while the spectral function A is like the density of states, it seems reasonable to expect that at *equilibrium*

$$G^n(E) = f_0(E)A(E), \quad G^p(E) = (1 - f_0(E))A(E) \tag{8.3.3}$$

This is indeed true. Using Eq.(8.3.2), it is easy to see that this solution satisfies Eq.(8.3.1) provided the scattering functions are given by

$$\Sigma^{in}(E) = f_0(E)\Gamma(E) \quad \text{and} \quad \Sigma^{out}(E) = (1 - f_0(E))\Gamma(E) \qquad (8.3.4)$$

If we are interested in calculating equilibrium quantities, the retarded Green's function (G^R) contains all the information we need. We can then calculate the spectral function $A = i[G^R - G^A]$ and the electron and hole correlation functions, G^n and G^p, are known automatically. It is only for non-equilibrium problems that we need to solve a kinetic equation like Eq.(8.3.1).

8.4 Calculating the self-energy

In general for non-equilibrium problems, the kinetic equations (see Eqs.(8.3.1))

$$G^n = G^R \Sigma^{in} G^A \quad \text{and} \quad G^p = G^R \Sigma^{out} G^A$$

have to be solved simultaneously with the equations for the Green's function (see Eqs.(8.2.1), (8.2.2))

$$G^R = \left[EI - H - \Sigma^R\right]^{-1} \quad G^A = \left[G^R\right]^+$$

To proceed, however, we need to know how to calculate the self-energy and scattering functions (Σ^R, Σ^{in} and Σ^{out}). That is what we will describe in this section.

We know that the self-energy and the scattering functions arise from two different sources, namely, (1) the interaction with the leads and (2) the phase-breaking interactions inside the conductor. Assuming that the two are independent of each other we can express these functions in the form

$$\Sigma^R = \Sigma^R_\varphi + \sum_p \Sigma^R_p$$

$$\Sigma^{in} = \left[\Sigma^{in}_\varphi + \sum_p \Sigma^{in}_p\right], \quad \Sigma^{out} = \left[\Sigma^{out}_\varphi + \sum_p \Sigma^{out}_p\right] \qquad (8.4.1)$$

The self-energy and scattering functions with subscript p arise from the interaction with lead p while those with subscript φ arise from phase-breaking interactions within the conductor. We have already derived the self-energy due to the leads in Chapter 3 (see Eq.(3.5.19)). We will summarize the results in the next section when we describe the overall solution procedure. In this section let us focus on the self-energy and scattering functions due to the interactions within the conductor.

The actual expressions for the self-energy and scattering functions Σ_φ^R, Σ_φ^{in} and Σ_φ^{out} depend on the type of interaction we wish to describe and the degree of approximation we want to use. These expressions are derived from perturbation theory and one can get increasingly complicated expressions as we go to higher orders. For a detailed description we refer the reader to the cited references. Here we will simply summarize the results for electron–electron interactions in the Hartree–Fock approximation and for electron–phonon interactions in the self-consistent Born approximation (SCBA).

Electron–electron interactions

In the Hartree–Fock approximation, electron–electron interactions do not give rise to any inscattering or outscattering functions:

$$\Sigma_\varphi^{in} = \Sigma_\varphi^{out} = 0 \quad \rightarrow \quad \Gamma_\varphi = \Sigma_\varphi^{in} + \Sigma_\varphi^{out} = 0$$

But it contributes to the self-energy function:

$$\Sigma_\varphi^R(\mathbf{r}, \mathbf{r'}; E) = U_H(\mathbf{r})\delta(\mathbf{r} - \mathbf{r'}) + \Sigma_F(\mathbf{r}, \mathbf{r'}) \qquad (8.4.2)$$

The first term is the Hartree potential:

$$U_H(\mathbf{r}) = \iint \frac{1}{2\pi} G^n(\mathbf{r}, \mathbf{r'}; E) \frac{e^2}{4\pi\varepsilon |\mathbf{r} - \mathbf{r'}|} d\mathbf{r'} dE \qquad (8.4.3)$$

This is easy to understand if we note that the electron density is related to the diagonal elements of the correlation function:

$$\bar{n}(\mathbf{r}) = \int \frac{1}{2\pi} \left[G^n(\mathbf{r}, \mathbf{r'}; E) \right]_{\mathbf{r'}=\mathbf{r}} dE$$

The second term is the exchange potential:

$$\Sigma_F^s(\mathbf{r}, \mathbf{r'}) = -\int G^{ns}(\mathbf{r}, \mathbf{r'}; E) \frac{e^2}{4\pi\varepsilon |\mathbf{r} - \mathbf{r'}|} dE \qquad (8.4.4)$$

The superscript 's' is added as a reminder that an electron only feels an exchange potential due to other electrons of the same spin.

The expression for the exchange term is not as obvious as the Hartree term but we will not discuss it further. Many-body perturbation theory provides a systematic method for evaluating the self-energy and scattering functions to any desired order (see Section 3 of Ref.[8.5] for example). The Hartree–Fock approximation represents the lowest order result. Higher order calculations lead to non-zero values of Σ_φ^{in} and Σ_φ^{out} as well.

Electron–phonon interactions

We will also discuss phonon scattering only in lowest order perturbation theory. In this approximation

$$\Sigma_\varphi^{in}(\mathbf{r},\mathbf{r}';E) = \int D(\mathbf{r},\mathbf{r}';\hbar\omega) G^n(\mathbf{r},\mathbf{r}';E - \hbar\omega) d(\hbar\omega) \qquad (8.4.5a)$$

$$\Sigma_\varphi^{out}(\mathbf{r},\mathbf{r}';E) = \int D(\mathbf{r},\mathbf{r}';\hbar\omega) G^p(\mathbf{r},\mathbf{r}';E + \hbar\omega) d(\hbar\omega) \qquad (8.4.5b)$$

where the function D describes the spatial correlation and energy spectrum of the phase-breaking scatterers ($\hbar\omega > 0$ corresponds to absorption and $\hbar\omega < 0$ to emission).

$$D(\mathbf{r},\mathbf{r}';\hbar\omega) = \sum_q |U_q|^2 \begin{Bmatrix} \exp[-iq.(\mathbf{r} - \mathbf{r}')] N_q \, \delta(\omega - \omega_q) \\ + \exp[+iq.(\mathbf{r} - \mathbf{r}')] (N_q + 1) \delta(\omega + \omega_q) \end{Bmatrix} \qquad (8.4.6)$$

where N_q is the number of phonons with wavevector q and frequency ω_q and U_q is the potential felt by an electron due to a single phonon with wavevector q. Assuming that the bath of phonons is always maintained in thermal equilibrium N_q is given by the Bose–Einstein function:

$$N_q = \frac{1}{\exp[\hbar\omega_q/k_B T] - 1}$$

This assumption is fairly good at low bias, though at high applied voltages the phonon bath too could deviate from equilibrium if heat-sinking arrangements are inadequate. To describe such 'hot phonon' effects one would have to solve a transport equation for the phonons self-consistently with that for the electrons. Relatively little work has been done on these lines.

We will discuss the expressions stated above (Eqs.(8.4.5a,b)) a little further. But before we do that let us state the expression for the self-energy function:

$$\Sigma_\varphi^R(\mathbf{r},\mathbf{r}';E) = -\Gamma_\varphi^H(\mathbf{r},\mathbf{r}';E) + \frac{i}{2}\Gamma_\varphi(\mathbf{r},\mathbf{r}';E) \qquad (8.4.7)$$

where
$$\Gamma_\varphi(\mathbf{r},\mathbf{r}';E) = \Sigma_\varphi^{in}(\mathbf{r},\mathbf{r}';E) + \Sigma_\varphi^{out}(\mathbf{r},\mathbf{r}';E)$$

and
$$\Gamma_\varphi^H(\mathbf{r},\mathbf{r}';E) = P\int\frac{\Gamma_\varphi(\mathbf{r},\mathbf{r}';E')}{E - E'}\, dE'$$

The 'P' in the last expression stands for principal part. The function $\Gamma_\varphi^H(E)$ is known as the Hilbert transform of the function $\Gamma_\varphi(E)$. This relationship between the two functions ensures that the Fourier transform of Σ^R is causal; that is, it vanishes for $t < 0$. To see this we write Eq.(8.4.7) in the form

$$\Sigma_\varphi^R(E) = \frac{i}{2}\Gamma_\varphi(E) \otimes \left[\delta(E) + 2i\,P\left(\frac{1}{E}\right)\right]$$

where the symbol $\otimes$ denotes convolution. On Fourier transforming the convolution becomes a product. Since the Fourier transform of the function

$$\delta(E) + 2i\,P\left(\frac{1}{E}\right)$$

is proportional to the step function $\vartheta(t)$ the Fourier transform of $\Sigma_\varphi^R(E)$ is causal:

$$\Sigma_\varphi^R(t) \sim \vartheta(t)\Gamma_\varphi(t)$$

Discussion of Eqs. (8.4.5a,b)

A formal derivation of Eqs.(8.4.5a,b) can be found in the cited references. Here we will just try to 'understand' these results and convince ourselves that they are reasonable. We will focus on Eq.(8.4.5a) for the inscattering function; the same arguments can be used to justify Eq.(8.4.5b) for the outscattering function by using holes instead of electrons. For simplicity let us consider a single phonon with a wavevector q and frequency ω_q, so that

$$D(\mathbf{r},\mathbf{r}';\hbar\omega) \to |U_q|^2 \begin{Bmatrix} \exp[-iq.(\mathbf{r}-\mathbf{r}')]N_q\,\delta(\omega-\omega_q) \\ +\exp[+iq.(\mathbf{r}-\mathbf{r}')](N_q+1)\delta(\omega+\omega_q) \end{Bmatrix}$$

and Eq.(8.4.5a) reduces to

$$\Sigma_\varphi^{in}(\mathbf{r},\mathbf{r}';E) = |U_q|^2 \left\{ \begin{array}{l} \exp[-iq.(\mathbf{r}-\mathbf{r}')]N_q\, G^n(\mathbf{r},\mathbf{r}';E-\hbar\omega_q) \\ +\exp[+iq.(\mathbf{r}-\mathbf{r}')](N_q+1)\,G^n(\mathbf{r},\mathbf{r}';E+\hbar\omega_q) \end{array} \right\}$$

This expression is interpreted as follows. The inscattering function at an energy E arises from two sources: due to electrons with energy $(E-\hbar\omega_q)$ absorbing a phonon and due to electrons with energy $(E+\hbar\omega_q)$ emitting a phonon. The first process (involving absorption) is proportional to N_q while the second process (involving emission) is proportional to (N_q+1), as we would expect.

We can get some insight into the meaning of Eq.(8.4.5a) by Fourier transforming to time domain. Since Eq.(8.4.5a) is a convolution integral, on Fourier transforming we obtain an ordinary product:

$$\Sigma_\varphi^{in}(\mathbf{r},\mathbf{r}';\tau) = \overline{D}(\mathbf{r},\mathbf{r}';\tau)G^n(\mathbf{r},\mathbf{r}';\tau) \tag{8.4.8}$$

The Fourier transformed function $\overline{D}$ describes the spatial and temporal correlation of the potential due to the phase-breaking scatterers. To see this let us write the potential energy felt by an electron due to a single phonon with wavevector q as

$$U(\mathbf{r},t) = U_q \exp[i(q.\mathbf{r}-\omega_q t)] + U_q^* \exp[-i(q.\mathbf{r}-\omega_q t)]$$

where U_q is a complex quantity with a random phase (reflecting the incoherent nature of the phonons). Then

$$D(\mathbf{r},\mathbf{r}';\tau) \rightarrow \langle U(\mathbf{r},t)U(\mathbf{r}',t+\tau) \rangle$$
$$= |U_q|^2 \left\{ \exp[-iq.(\mathbf{r}-\mathbf{r}')-i\omega_q\tau] + \exp[+iq.(\mathbf{r}-\mathbf{r}')+i\omega_q\tau] \right\}$$

This is consistent with the definition of D (see Eq.(8.4.6)) if we now multiply the first term (absorption) by the number of phonons N_q and the second term (stimulated + spontaneous emission) by N_q+1.

To understand Eq.(8.4.8), recall that the scattering function Σ^{in} describes the correlations of the source term $S(r)$ (see derivation of Eq.(8.3.1)):

$$[E-H_C]\Psi(\mathbf{r}) - \int \Sigma^R(\mathbf{r},\mathbf{r}_1)\Psi(\mathbf{r}_1)d\mathbf{r}_1 = S(\mathbf{r})$$

Now the source term is given by $U\Psi$ if there is a scattering potential U. Hence we can write

$$\Sigma_{\varphi}^{\text{in}}(\mathbf{r}, \mathbf{r}'; \tau) = \langle S(\mathbf{r}, t) S(\mathbf{r}', t + \tau) \rangle$$

$$= \langle U(\mathbf{r}, t) \Psi(\mathbf{r}, t) U(\mathbf{r}', t + \tau) \Psi^*(\mathbf{r}', t + \tau) \rangle$$

As a first-order approximation we can write

$$\Sigma_{\varphi}^{\text{in}}(\mathbf{r}, \mathbf{r}'; \tau) \approx \langle U(\mathbf{r}, t) U(\mathbf{r}, t + \tau) \rangle \langle \Psi(\mathbf{r}', t) \Psi^*(\mathbf{r}', t + \tau) \rangle$$

$$\rightarrow \overline{D}(\mathbf{r}, \mathbf{r}'; \tau) G^{\text{n}}(\mathbf{r}, \mathbf{r}'; \tau)$$

since $\overline{D}$ represents the correlations of the scattering potential U and G^{n} represents the correlations of the wavefunction Ψ.

8.5 Summary of solution procedure

We have discussed the individual equations that need to be solved simultaneously in order to calculate the correlation function. The purpose of this section is to put the individual pieces together and present an overall procedure. The method we will describe is based on the tight-binding model, which allows us to incorporate the effect of the leads in a straightforward manner as discussed in Section 3.5. This approach is basically an extension of that described by C. Caroli *et al.* (1971) in *J. Phys. C: Solid State Physics*, **4**, 916. Alternative approaches based on a continuum representation have also been described in the literature (see for example, T. E. Feuchtwang (1976), *Phys. Rev. B*, **13**, 517 and G. B. Arnold (1985), *J. Low Temp. Physics*, **59**, 143).

In the tight-binding model, we start by choosing a discrete lattice in real space spanning the conductor (see Fig. 8.5.1). If there are N points on the lattice then each of the quantities of interest (like G^{n}, G^{R}, Σ^{n}, Σ^{R} etc.) is a matrix of dimensions $(N \times N)$ for *each energy*. We now proceed as follows:

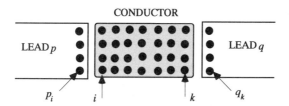

Fig. 8.5.1. A conductor connected to two leads p and q. A point in lead p is labeled p_i if it is adjacent to point i inside the conductor.

Step 1: First we calculate the self-energy function due to the leads using the results obtained in Section 3.5 (see Eq.(3.5.19)):

$$\Sigma_P^R(i,j;E) = t^2 g_p^R(p_i, p_j)$$
$$= -t \sum_{m \in p} \chi_m(p_i) \exp[+ik_m a] \chi_m(p_j) \qquad (8.5.1a)$$

The parameter $t \equiv \hbar^2/2ma^2$ is related to the spacing a between the discrete lattice sites which are labeled by the indices i and j. A site in lead p is labeled p_i if it is adjacent to site i in the conductor. Using Eq.(8.2.8)

$$\Gamma_P(i,j;E) = \sum_{m \in p} \chi_m(p_i) \frac{\hbar v_m}{a} \chi_m(p_j) \qquad (8.5.1b)$$

Note that the wavenumber k_m and the velocity v_m for mode m are related to the energy E through the tight-binding dispersion relation (see Eqs.(3.5.8a,b)):

$$E = \varepsilon_m + 2t(1 - \cos(k_m a)) \quad \text{and} \quad \hbar v_m \equiv \partial E/\partial k_m = 2at \sin(k_m a)$$

where ε_m is the cut-off energy for mode m.

Next we need the inscattering and outscattering functions. Their sum is equal to Γ (Eq.(8.2.8)). To obtain these functions individually, we assume that each lead p is maintained in local equilibrium with some Fermi distribution $f_p(E)$. This allows us to write down the inscattering and outscattering functions using the equilibrium solution from Eq.(8.3.4):

$$\Sigma_P^{in}(i,j;E) = f_p(E)\Gamma_P(i,j;E) \qquad (8.5.1c)$$

$$\Sigma_P^{out}(i,j;E) = (1 - f_p(E))\Gamma_P(i,j;E) \qquad (8.5.1d)$$

Step 2: Next we calculate the Green's function (see Eq.(8.2.1))

$$G^R = \left[EI - H_C - \Sigma^R\right]^{-1}, \quad G^A = \left[G^R\right]^+ \qquad (8.5.2a)$$

To perform this step we need to use the matrix representation for the Hamiltonian operator H_C (see Eq.(8.2.5)) that was developed in Chapter 3 using the method of finite differences. The result is repeated here for convenience (see Eqs.(3.5.9a,b)):

$$\left[EI - H_C\right]_{ij} = E - U(\mathbf{r}_i) - zt \quad \text{if } i = j$$

$$= \tilde{t}_{ij} \quad \text{if } i \text{ and } j \text{ are nearest neighbors} \qquad (8.5.2b)$$

$$= 0 \quad \text{otherwise}$$

where (1) z is the number of nearest neighbors ($z = 2$ for a linear chain and $z = 4$ for a square lattice), (2) $\mathbf{r}_i$ is the position vector for lattice site i. The nearest neighbor coupling is given by

$$\tilde{t}_{ij} = t \exp\left[ie\mathbf{A}.(\mathbf{r}_i - \mathbf{r}_j)/\hbar\right] \qquad (8.5.2c)$$

The vector potential $\mathbf{A}$ is evaluated at a point halfway between sites i and j, that is, at $(\mathbf{r}_i + \mathbf{r}_j)/2$.

Step 3 is to calculate the correlation functions (see Eqs.(8.3.1))

$$G^n = G^R \Sigma^{in} G^A \quad \text{and} \quad G^p = G^R \Sigma^{out} G^A \qquad (8.5.3)$$

Note that the the scattering functions are obtained by summing those due to the interactions and those due to the leads.

$$\Sigma^{in} = \left[\Sigma^{in}_\varphi + \sum_p \Sigma^{in}_p\right], \quad \Sigma^{out} = \left[\Sigma^{out}_\varphi + \sum_p \Sigma^{out}_p\right]$$

The component due to the interactions gets updated from one iteration to the next.

Step 4 is to calculate the self-energy and scattering functions arising from the interactions, using the appropriate expressions from Eqs.(8.4.6)–(8.4.8). The steps are summarized in Fig. 8.5.2. Once we complete step 4, we check to see if the self-energy and scattering functions due to the interactions have changed since the last iteration. If so we repeat steps 2, 3 and 4 till these functions converge. Note that at each iteration we are required to invert ($N \times N$) matrices for each energy channel. The number of energy channels depends on the temperature, the bias and also the correlation energy (that is, the energy range over which the Green's function can be assumed to be nearly constant).

Step 5: After we have obtained a converged solution for the correlation function we can use it to calculate any quantity of interest such as the *terminal current*. The current at lead p is given by

$$I_p = 2 \text{ (for spin)} \times \int i_p(E) dE$$

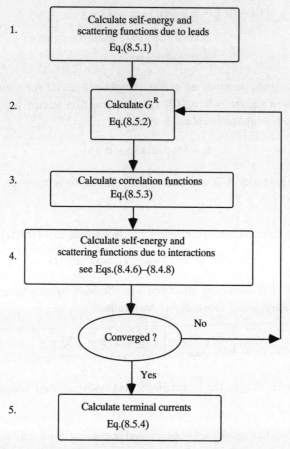

Fig. 8.5.2. Block diagram illustrating the iterative procedure to be followed in applying the NEGF formalism. No iteration is necessary if we neglect interactions. We can go directly from step 3 to step 5.

where $$i_p = \frac{e}{h}\mathrm{Tr}\left[\Sigma_p^{\mathrm{in}}G^{\mathrm{p}} - \Sigma_p^{\mathrm{out}}G^{\mathrm{n}}\right] = \frac{e}{h}\mathrm{Tr}\left[\Sigma_p^{\mathrm{in}}A - \Gamma_p G^{\mathrm{n}}\right] \qquad (8.5.4)$$

The two spin contributions we assume to be degenerate, as we have done throughout this book. Note, however, that the factor of 2 is not included in $i_p(E)$ as we did earlier (cf. Eq.(2.5.7))

A proper derivation of Eq.(8.5.4) requires an extended discussion which we will take up in the next section. But it is easy to see that the result is quite reasonable using a simple heuristic argument. Let us assume we are using a representation $|\alpha\rangle$ in which the correlation functions are diagonal (the trace of course is the same in any representation). In this representation we can write

$$i_p = \sum_\alpha \left[\Sigma_p^{in}(\alpha,\alpha) G^p(\alpha,\alpha) - \Sigma_p^{out}(\alpha,\alpha) G^n(\alpha,\alpha) \right]$$

This is quite reasonable if we note that $G^p(\alpha,\alpha)$ (or $G^n(\alpha,\alpha)$) represent the probability that state $|\alpha\rangle$ is empty (or occupied) while $\Sigma_p^{in}(\alpha,\alpha)$ (or $\Sigma_p^{out}(\alpha,\alpha)$) represent the rate at which electrons are scattered from lead p into (or out of) state $|\alpha\rangle$ if empty (or occupied).

8.6 Current flow and energy exchange

In the last section we have seen how we can calculate the correlation function $G^n(\mathbf{r},\mathbf{r}';E)$. Once we know the correlation function, we can calculate all other quantities of interest. For example we have already seen that the electron density is given by (see Eq.(8.1.3))

$$2\pi n(\mathbf{r};E) = G^n(\mathbf{r},\mathbf{r};E), \quad \underline{n}(\mathbf{r}) = 2\int n(\mathbf{r};E)dE \qquad (8.6.1)$$

In this section we will show how the correlation function can be used to calculate the current density, the terminal current, the energy current and the energy exchanged with the reservoir.

Eq.(8.6.1) follows quite simply if we write the electron density in terms of the wavefunction as $n(\mathbf{r};E) = \Psi(\mathbf{r})\Psi^*(\mathbf{r})$ and replace

$$\Psi(\mathbf{r})\Psi^*(\mathbf{r}) \quad \text{with} \quad G^n(\mathbf{r},\mathbf{r}';E)/2\pi$$

This is the basic approach we will use to obtain expressions for the current density etc. in terms of the correlation function. We will first express the quantity of interest in terms of the one-particle wavefunction and then replace $\Psi(\mathbf{r})\Psi^*(\mathbf{r})$ with $G^n(\mathbf{r},\mathbf{r}';E)/2\pi$. For a more rigorous discussion we refer the reader to [8.7].

Current density

The current density is given by (see Eq.(2.6.6))

$$\mathbf{J} = \frac{e}{2m}\left(\Psi[(\mathbf{p} - e\mathbf{A})\Psi]^* - \Psi^*[(\mathbf{p} - e\mathbf{A})\Psi] \right)$$

which can be written in the form

$$\mathbf{J}(\mathbf{r};E) = \frac{e}{2m}\Big[(\mathbf{p} - \mathbf{p}')\Psi(\mathbf{r})\Psi^*(\mathbf{r}')\Big]_{\mathbf{r}'=\mathbf{r}} - \frac{e^2}{m}A(\mathbf{r})\Big[\Psi(\mathbf{r})\Psi^*(\mathbf{r}')\Big]_{\mathbf{r}'=\mathbf{r}}$$

where $\mathbf{p} \equiv -i\hbar\nabla$ and $\mathbf{p}' \equiv -i\hbar\nabla'$, where the gradient operators ∇ and ∇' operate on $\mathbf{r}$ and $\mathbf{r}'$ respectively. Finally we replace $\Psi(\mathbf{r})\Psi^*(\mathbf{r}')$ with $G^n(\mathbf{r},\mathbf{r}';E)/2\pi$ as before to obtain

$$2\pi\mathbf{J}(\mathbf{r};E) = \left[-\frac{ie\hbar}{2m}(\nabla - \nabla')G^n(\mathbf{r},\mathbf{r}';E) - \frac{e^2}{m}A(\mathbf{r})G^n(\mathbf{r},\mathbf{r}';E)\right]_{\mathbf{r}'=\mathbf{r}}$$

$$\underline{\mathbf{J}}(\mathbf{r}) = 2 \text{ (for spin)} \times \int \mathbf{J}(\mathbf{r};E)dE \tag{8.6.2}$$

Once we have solved for G^n, we can calculate the current density $\mathbf{J}(\mathbf{r};E)$ throughout the conductor using Eq.(8.6.2). The terminal current per unit energy can then be obtained by integrating $\mathbf{J}(\mathbf{r};E)$ over the cross-section of the corresponding contact:

$$i_p(E) = \int \mathbf{J}(\mathbf{r};E).d\mathbf{S}_p$$

where $\mathbf{S}_p$ is the surface separating the conductor from the lead p. However, if we are not interested in the detailed current flow pattern then the terminal current can be obtained directly without calculating the current density, as we will now show.

Current operator

To obtain the terminal currents directly, we need to identify a quantity that will tell us the rate at which electrons are lost from the system into the leads or due to various scattering processes. The divergence of the current density: $\nabla.\mathbf{J}(\mathbf{r};E)$ represents such a quantity, since it tells us the rate at which particles are lost from the surrounding volume. As we have discussed earlier the divergence of the current can be written in terms of the wavefunction as (see Eq.(8.2.7))

$$i\hbar\nabla.\mathbf{J} = \big[H_C\Psi\big]^*\Psi - \Psi^*\big[H_C\Psi\big]$$

We rewrite this in the form

$$i\hbar\nabla.\mathbf{J}(\mathbf{r};E) = \frac{e}{2m}\Big(\big[H_C(\mathbf{r}) - H_C(\mathbf{r}')\big]\Psi(\mathbf{r})\Psi^*(\mathbf{r}')\Big)_{\mathbf{r}'=\mathbf{r}}$$

and replace $\Psi(\mathbf{r})\Psi^*(\mathbf{r}')$ with $G^n(\mathbf{r},\mathbf{r}';E)/2\pi$ as before to obtain

$$i2\pi\hbar\nabla.\mathbf{J}(\mathbf{r};E) = \frac{e}{2m}\left(\left[H_\mathrm{C}(\mathbf{r}) - H_\mathrm{C}(\mathbf{r'})\right]G^\mathrm{n}(\mathbf{r},\mathbf{r'})\right)_{\mathbf{r'}=\mathbf{r}}$$

It is convenient to define a current operator

$$I_{\mathrm{op}}(E) \equiv \frac{ie}{h}\left[H_\mathrm{C}G^\mathrm{n} - G^\mathrm{n}H_\mathrm{C}\right] \tag{8.6.3}$$

whose diagonal elements are equal to the divergence of the current density:

$$I_{\mathrm{op}}(\mathbf{r},\mathbf{r};E) = \nabla.\mathbf{J}(\mathbf{r};E) \tag{8.6.4}$$

Making use of Eqs.(8.6.3), (8.5.3) and (8.5.2) it is straightforward to show that (see Exercise E.8.2 at the end of this chapter)

$$I_{\mathrm{op}}(E) \equiv \frac{ie}{h}\left[\Sigma^{\mathrm{in}}G^\mathrm{R} - G^\mathrm{A}\Sigma^{\mathrm{in}} - \Sigma^\mathrm{R}G^\mathrm{n} + G^\mathrm{n}\Sigma^\mathrm{A}\right] \tag{8.6.5}$$

Terminal current

The trace of the current operator represents the net outflow per unit energy across an imaginary surface S enclosing the conductor (see Fig. 8.6.1):

$$\mathrm{Tr}\left[I_{\mathrm{op}}\right] \equiv \int I_{\mathrm{op}}(\mathbf{r},\mathbf{r};E)\mathrm{d}\mathbf{r} = \int \nabla.\mathbf{J}(\mathbf{r};E)\mathrm{d}\mathbf{r}$$

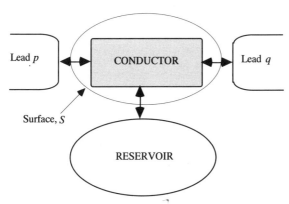

Fig. 8.6.1. Interactions can be viewed as processes involving exchange of particles with a reservoir. Particles enter the reservoir with an energy E and are returned with a different energy E'. There is no net exchange of particles though there is a net exchange of energy.

Using Eq.(8.6.5) we can write

$$\text{Tr}\left[I_{\text{op}}\right] = \frac{ie}{h}\text{Tr}\left[\Sigma^{\text{in}}(G^R - G^A) - (\Sigma^R - \Sigma^A)G^n\right]$$

Making use of Eqs.(8.2.3) and (8.2.8), we obtain

$$\text{Tr}\left[I_{\text{op}}\right] = \frac{e}{h}\text{Tr}\left[\Sigma^{\text{in}}A - \Gamma G^n\right]$$

$$= \frac{e}{h}\text{Tr}\left[\Sigma^{\text{in}}G^p - \Sigma^{\text{out}}G^n\right] \tag{8.6.6}$$

This last expression is easily understood following the heuristic argument we used earlier to justify Eq.(8.5.4). Σ^{in} is the rate of inscattering into a state if it is empty, while G^p is the density of empty states, so that the first term $\Sigma^{\text{in}}G^p$ represents the actual rate of inscattering. Similarly the second term $\Sigma^{\text{out}}G^n$ represents the rate of outscattering.

We know that the inscattering and outscattering functions are given by the sum of individual contributions from the leads and from the interactions (see Eq.(8.4.1)):

$$\Sigma^{\text{in}} = \left[\Sigma_\varphi^{\text{in}} + \sum_p \Sigma_p^{\text{in}}\right], \quad \Sigma^{\text{out}} = \left[\Sigma_\varphi^{\text{out}} + \sum_p \Sigma_p^{\text{out}}\right]$$

This suggests that we split up the net outflow given by Eq.(8.6.6) into individual components:

$$\text{Tr}\left[I_{\text{op}}(E)\right] = i_\varphi(E) + \sum_p i_p(E) \tag{8.6.7}$$

where

$$i_\varphi(E) = \frac{e}{h}\text{Tr}\left[\Sigma_\varphi^{\text{in}}G^p - \Sigma_\varphi^{\text{out}}G^n\right]$$

and

$$i_p(E) = \frac{e}{h}\text{Tr}\left[\Sigma_p^{\text{in}}G^p - \Sigma_p^{\text{out}}G^n\right]$$

The term $i_p(E)$ gives us the influx from the lead p, which can be integrated over energy to give the corresponding terminal current as stated earlier in Eq.(8.5.4). Interactions can be viewed as processes involving exchange of particles with a conceptual reservoir. From this point of view, the term $i_\varphi(E)$ represents the flow of particles into this conceptual reservoir (see Fig. 8.6.1).

Energy exchange

Interactions only reshuffle electrons around in energy. There is no net loss or gain of electrons in the process. Particles enter the reservoir with an energy E and are returned with a different energy E'. If $i_\varphi(E)$ is positive at some energy it has to be negative at some other energy such that

$$\int i_\varphi(E)\mathrm{d}E = 0$$

This means that regardless of the detailed nature of the interactions, the following relation must be satisfied:

$$\int \mathrm{Tr}\Big[\Sigma_\varphi^{\mathrm{in}}\,G^{\mathrm{p}}\Big]\mathrm{d}E = \int \mathrm{Tr}\Big[\Sigma_\varphi^{\mathrm{out}}\,G^{\mathrm{n}}\Big]\mathrm{d}E \qquad (8.6.8)$$

Note, however, that $i_\varphi(E)$ is not zero at all energies unless the interactions are purely non-dissipative. The power dissipated can be calculated by integrating the outflux of energy into the reservoir over all energies:

$$\text{Power dissipated} = -2\ (\text{for spin}) \times \frac{1}{e}\int E i_\varphi(E)\mathrm{d}E$$

Thus the NEGF formalism can be used to calculate the power dissipated inside the conductor (see R. K. Lake and S. Datta (1992), *Phys. Rev. B*, **46**, 4757). This need not equal the applied voltage times the current, since part of the energy could be dissipated inside the contacts. Indeed if inelastic processes were completely absent inside the conductor, all the dissipation would occur in the contacts.

8.7 Relation to the Landauer–Büttiker formalism

Superficially, the NEGF formalism looks very different from the Landauer–Büttiker formalism. The NEGF formalism focuses on the internal state of the conductor. Even the scattering functions due to a lead p ($\Sigma_p^{\mathrm{in}}(\mathrm{r,r'})$ and $\Sigma_p^{\mathrm{out}}(\mathrm{r,r'})$) are defined at points $(\mathrm{r,r'})$ located inside the conductor. By contrast, in the Landauer approach the central quantity is the transmission function from one contact to another. The internal state of the conductor usually never appears in the discussion. However, we have seen in Section 3.5 that the transmission function can be expressed in terms of internal quantities (see Eq.(3.5.20)). We will show in this section that we obtain precisely this result from the NEGF formalism as well when we consider coherent transport (without any electron–phonon or electron–electron interactions). We will also show the approximations inherent in the phenomenological model used to incorporate phase-breaking into the Landauer–Büttiker formalism (see discussion in Section 2.6).

Combining Eqs.(8.5.3) and (8.5.4) and making use of Eq.(8.2.8) we can write

$$
\begin{aligned}
i_p &= \frac{e}{h} \mathrm{Tr}\left[\Sigma_p^{\mathrm{in}} G^{\mathrm{R}} \Sigma^{\mathrm{out}} G^{\mathrm{A}} - \Sigma_p^{\mathrm{out}} G^{\mathrm{R}} \Sigma^{\mathrm{in}} G^{\mathrm{A}} \right] \\
&= \frac{e}{h} \mathrm{Tr}\left[\Sigma_p^{\mathrm{in}} G^{\mathrm{R}} \Gamma G^{\mathrm{A}} - \Gamma_p G^{\mathrm{R}} \Sigma^{\mathrm{in}} G^{\mathrm{A}} \right]
\end{aligned}
$$

Writing the scattering functions in terms of its components (Eq.(8.4.1))

$$
\Sigma^{\mathrm{in}} = \left[\Sigma_\varphi^{\mathrm{in}} + \sum_q \Sigma_q^{\mathrm{in}} \right], \quad \Gamma = \left[\Gamma_\varphi + \sum_q \Gamma_q \right]
$$

we can separate out the total current into its coherent and non-coherent components as follows:

$$
i_p = \left[i_p \right]_{\mathrm{coherent}} + \left[i_p \right]_{\mathrm{non\text{-}coherent}} \tag{8.7.1}
$$

$$
\left[i_p \right]_{\mathrm{coherent}} = \frac{e}{h} \sum_q \mathrm{Tr}\left[\Sigma_p^{\mathrm{in}} G^{\mathrm{R}} \Gamma_q G^{\mathrm{A}} - \Gamma_p G^{\mathrm{R}} \Sigma_q^{\mathrm{in}} G^{\mathrm{A}} \right] \tag{8.7.2}
$$

$$
\left[i_p \right]_{\mathrm{non\text{-}coherent}} = \frac{e}{h} \mathrm{Tr}\left[\Sigma_p^{\mathrm{in}} G^{\mathrm{R}} \Gamma_\varphi G^{\mathrm{A}} - \Gamma_p G^{\mathrm{R}} \Sigma_\varphi^{\mathrm{in}} G^{\mathrm{A}} \right] \tag{8.7.3}
$$

It is interesting to note that if the scattering functions $\Sigma_\varphi^{\mathrm{in}}, \Sigma_\varphi^{\mathrm{out}}$ are zero then the non-coherent component of the current is zero. This means that even when we take the Hartree–Fock interaction into account (see Eqs.(8.4.2)–(8.4.4)) transport remains coherent. It affects the dynamical properties of the electrons through the self-energy function $\Sigma_\varphi^{\mathrm{R}}$ but since there is no inscattering or outscattering the current does not have any non-coherent component.

Coherent component

Making use of Eq.(8.5.1c) we can rewrite the coherent component as

$$
\left[i_p \right]_{\mathrm{coherent}} = \frac{e}{h} \sum_q \mathrm{Tr}\left[\Gamma_p G^{\mathrm{R}} \Gamma_q G^{\mathrm{A}} \right] (f_p - f_q)
$$

so that the terminal current can be expressed in the familiar form (see Eq.(2.5.7)):

$$\left[I_p\right]_{\text{coherent}} = 2 \text{ (for spin)} \times \int i_p(E)\mathrm{d}E$$

$$= \frac{2e}{h} \int \sum_q \overline{T}_{pq}\left[f_p - f_q\right]\mathrm{d}E$$

where
$$\overline{T}_{pq} = \mathrm{Tr}\left[\Gamma_p\, G^{\mathrm{R}}\,\Gamma_q\, G^{\mathrm{A}}\right] \tag{8.7.4}$$

in agreement with the result obtained earlier (see Eq.(3.5.20)).

Non-coherent component

It will be recalled that in Section 2.6 we described a phenomenological model using a floating voltage probe to simulate non-coherent processes. It is apparent from the above description that the effect of interactions is just like that of an extra lead attached to the conductor. The current $i_\varphi(E)$ at this fictitious lead integrated over energy is equal to zero just like a floating probe (see Eq.(8.6.8)). If we could express the inscattering and outscattering functions in the form

$$\left[\Sigma_\varphi^{\text{in}}\right] = f_\varphi\left[\Gamma_\varphi\right] \quad \text{and} \quad \left[\Sigma_\varphi^{\text{out}}\right] = (1 - f_\varphi)\left[\Gamma_\varphi\right] \tag{8.7.5}$$

then we could interpret $f_\varphi(E)$ as the distribution function at a fictitious voltage probe and express the non-coherent current in the form

$$\left[I_p\right]_{\text{non-coherent}} = \frac{2e}{h} \int \overline{T}_{p\varphi}\left[f_p - f_\varphi\right]\mathrm{d}E$$

where
$$\overline{T}_{p\varphi} = \mathrm{Tr}\left[\Gamma_p\, G^{\mathrm{R}}\,\Gamma_\varphi\, G^{\mathrm{A}}\right] \tag{8.7.6}$$

One still runs into difficulties in calculating the distribution function $f_\varphi(E)$, unless vertical flow is neglected as we discussed in Section 2.6.

But it should be noted that in general it may not even be accurate to describe interactions in terms of a simple probe distribution function $f_\varphi(E)$. This is because we cannot always express the inscattering and outscattering functions in the form shown in Eq.(8.7.5). We need a distribution function $[F_\varphi]$ that is a matrix and not just a number ($[I]$: identity matrix):

$$\left[\Sigma_\varphi^{\text{in}}\right] = \left[F_\varphi\right]\left[\Gamma_\varphi\right] \quad \text{and} \quad \left[\Sigma_\varphi^{\text{out}}\right] = \left[I - F_\varphi\right]\left[\Gamma_\varphi\right] \tag{8.7.7}$$

Unless $[F_\varphi]$ is diagonal the current cannot be expressed in the form shown in Eq.(8.7.6). It is still possible to express the current in the linear response regime in terms of effective transmission functions but the resulting expressions for the transmission are far more complicated (see S. Datta (1992), *Phys. Rev. B*, **46**, 9493).

8.8 Relation to the Boltzmann formalism

A transport theory is based on some dynamical equation like Newton's law or the Schrödinger equation. However, dynamics alone is not enough, whether it is classical or quantum. This is because dynamical equations are reversible, while transport processes are dissipative and irreversible. Transport theories usually introduce this irreversibility by assuming that the system of interest is in contact with a vast 'reservoir' that is always in thermal equilibrium and continually tries to restore the system to equilibrium through random interactions. This is particularly clear in the Landauer approach where the system and the reservoir are spatially distinct entities. The system is the mesoscopic conductor which is assumed to be non-dissipative and as such can be described by a dynamical equation. The reservoirs are the contacts which accept the non-equilibrium distribution of electrons emerging from the conductor and reinject a fully thermalized distribution back into the conductor. This separation into a dissipative and a non-dissipative segment is an idealization, though it does seem to capture much of the physics of mesoscopic transport.

In general dynamics and dissipation are completely intertwined and for a quantitative description we need a transport equation like the Boltzmann equation which combines Newton's law with a probabilistic description of random scattering forces. We could use the Boltzmann equation to describe transport in mesoscopic systems including dissipative processes inside the conductor, as long as quantum interference effects do not play a significant role. This is the approach followed by device engineers for the simulation of small electronic devices having dimensions comparable to the mean free path of the electrons (see, for example, M. S. Lundstrom (1990), *Fundamentals of Carrier Transport*, Modular Series on Solid-state Devices, vol. X, eds. R. F. Pierret and G. W. Neudeck, (New York, Addison–Wesley)). The NEGF formalism provides a general framework for describing quantum transport, much the same way that the Boltzmann formalism provides a general framework for semiclassical transport. Both formalisms combine a dynamical equation with a probabilistic description of scattering processes:

Boltzmann $\Rightarrow$ Semiclassical dynamics + Random scattering

NEGF $\Rightarrow$ Quantum dynamics + Random scattering

In this section we will try to identify the corresponding concepts in the two formalisms. For this comparison we will ignore the leads and assume that we are discussing current flow in the interior of a large conductor.

Basic equations

The central quantity in semiclassical transport theory is the distribution function $f(\mathbf{r},\mathbf{k})$ which is described by the Boltzmann equation. For steady-state transport

$$v.\nabla f + (eE/\hbar).\nabla_k f + S^{\text{out}}(\mathbf{r},\mathbf{k})f(\mathbf{r},\mathbf{k}) = S^{\text{in}}(\mathbf{r},\mathbf{k})(1 - f(\mathbf{r},\mathbf{k}))$$

where S^{out} and S^{in} are the outscattering and the inscattering functions respectively. We rewrite this equation in the form (suppressing the arguments $\mathbf{r}$, $\mathbf{k}$ for clarity)

$$v.\nabla f + (eE/\hbar).\nabla_k f + (S^{\text{out}} + S^{\text{in}})f = S^{\text{in}} \tag{8.8.1}$$

and convert it into an integral equation:

$$f(\mathbf{r},\mathbf{k}) = \iint F(\mathbf{r},\mathbf{k};\mathbf{r}_1,\mathbf{k}_1) S^{\text{in}}(\mathbf{r}_1,\mathbf{k}_1)d\mathbf{r}_1 d\mathbf{k}_1 \tag{8.8.2a}$$

where F is the Green's function or the 'impulse response' of the Boltzmann equation:

$$v.\nabla F + (eE/\hbar).\nabla_k F + (S^{\text{out}} + S^{\text{in}})F = \delta(\mathbf{r} - \mathbf{r}_1)\delta(\mathbf{k} - \mathbf{k}_1) \tag{8.8.2b}$$

Eq.(8.8.2a) looks formally somewhat similar to the central equation in the NEGF formalism ($G^n = G^R \Sigma^{\text{in}} G^A$) which can be written as

$$G^n(\mathbf{r},\mathbf{r}';E) = \iint F(\mathbf{r},\mathbf{r}';\mathbf{r}_1,\mathbf{r}_1';E)\Sigma^{\text{in}}(\mathbf{r}_1,\mathbf{r}_1';E)d\mathbf{r}_1 d\mathbf{r}_1' \tag{8.8.3}$$

where we have defined

$$F(\mathbf{r},\mathbf{r}';\mathbf{r}_1,\mathbf{r}_1';E) \equiv G^R(\mathbf{r},\mathbf{r}_1;E)G^A(\mathbf{r}_1',\mathbf{r}';E) = G^R(\mathbf{r},\mathbf{r}_1;E)G^R(\mathbf{r}',\mathbf{r}_1';E)*$$

G^R being the Green's function (or impulse response) of the Schrödinger equation including the self-energy functions:

$$\left[E - H - \sigma + \frac{i}{2}(\Sigma^{\text{in}} + \Sigma^{\text{out}})\right]G^R(\mathbf{r},\mathbf{r}') = \delta(\mathbf{r} - \mathbf{r}')$$

Analogous concepts

Comparing Eqs.(8.8.2) and (8.8.3) it is easy to see the similarity in the formal structure of the two formalisms and pick out the concepts that are analogous:

NEGF	Boltzmann
$G^n(\mathbf{r},\mathbf{r}';E)$	$f(\mathbf{r},\mathbf{k})$
$G^p(\mathbf{r},\mathbf{r}';E)$	$1 - f(\mathbf{r},\mathbf{k})$
$\Sigma^{out}(\mathbf{r},\mathbf{r}';E)$	$S^{out}(\mathbf{r},\mathbf{k})$
$\Sigma^{in}(\mathbf{r},\mathbf{r}';E)$	$S^{in}(\mathbf{r},\mathbf{k})$

In the NEGF formalism the 'impulse response' $F(\mathbf{r},\mathbf{r}';\mathbf{r}_1,\mathbf{r}_1';E)$ is obtained from a Schrödinger-like equation including an imaginary potential ($i(\Sigma^{in} + \Sigma^{out})/2$) that leads to the disappearance of particles. Of course, the particles do not really disappear. They are reinjected through the source term Σ^{in} in the equation $G^n = G^R \Sigma^{in} G^A$. Thus there are two aspects to the problem described by two different equations which have to be solved simultaneously. A Schrödinger-like equation describes the dynamics of quasi-particles as they propagate and 'decay' via scattering processes while a transport equation ensures that the particles that 'decay' are reinjected. In the usual form of the Boltzmann equation these two aspects of the problem are intertwined in the same equation, Eq.(8.8.1). But it is possible to separate out the two aspects as we have done in Eq.(8.8.2) and thereby exhibit the similarity with the NEGF formalism.

The self-energy functions Σ^{in} and Σ^{out} in the NEGF formalism play roles analogous to the inscattering and outscattering functions (S^{in} and S^{out}) in the Boltzmann formalism. In both formalisms irreversibility is introduced through the assumption that the scattering functions can be evaluated assuming that the reservoir is always in thermal equilibrium with no memory of past interactions.

Earlier in Section 8.2 we pointed out that the correct phase-breaking rate is not just equal to the outscattering function, but is obtained by adding the inscattering and outscattering functions $\Gamma = \Sigma^{in} + \Sigma^{out}$ (see Eq.(8.2.8)). Eqs.(8.8.2a,b) have been written in a form that corresponds to this viewpoint. But instead of Eq.(8.8.2a) we could just as well have written

$$f(\mathbf{r},\mathbf{k}) = \iint F'(\mathbf{r},\mathbf{k};\mathbf{r}_1,\mathbf{k}_1) S^{in}(\mathbf{r}_1,\mathbf{k}_1)(1 - f(\mathbf{r}_1,\mathbf{k}_1))d\mathbf{r}_1 d\mathbf{k}_1$$

with F' defined by a slightly different equation (cf.Eq.(8.8.2b))

$$v.\nabla F' + (eE/\hbar).\nabla_k F' + S^{out} F' = \delta(\mathbf{r} - \mathbf{r}_1)\delta(\mathbf{k} - \mathbf{k}_1)$$

A transition that is blocked due to the exclusion principle breaks the phase and should be viewed as a 'self-scattering event' which scatters an electron back into the same state with its phase randomized. Such self-scattering, however, makes no difference to the Boltzmann formalism which does not keep track of the phase anyway. We can write the Boltzmann equation in a form that includes the self-scattering (as we have done in Eqs.(8.8.2a,b)) and makes it look more like the NEGF, but this is purely an algebraic manipulation.

Distribution function

The correlation function G^n is analogous to the distribution function f used in the Boltzmann theory. In the Boltzmann formalism all one-particle quantities of interest like the electron density (n), the particle current density (J_P) or the energy density (U) can be calculated once we know the distribution function:

$$n(\mathbf{r}) = \int f(\mathbf{r}, \mathbf{k}) d\mathbf{k}, \quad J(\mathbf{r}) = \int v(\mathbf{k}) f(\mathbf{r}, \mathbf{k}) d\mathbf{k}, \quad U(\mathbf{r}) = \int E(\mathbf{k}) f(\mathbf{r}, \mathbf{k}) d\mathbf{k} \quad (8.8.4)$$

The same is true of the correlation function in the NEGF formalism. All one-particle quantities of interest can be obtained once the correlation function is known (see Section 8.6):

$$n(\mathbf{r}) = \int \frac{1}{2\pi} \left[G^n(\mathbf{r}, \mathbf{r}'; E) \right]_{\mathbf{r}'=\mathbf{r}} dE$$

$$J(\mathbf{r}) = \int \frac{1}{2\pi} \left[-\frac{ie\hbar}{2m}(\nabla - \nabla') G^n(\mathbf{r}, \mathbf{r}'; E) - \frac{e^2}{m} A(\mathbf{r}) G^n(\mathbf{r}, \mathbf{r}'; E) \right]_{\mathbf{r}'=\mathbf{r}} dE$$

$$U(\mathbf{r}) = \int \frac{1}{2\pi} E \left[G^n(\mathbf{r}, \mathbf{r}'; E) \right]_{\mathbf{r}'=\mathbf{r}} dE \quad (8.8.5)$$

Wigner function

It is possible to transform variables to define a function $G^W(\mathbf{R}, \mathbf{k}; E)$ related to the correlation function $G^n(\mathbf{r}, \mathbf{r}'; E)$ such that Eqs.(8.8.5) look just like Eqs.(8.8.4). The transformation proceeds as follows. We first transform $(\mathbf{r},\mathbf{r}')$ to center of mass and relative coordinates $(\mathbf{R},\boldsymbol{\rho})$:

$$\mathbf{R} = (\mathbf{r} + \mathbf{r}')/2 \text{ and } \boldsymbol{\rho} = (\mathbf{r} - \mathbf{r}')$$

to write the correlation function in the form $G^n(\mathbf{R},\boldsymbol{\rho};E)$. Next we Fourier transform the relative coordinate $\boldsymbol{\rho}$ to obtain the Wigner function:

$$G^W(\mathbf{R},\mathbf{k};E) \sim \int \exp[-i\mathbf{k}.\boldsymbol{\rho}]G^n(\mathbf{R},\boldsymbol{\rho};E)d\boldsymbol{\rho} \tag{8.8.6}$$

The appeal of the Wigner function arises from the fact that if we rewrite the expressions for the electron density, current density etc. in terms of $G^W(\mathbf{R},\mathbf{k})$ they take on a form just like the semiclassical expressions in Eqs.(8.8.4):

$$n(\mathbf{R}) \sim \iint G^W(\mathbf{R},\mathbf{k};E)dEd\mathbf{k}, \quad J(\mathbf{R}) \sim \iint v(\mathbf{k})G^W(\mathbf{R},\mathbf{k};E)dEd\mathbf{k}$$

$$U(\mathbf{R}) \sim \iint E G^W(\mathbf{R},\mathbf{k};E)dEd\mathbf{k} \tag{8.8.7}$$

where $v(\mathbf{k}) = (\hbar\mathbf{k} - eA)/m$. Starting from Eq.(8.8.5) we can obtain the expression for the electron density (n) as follows:

$$n(\mathbf{R}) \sim \int \left[G^n(\mathbf{r},\mathbf{r}';E)\right]_{\mathbf{r}'=\mathbf{r}} dE \sim \int \left[G^n(\mathbf{R},\boldsymbol{\rho};E)\right]_{\boldsymbol{\rho}=0} dE$$

$$\sim \int \left[\int \exp[+i\mathbf{k}\boldsymbol{\rho}]G^W(\mathbf{R},\mathbf{k};E)d\mathbf{k}\right]_{\boldsymbol{\rho}=0} dE \sim \iint G^W(\mathbf{R},\mathbf{k};E)dEd\mathbf{k}$$

The expression for the energy density (U) can be obtained along similar lines. The expression for the current density is also obtained similarly, once we note that the operator $(\nabla - \nabla')$ in real space corresponds to multiplying by 'i$\mathbf{k}$' in the transformed space:

$$J(\mathbf{R}) = \int \left[-\frac{ie\hbar}{2m}\nabla_\rho G^n(\mathbf{R},\boldsymbol{\rho};E) - \frac{e^2}{m}A(\mathbf{r})G^n(\mathbf{R},\boldsymbol{\rho};E)\right]_{\boldsymbol{\rho}=0} dE$$

$$= \int \left\{\int \exp(i\mathbf{k}\boldsymbol{\rho})\left[\frac{e\hbar\mathbf{k}}{2m}G^W(\mathbf{R},\mathbf{k};E) - \frac{e^2}{m}A(\mathbf{R})G^W(\mathbf{R},\mathbf{k};E)\right]d\mathbf{k}\right\}_{\boldsymbol{\rho}=0} dE$$

$$= \iint \left[\frac{e\hbar\mathbf{k}}{m} - \frac{e^2A}{m}\right]G^W(\mathbf{R},\mathbf{k};E)dEd\mathbf{k}$$

One difference between the semiclassical distribution function $f(\mathbf{r},\mathbf{k})$ and the Wigner function $G^W(\mathbf{R},\mathbf{k};E)$ is that in the semiclassical picture the energy E is related to the magnitude of the wavevector $\mathbf{k}$ (and thus does not appear explicitly) while in the quantum mechanical picture there is no relationship in general since plane waves are not necessarily energy eigenstates.

There is another very important difference between the Wigner function and semiclassical distribution functions. The semiclassical distribution function $f(\mathbf{r},\mathbf{k})$ has a simple physical meaning: it is the number of particles at $\mathbf{r}$ with momentum $\hbar\mathbf{k}$. But we cannot in general interpret the Wigner function in the same way because it is not positive definite. To clarify this point, let us look at a simple example. Suppose we have electrons with a single energy E which are reflected from a barrier so that the wavefunction is composed of incident and reflected waves as follows:

$$\Psi(z) = \exp(i\beta z) + r\exp(-i\beta z)$$

where r is the reflection coefficient and $\beta = \sqrt{2mE}/\hbar$. The correlation function is given by

$$
\begin{aligned}
G^{n}(z,z') &= \Psi(z)\Psi^{*}(z') \\
&= \exp\left[i\beta(z-z')\right] + rr^{*}\exp\left[-i\beta(z-z')\right] \\
&\quad + r^{*}\exp\left[i\beta(z+z')\right] + r\exp\left[-i\beta(z+z')\right]
\end{aligned}
$$

Transforming to center of mass and relative coordinates ($R = (z + z')/2$ and $\rho = (z - z')$)

$$G^{n}(R,\rho) = \exp\left[i\beta\rho\right] + rr^{*}\exp\left[-i\beta\rho\right] + r^{*}\exp\left[i2\beta R\right] + r\exp\left[-i2\beta R\right]$$

Hence the Wigner function is given by

$$G^{W}(R,k=0) = r^{*}\exp\left[i2\beta R\right] + r\exp\left[-i2\beta R\right]$$

$$G^{W}(R,k=+\beta) = 1$$

$$G^{W}(R,k=-\beta) = rr^{*}$$

The components of the Wigner function at $k = +\beta$ and at $k = -\beta$ make perfect sense. The number of particles with $k = +\beta$ is 1 while the number with $k = -\beta$ is equal to the reflection probability rr^{*}.

But what is the meaning of the component at $k = 0$? It oscillates in space between $\pm 2|r|$ and defies any simple classical interpretation since it takes on negative values as well. The basic reason for this difficulty is that by trying to define R and k simultaneously with arbitrary precision we have violated the uncertainty relation. The price we have to pay is the appearance of negative probabilities. This is a generic problem that arises whenever we try to describe quantum correlations in terms of classical concepts. One way to get around this is to define a new function (known as the Husimi function) obtained by averaging the Wigner function over a spatial region greater than a wavelength. The unphysical component $G^W(R,k = 0)$ then vanishes. By giving up on spatial precision we satisfy the uncertainty principle.

The formal similarity between Eqs.(8.8.4) and (8.8.7) is rather attractive and several authors have used the Wigner function or the Husimi function to describe quantum transport. Other types of distribution functions have also been described in the literature which are suitable for specific classes of problems (see for example the 'quasi-classical approximation' discussed in Section III of Ref.[8.6]).

8.9 Strongly interacting systems

Our discussion in this chapter has been based on a simple one-particle picture, assuming that an individual particle sees an effective potential (the self-energy) due to its interaction with the surroundings. As meso-scopic conductors get smaller and begin to resemble large molecules, this concept of an effective potential is breaking down as we saw in Section 6.3 in our discussion of single-electron tunneling. This is known as the regime of strongly correlated transport and it is gaining increasing atten-tion. Note that the word 'correlation' here is used to denote something very different from the phase correlations of the one-particle wavefunction that we have been talking about so far.

The NEGF formalism is usually derived from a many-body approach [8.1–8.8] using perturbation theory to treat the interactions. This is similar in spirit to what we did in Section 5.5 to derive a self-energy function for impurity scattering (see Fig. 5.5.4). Of course, the details are much more complicated since it involves many-body perturbation theory and is be-yond the scope of this book. The point is that this approach provides a systematic method for evaluating the self-energy and scattering functions taking the interactions into account up to any desired order. As such it might seem that we should be able to use this approach even for very strong electron–phonon or electron–electron interactions. We just need to go to higher orders. But this is not correct. For sufficiently strong interac-tions the perturbation expansion breaks down. This is exactly what would happen even with the single-particle perturbation theory described in

Section 5.5, if the impurity potential were strong enough to form localized atomic-like states.

Is the NEGF formalism applicable to the strongly correlated transport regime? The answer is that we cannot use

$$G^R = \left[EI - H - \Sigma^R \right]^{-1} \tag{8.5.2}$$

or

$$G^n = G^R \, \Sigma^{in} \, G^A \tag{8.5.3}$$

since we cannot describe the interactions by self-energy and scattering functions like Σ_φ^R or $\Sigma_\varphi^{in}, \Sigma_\varphi^{out}$. But if we could use some other non-perturbative technique to calculate the Green's functions and the correlation functions, then the current can be calculated from the relation

$$I_p = \frac{2e}{h} \int \mathrm{Tr} \left[\Sigma_p^{in} A - \Gamma_p G^n \right] dE \tag{8.5.4}$$

This may not be obvious since we made use of Eqs.(8.5.2) and (8.5.3) in deriving Eq.(8.5.4). However, this relation has been derived without making use of Eqs.(8.5.2) and (8.5.3) thus proving its applicability to strongly interacting systems even when Eqs.(8.5.2) and (8.5.3) may not be applicable (see Y. Meir and N. S. Wingreen (1992), *Phys. Rev. Lett.* **68**, 2512).

An interesting result pointed out by Meir and Wingreen is that for a two-terminal conductor with 'proportionate' coupling to the two leads, that is, with $\Gamma_1 = \lambda \Gamma_2$ where λ is a constant, we can write the current as

$$I = I_1 = -I_2 = \frac{2e}{h} \int \mathrm{Tr} \left[\gamma A \right] \left[f_1 - f_2 \right] dE \quad \text{where} \quad \gamma \equiv \frac{1}{1+\lambda} \Gamma_1$$

This result is obtained from Eq.(8.5.4) by noting that since I_1 must be equal to $-I_2$, we can write the terminal current as $(I_1 - \lambda I_2)/(1 + \lambda)$. Thus for this limited class of conductors, the current is given by the same expression as in the non-interacting case (cf. Eq.(2.5.1)), with the transmission function given by $\mathrm{Tr}[\gamma A]$. Of course, for strongly interacting systems we still need suitable non-perturbative techniques for evaluating the spectral function A (see for example, S. Hershfield *et al.* (1991), *Phys. Rev. Lett.*, **67**, 3720, A. L. Yeyati *et al.* (1993), *Phys. Rev. Lett.*, **71**, 2991 and P. A. Lee (1993), *Physica B*, **189**, 1).

Another special case where the current can be expressed in this form is when the conductor is strongly coupled to one of the leads, say lead q (see Fig. 3.7.4). We can then assume that the conductor is essentially in equilibrium with lead q so that $G^n \approx f_q A$ (see Eq.(8.3.3)). Since $\Sigma_p^{in} = f_p \Gamma_p$ (see Eq.(8.5.1c)), we can write

$$I_p \approx \frac{2e}{h} \int \text{Tr}[\Gamma_p A][f_p - f_q] dE$$

Transport in strongly correlated systems is a rapidly evolving field of research and we will not discuss it further in this book. It is interesting to note that similar issues arise even for classical transport in the presence of strong correlations. The Boltzmann equation which is widely used to describe classical transport in dilute systems is inadequate once the potential energy due to interparticle interactions exceeds the kinetic energy. One then has to worry about higher order distribution functions (two-particle, three-particle etc.) described by equations higher up in the BBGKY hierarchy (see for example S. Ichimaru (1973), *Basic Principles of Plasma Physics*, Frontiers in Physics, Benjamin/Cummings). Similarly in the NEGF formalism one needs higher order Green's functions to describe strongly correlated systems.

8.10 An example: resonant tunneling with phonon scattering

We end this chapter with a simple example (adapted from R. Lake *et al.* (1992), *Phys. Rev. B*, **47**, 6427) that can be worked out analytically by making suitable approximations. In general, to apply this formalism to actual conductors, we need to resort to numerical calculations. But this simple example serves to illustrate how the formalism is applied to concrete problems. The problem we will address is the effect of phonon scattering on the current flow through a resonant tunneling (RT) diode. The basic physics has already been discussed in Section 6.2. Here we will just outline the key results before we proceed to derive them.

First we consider an RT diode biased such that the incident electrons from the left lead can tunnel through the resonant level in the well (see Fig. 8.10.1a). This is the condition for peak current. We have seen in Chapter 6 that the current (per transverse mode) is given by

$$I_P = \frac{2e}{\hbar} \frac{\Gamma_1 \Gamma_2}{\Gamma_1 + \Gamma_2} \tag{8.10.1a}$$

where Γ_1 and Γ_2 are the energy broadening of the resonant level due to the coupling to leads 1 and 2 respectively. As we discussed in Chapter 6 this result is unaffected by phase-breaking processes. The current is the

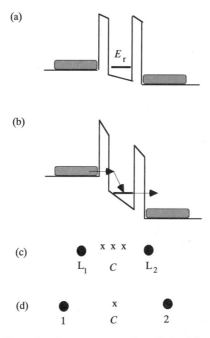

Fig. 8.10.1. A one-dimensional resonant tunneling diode: (*a*) potential energy diagram at a bias corresponding to main peak; (*b*) potential energy diagram at a bias corresponding to the phonon peak; (*c*) a discrete model with one lattice point each in the two barriers and in the well; (*d*) a simpler approximate model which accounts for the barriers through a reduced coupling between the leads and the well.

same for both coherent and sequential tunneling. We will derive this result using the NEGF formalism.

The second question we address involves 'vertical' flow of carriers from one energy to another. We consider a resonant tunneling diode biased such that the resonant energy level lies below the energy of the incident electrons (see Fig. 8.10.1b). This condition corresponds to the valley current where very little current flows in the absence of scattering. Scattering processes, however, induce vertical transitions and can lead to a 'phonon peak' in the valley current when the resonant energy level lies one optical phonon energy $\hbar\omega_0$ below the energy of the incident electrons. We will derive the following expression for the current per transverse mode at the phonon peak:

$$I_{PP} = \frac{2e}{\hbar} \frac{g\Gamma_1\Gamma_2}{g\Gamma_1 + \Gamma_2} \tag{8.10.1b}$$

where g is a dimensionless constant ($<<1$) describing the strength of the electron–phonon coupling.

Comparing Eqs.(8.10.1a) and (8.10.1b) for the main peak and the phonon peak we note that the only difference is that Γ_1 has been replaced by $g\Gamma_1$. This result can be understood intuitively. At the main peak an electron can tunnel directly into the resonant level. But at the phonon peak, an electron has to tunnel in from the left and emit a phonon in order to get into the resonant level. Consequently the effective tunneling rate is reduced by a factor g.

We will now derive Eqs.(8.10.1a,b) from the NEGF formalism, following the steps summarized in Fig. 8.5.2. Note that the basic difference between the main peak and the phonon peak (that is, between Eq.(8.10.1a) and (8.10.1b)) arises in step 4 where we evaluate the scattering functions due to the interaction with phonons.

Discrete lattice model

If we use one lattice point in each of the two barriers and one point in the well (see Fig. 8.10.1c) then we would have to work with (3×3) matrices. We can simplify the algebra with an approximation that captures much of the physics if the tunneling probability through the barriers is very small. In this approximation we get rid of the lattice points inside the barriers and reduce the coupling of the device to the leads to account for the barriers. We then have only one point inside the conductor so that we have to work with (1×1) matrices.

Let us now proceed with the steps shown in Fig. 8.5.2.

Step 1 Self-energy and scattering functions due to leads, Eq.(8.5.1)

$$\Sigma_p^R(C,C) = -te^{ik_p a}\alpha_p^2$$

where C denotes the single point inside the conductor and α_p is a number much less than one representing the reduced coupling of the well to lead p due to the barriers ($p = 1,2$). The entire structure including the leads is assumed to be single-moded. Since we are dealing with (1×1) matrices we will drop the argument (C,C) from here on:

$$\Sigma_p^R = \sigma_p - \frac{i}{2}\Gamma_p \qquad (8.10.2a)$$

$$\sigma_p \equiv t\alpha_p^2 \cos(k_p a) \quad \text{and} \quad \Gamma_p \equiv \alpha_p^2 \frac{\hbar v_p}{a}$$

We assume that in our energy range of interest $f_1(E) = 1$ and $f_2(E) = 0$. Hence

$$\Sigma_1^{in} = \Gamma_1, \quad \Sigma_1^{out} = 0$$
$$\Sigma_2^{in} = 0, \quad \Sigma_2^{out} = \Gamma_2$$

(8.10.2b)

Step 2 Green's function, Eq.(8.5.2)

$$G^R = \left[E - H - \Sigma^R \right]^{-1}$$
$$= \left((E - E_r') + \frac{i}{2}(\Gamma_\varphi + \Gamma_1 + \Gamma_2) \right)^{-1} = \frac{1}{\varepsilon}$$

(8.10.3)

where

$$\varepsilon \equiv (E - E_r') + \frac{i}{2}(\Gamma_\varphi + \Gamma_1 + \Gamma_2)$$
$$E_r' = E_r - \alpha_1^2 t \cos(k_1 a) - \alpha_2^2 t \cos(k_2 a) \approx E_r$$

Γ_φ represents the sum of the inscattering and outscattering functions arising from the interaction with the phonons:

$$\Gamma_\varphi = \Sigma_\varphi^{in} + \Sigma_\varphi^{out}$$

Step 3 Correlation functions, Eq.(8.5.3)

$$G^n = \Sigma^{in} \left| G^R \right|^2 = \frac{\Sigma_\varphi^{in} + \Gamma_1}{\left| \varepsilon \right|^2}$$
$$G^p = \Sigma^{out} \left| G^R \right|^2 = \frac{\Sigma_\varphi^{out} + \Gamma_2}{\left| \varepsilon \right|^2}$$

(8.10.4)

Let us skip *step 4* for the moment. We will return to it shortly.

Step 5 Terminal currents, Eq.(8.5.4):

$$I_1 = \frac{2e}{h} \int \Sigma_1^{in} G^p dE = \frac{2e}{h} \int \left[\frac{\Gamma_1(\Sigma_\varphi^{out} + \Gamma_2)}{\left| \varepsilon \right|^2} \right] dE$$

$$I_2 = -\frac{2e}{h} \int \Sigma_2^{out} G^n dE = -\frac{2e}{h} \int \left[\frac{\Gamma_2(\Sigma_\varphi^{in} + \Gamma_1)}{\left| \varepsilon \right|^2} \right] dE$$

(8.10.5)

$$I_\varphi = \frac{2e}{h} \int \left[\Sigma_\varphi^{in} G^p - \Sigma_\varphi^{out} G^n \right] dE = \frac{2e}{h} \int \left[\frac{\Sigma_\varphi^{in} \Gamma_2 - \Sigma_\varphi^{out} \Gamma_1}{\left| \varepsilon \right|^2} \right] dE$$

where we have made use of Eq.(8.10.4). If we are interested in coherent transport without interactions *then we can stop right here.* Since $\Sigma_\varphi^{in} = 0$ and $\Sigma_\varphi^{out} = 0$, we obtain

$$I_1 = -I_2 = \frac{2e}{h} \int \frac{\Gamma_1 \Gamma_2}{|\varepsilon|^2} \, dE \quad \text{and} \quad I_\varphi = 0$$

But the real value of the NEGF formalism is that it allows us to include interactions. To do this, however, we need to perform *step 4* which requires us to evaluate Σ_φ^{in} and Σ_φ^{out} in terms of the correlation functions G^n and G^p inside the well. This step has to be handled differently for the main peak and for the phonon peak. We will take these up one by one.

Step 4 Main peak

Next we need to make use of the relation between the correlation functions and the scattering functions (see Eqs.(8.4.5a,b)):

$$\Sigma_\varphi^{in}(E) = \int D(\hbar\omega) G^n(E - \hbar\omega) \, d(\hbar\omega)$$

$$\Sigma_\varphi^{out}(E) = \int D(\hbar\omega) G^p(E + \hbar\omega) \, d(\hbar\omega)$$

Since both G^n and G^p are sharply peaked around the resonance energy we can assume that the function D is independent of energy over this small energy range of interest. We then obtain using Eq.(8.10.4)

$$\Sigma_\varphi^{in} = D \int G^n(E') dE' = 2\pi D \frac{\Sigma_\varphi^{in} + \Gamma_1}{\Gamma_\varphi + \Gamma_1 + \Gamma_2} \qquad (8.10.6a)$$

and

$$\Sigma_\varphi^{out} = 2\pi D \frac{\Sigma_\varphi^{out} + \Gamma_2}{\Gamma_\varphi + \Gamma_1 + \Gamma_2} \qquad (8.10.6b)$$

where we have made use of the integral

$$\int \frac{dE}{|\varepsilon|^2} = \int \frac{1}{(E - E_r)^2 + \left(\dfrac{\Gamma_\varphi + \Gamma_1 + \Gamma_2}{2}\right)^2} \, dE \approx \frac{2\pi}{\Gamma_\varphi + \Gamma_1 + \Gamma_2} \qquad (8.10.7)$$

Adding Eqs.(8.10.6a) and (8.10.6b) we obtain

$$\Gamma_\varphi = 2\pi D \qquad (8.10.8a)$$

Also,
$$\frac{\Sigma_\varphi^{in}}{\Sigma_\varphi^{out}} = \frac{\Sigma_\varphi^{in} + \Gamma_1}{\Sigma_\varphi^{out} + \Gamma_2} = \frac{\Gamma_1}{\Gamma_2} \quad \Rightarrow \quad \frac{\Sigma_\varphi^{in}}{\Gamma_\varphi} = \frac{\Gamma_1}{\Gamma_1 + \Gamma_2} \qquad (8.10.8b)$$

Step 4 Phonon peak

The basic difference between the main peak and the phonon peak is that at the main peak transport takes place at one energy while at the phonon peak transport takes place around two distinct energies which we will denote by E_a and E_b, where $E_b \approx E_r$ and $E_a \approx E_r + \hbar\omega_0$. Assuming that $\hbar\omega_0 \gg k_B T$ we can neglect phonon absorption so that there is no inscattering at the upper energy E_a and no outscattering at the lower energy E_b:

$$\Sigma_\varphi^{in}(E_a) = 0, \quad \Sigma_\varphi^{out}(E_b) = 0$$

Next we need to evaluate the outscattering function at the upper energy E_a and the inscattering function at the lower energy E_b. Assuming a single frequency phonon spectrum we can write

$$D(\hbar\omega) \approx D_0 \delta(\hbar\omega + \hbar\omega_0)$$

so that

$$\Sigma_\varphi^{out}(E_a) \approx D_0 G^p(E_a - \hbar\omega_0)$$

and

$$\Sigma_\varphi^{in}(E_b) \approx D_0 G^n(E_b + \hbar\omega_0)$$

Thus we need the electron correlation function at the upper energy and the hole correlation function at the lower energy. From Eq.(8.10.4) we have

$$G^n(E_a) \approx \frac{\Gamma_1^a}{(\hbar\omega_0)^2} \quad \text{and} \quad G^p(E_b) = \frac{\Gamma_2^b}{(E_b - E_r)^2 + \left(\dfrac{\Gamma_2^b + \Gamma_\varphi^b}{2}\right)^2}$$

Defining

$$g \equiv \frac{D_0}{(\hbar\omega_0)^2}$$

we can write

$$\Sigma_\varphi^{in}(E_b) = \frac{D_0}{(\hbar\omega_0)^2}\Gamma_1^a = g\Gamma_1^a \tag{8.10.9}$$

and

$$\Sigma_\varphi^{out}(E_a) = D_0 \frac{\Gamma_2^b}{(E_a - E_r - \hbar\omega_0)^2 + \left(\dfrac{\Gamma_2^b + g\Gamma_1^a}{2}\right)^2}$$

We have added the superscripts 'a' and 'b' to Γ_1 and Γ_2 to denote the energies at which these are evaluated and made use of the fact that $\Gamma_1^b = 0$.

Also, we have assumed that $E_a - E_r \approx \hbar \omega_0$ is much larger than the scattering functions.

Step 5 Main peak

From Eqs.(8.10.5) and (8.10.8b), we can write the current in lead 1

$$I_1 = \frac{2e}{h} \int \left[\frac{\Gamma_1 \Gamma_2}{|\varepsilon|^2} + \frac{\Gamma_1 \Gamma_\varphi}{|\varepsilon|^2} \frac{\Gamma_2}{\Gamma_1 + \Gamma_2} \right] dE = I_{coh} + I_{seq}$$

Making use of the integral in Eq.(8.10.7) we obtain

$$I_{coh} = \frac{2e}{h} \int \frac{\Gamma_1 \Gamma_2}{|\varepsilon|^2} \, dE = \frac{2e}{\hbar} \frac{\Gamma_1 \Gamma_2}{\Gamma_1 + \Gamma_2 + \Gamma_\varphi}$$

$$I_{seq} = \frac{2e}{h} \int \frac{\Gamma_1 \Gamma_\varphi}{|\varepsilon|^2} \frac{\Gamma_2}{\Gamma_1 + \Gamma_2} \, dE$$

$$= \frac{2e}{\hbar} \frac{\Gamma_1 \Gamma_2}{\Gamma_1 + \Gamma_2 + \Gamma_\varphi} \frac{\Gamma_\varphi}{\Gamma_1 + \Gamma_2} = \frac{\Gamma_\varphi}{\Gamma_1 + \Gamma_2} I_{coh}$$

The two terms can be identified as the coherent and the sequential components of the current, in agreement with the results obtained in Chapter 6 using heuristic arguments (see Eqs.(6.2.4) and (6.2.6)). The total current is independent of the phase-breaking rate (Γ_φ)

$$I_{coh} + I_{seq} = \frac{2e}{\hbar} \frac{\Gamma_1 \Gamma_2}{\Gamma_1 + \Gamma_2}$$

as stated earlier (Eq.(8.10.1a)). Using the expressions for the scattering functions derived above (Eq.(8.10.8)), it is easy to see from Eq.(8.10.5) that the current in lead 2 is simply the negative of the current in lead 1: $I_2 = -I_1$.

Note that from Eq.(8.10.5) the current due to scattering is zero at all energies: $I_\varphi = 0$. This is basically the reason the simple heuristic argument in Section 6.2 works so well. *There is no net 'vertical flow' of electrons from one energy to another, in spite of the presence of interactions.* All the current flows at the same energy as sketched in Fig. 8.10.2a.

Next we will discuss the phonon peak where we will see that the sequential current flows 'vertically' from one energy to another as shown in Fig. 8.10.2b.

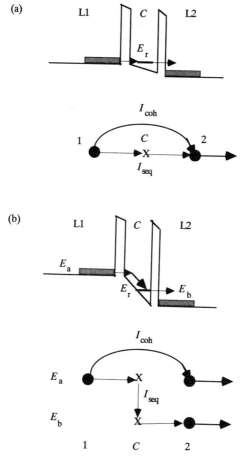

Fig. 8.10.2. Schematic representation of the coherent and sequential components of the current at the (*a*) main peak and at the (*b*) phonon peak.

Step 5 Phonon peak

Now we can calculate the terminal currents from Eq.(8.10.5) using the in-scattering and outscattering functions from Eq.(8.10.9). Making use of the integral in Eq.(8.10.7) we obtain from Eq.(8.10.5)

$$I_1^a = \frac{2e}{h} \int i_1(E_a) dE_a = I_{coh} + I_{seq} \qquad (8.10.10)$$

where
$$I_{\text{coh}} \equiv \frac{2e}{h} \Delta E \frac{\Gamma_1 \Gamma_2}{(\hbar\omega_0)^2}$$

$$I_{\text{seq}} \equiv \frac{2e}{\hbar} \frac{g\Gamma_1^{\text{a}} \Gamma_2^{\text{b}}}{g\Gamma_1^{\text{a}} + \Gamma_2^{\text{b}}}$$

The first term is the coherent term (ΔE is the energy spread of the incident electrons) which is usually very small. The second term is the sequential term which has the form stated earlier in Eq.(8.10.1b). There is no current at lead 1 around the energy E_{b}. Similarly for lead 2 we obtain

$$I_2^{\text{a}} = -I_{\text{coh}}, \quad I_2^{\text{b}} = -I_{\text{seq}}$$

while the current due to scattering is given by

$$I_\varphi^{\text{a}} = -I_{\text{seq}}, \quad I_\varphi^{\text{b}} = +I_{\text{seq}}$$

This illustrates what we have referred to as vertical flow. The sequential current flows in at energy E_{a} and out at energy E_{b}, leading to the current flow pattern shown in Fig. 8.10.2b.

Summary

For a proper description of quantum dynamics we need to keep track of phase-correlations. This requires a number of new concepts like the correlation function ($G^<$, $G^>$), the scattering function ($\Sigma^<$, $\Sigma^>$), the Green's function (G^{R}) and the self-energy (Σ^{R}). These concepts are introduced in Sections 8.1–8.2 and the basic equations are introduced in Sections 8.3–8.4. The overall approach for applying the NEGF formalism to specific problems is summarized in Section 8.5. Once we have calculated the correlation function for a particular structure, all other quantities of interest like the terminal current can be obtained as described in Section 8.6. Next we discuss the relationship of this formalism to the Landauer–Büttiker formalism (Section 8.7) and the Boltzmann formalism (Section 8.8). In Section 8.9 we discuss the difficulties of applying this formalism to strongly interacting systems where the perturbation theoretic treatment of interactions is not valid. We end in Section 8.10 with a simple illustrative example which permits an analytical solution.

We have developed the basic concepts of the NEGF formalism using elementary kinetic arguments without using the formalism of second quantization. One aspect that we have not done justice to is the systematic calculation of the self-energy functions for different types of interactions. This requires many-body perturbation theory beyond the scope of this book and we refer interested readers to Refs.[8.1]–[8.8]. However, the structure of the NEGF formalism can be appreciated without going into the details of how the self-energy functions are calculated, just as one can appreciate the Boltzmann formalism without knowing how the scattering functions are calculated.

Exercises

E.8.1 Show that the divergence of the probability current density (see Eq.(2.6.6))

$$\mathbf{J}(\mathbf{r}) = \frac{1}{2m}\left(\left[(\mathbf{p}-e\mathbf{A})\Psi\right]^*\Psi - \Psi^*\left[(\mathbf{p}-e\mathbf{A})\Psi\right]\right)$$

can be written as

$$i\hbar\nabla.\mathbf{J}(\mathbf{r}) = \left[H_C\Psi\right]^*\Psi - \Psi^*\left[H_C\Psi\right]$$

E.8.2 Making use of Eqs.(8.5.2) and (8.5.3) show that

$$H_C G^n - G^n H_C = G^R \Sigma^{in} - \Sigma^{in} G^A - \Sigma^R G^n + G^n \Sigma^A$$

E.8.3. Show that the scattering functions given in Section 8.4

$$\Sigma_\varphi^{in}(\mathbf{r},\mathbf{r}';E) = \int D(\mathbf{r},\mathbf{r}';\hbar\omega)G^n(\mathbf{r},\mathbf{r}';E-\hbar\omega)\mathrm{d}(\hbar\omega) \qquad (8.4.5a)$$

$$\Sigma_\varphi^{out}(\mathbf{r},\mathbf{r}';E) = \int D(\mathbf{r},\mathbf{r}';\hbar\omega)G^p(\mathbf{r},\mathbf{r}';E+\hbar\omega)\mathrm{d}(\hbar\omega) \qquad (8.4.5b)$$

obey the relation

$$\int \mathrm{Tr}\left[\Sigma_\varphi^{in} G^p\right]\mathrm{d}E = \int \mathrm{Tr}\left[\Sigma_\varphi^{out} G^n\right]\mathrm{d}E \qquad (8.6.8)$$

required for current conservation. Note that $D(\mathbf{r},\mathbf{r}';\hbar\omega) = D(\mathbf{r}',\mathbf{r};\hbar\omega)$.

E.8.4 Show that Eqs.(8.5.2) and (8.5.3) can be recast in terms of the time ordered and anti-time ordered functions (see Eq.(8.1.5)) as

$$\begin{bmatrix} E-H_C-\Sigma^T & +i\Sigma^{in} \\ +i\Sigma^{out} & E-H_C+\Sigma^{\overline{T}} \end{bmatrix}\begin{bmatrix} G^T & -iG^n \\ -iG^p & -G^{\overline{T}} \end{bmatrix} = \begin{bmatrix} I & 0 \\ 0 & I \end{bmatrix}$$

This is the form commonly found in the literature. It can be written compactly in the form

$$\left[(E - H_C)\underline{I} - \underline{\Sigma}\right]\left[\underline{G}\right] = \left[\underline{I}\right]$$

where we have defined

$$\underline{G} \equiv \begin{bmatrix} G^T & -iG^n \\ -iG^p & -G^{\bar{T}} \end{bmatrix}, \quad \underline{\Sigma} \equiv \begin{bmatrix} \Sigma^T & -i\Sigma^{in} \\ -i\Sigma^{out} & -\Sigma^{\bar{T}} \end{bmatrix}, \quad \underline{I} \equiv \begin{bmatrix} I & 0 \\ 0 & I \end{bmatrix}$$

This result is usually derived from a many-body approach using a perturbation expansion for the interactions. This yields a systematic diagrammatic method for evaluating the self-energy functions up to any desired order in the interaction.

Further reading

A few standard references and review articles are listed below. The list is far from exhaustive.

[8.1]   Martin, P. C. and Schwinger, J. (1959). 'Theory of many-particle systems.I', *Phys. Rev.* **115**, 1342.

[8.2]   Kadanoff, L. P. and Baym, G. (1962). *Quantum Statistical Mechanics*, Frontiers in Physics Lecture Note Series, Benjamin/Cummings.

[8.3]   Keldysh, L. V. (1965). 'Diagram technique for non-equilibrium processes', *Sov. Phys. JETP* **20**, 1018.

[8.4]   Langreth, D. C. (1976). In *Linear and Non-linear Electron Transport in Solids*, eds. J. T. Devreese and E. van Doren, NATO Advanced Study Institute Series B, vol.17, p.3, Plenum, New York.

[8.5]   Danielewicz, P. (1984). 'Quantum theory of non-equilibrium processes', *Ann. Phys.* **152**, 239.

[8.6]   Rammer, J. and Smith, H. (1986). 'Quantum field-theoretical methods in transport theory of metals', *Rev. Mod. Phys.* **58**, 323.

[8.7]   Mahan, G. D. (1987). 'Quantum transport equation for electric and magnetic fields' *Phys. Rep.* **145**, 251.

[8.8]   Khan, F. S., Davies, J. H. and Wilkins, J. W. (1987). 'Quantum transport equations for high electric fields' *Phys. Rev. B*, **36**, 2578.

The above references largely focus on infinite homogeneous media. Many authors have applied the NEGF formalism to problems involving finite structures. Since we are not aware of any review articles covering this aspect we cite a few papers below. Once again the list is far from exhaustive.

Tunneling

[8.9] Caroli, C., Combescot, R., Nozieres, P. and Saint-James, D. (1972). 'A direct calculation of the tunneling current: IV. Electron–phonon interaction effects', *J. Phys. C: Solid State Physics*, **5**, 21.

[8.10] Feuchtwang, T. E. (1976). 'Tunneling theory without the transfer Hamiltonian formalism', *Phys. Rev. B*, **13**, 517.

Superconductors

[8.11] Khlus, V. A. (1987). 'Current and voltage fluctuations in microjunctions between normal metals and superconductors', *Sov. Phys. JETP*, **66**, 1243.

[8.12] Arnold, G. B. (1985). 'Superconducting tunneling without the tunneling Hamiltonian', *J. Low Temp. Physics*, **59**, 143.

Relation to the Landauer–Büttiker formalism

[8.13] McLennan, M. J., Lee, Y. and Datta, S. (1991). 'Voltage drop in mesoscopic systems', *Phys. Rev. B*, **43**, 13 846.

[8.14] Pastawski, H. M. (1992). 'Classical and quantum transport from generalized Landauer–Büttiker equations. II. Time-dependent resonant tunneling', *Phys. Rev. B*, **46**, 4053.

Resonant tunneling

[8.15] Anda, E. V. and Flores, F. (1991). 'The role of inelastic scattering in resonant tunneling heterostructures' *J. Phys. Cond. Matter*, **2**, 8023.

[8.16] Runge, E. and Ehrenreich, H. (1992). 'Non-equilibrium transport in alloy-based resonant tunneling systems' *Annals of Physics*, **219**, 55.

[8.17] Lake, R. and Datta, S. (1993). 'Rate equations from the Keldysh formalism applied to the phonon peak in resonant tunneling diodes', *Phys. Rev. B*, **47**, 6427.

Transient transport

[8.18] Jauho, A. P., Wingreen, N. S. and Meir, Y. (1994). 'Time-dependent transport in interacting and non-interacting resonant-tunneling systems', *Phys. Rev. B*, **50**, 5528 and references therein.

More citations on quantum kinetic equations can be found in device-oriented review articles such as

[8.19] Ferry, D. K. and Grubin, H. L. 'Modeling of quantum transport in semiconductor devices' *Solid State Physics*, Academic Press, New York (to be published).

[8.20] Buot, F. (1994). 'Mesoscopic physics and nanoelectronics: Nanoscience and nanotechnology' *Physics Reports*, **234**, 73.

Concluding remarks

The 1980s were a very exciting time for mesoscopic physics characterized by a fruitful interplay between theory and experiment. What emerged in the process is a conceptual framework for describing current flow on length scales shorter than a mean free path. This conceptual framework is what we have tried to convey in this book. The activity in this field has expanded so much over the last few years that we have inevitably missed many interesting topics, such as persistent currents in normal metal rings, quantum chaos in microstructures, etc.

The development of the field is far from complete. So far both the theoretical and the experimental work has been almost entirely in the area of steady-state transport and many basic concepts remain to be clarified in the area of time-varying current flow as well as current fluctuations. Another emerging direction seems to be the study of mesoscopic conductors involving superconducting components. Finally, as we study current flow in smaller and smaller structures it seems clear that electron–electron interactions will play an increasingly significant role. As a result it will be necessary to go beyond the one-particle picture that is generally used in mesoscopic physics. Single-electron tunneling is a good example of this and it is likely that there will be many more developments involving current flow in strongly correlated systems.

It thus seems quite likely that the study of current flow in small conductors will continue to produce exciting new physics in the coming years. But will mesoscopic physics ever have any application? Device engineers of today involved in the design of sub-micron microelectronic devices are finding it necessary to go beyond the traditional drift–diffusion theory. The foundation for this work was laid by the basic physics-oriented research of the 1960s and 1970s. It is only reasonable to expect that the basic physics oriented research of today will eventually find use in describing microelectronic devices of the future. Some of the effects we have discussed here may play a more significant role in the operation of conventional devices, intentionally or otherwise. Also, non-conventional devices that rely on quantum effects (like resonant tunneling and single-electron tunneling) may find useful niches in specialised applications.

But will non-conventional mesoscopic devices ever have a major revolutionary impact on the electronics industry? At this point in time there is really no way even to guess at an answer to this question, though there is no dearth of opinions and prejudices. Microelectronic devices have been continually shrinking in size ever since their inception 40 years ago; currently they have a minimum feature size of about 500 nm. It is generally agreed that there are major barriers to be overcome if the feature size were to shrink beyond ~ 50 nm. But experts say that on this scale the real problem is not to build smaller transistors but to develop a scheme for interconnecting them. It is likely that a suitable scheme will emerge in due time but there is really no reason to believe that this new architecture will require revolutionary rather than evolutionary devices for its implementation. The current status of mesoscopic physics thus seems somewhat like that of semiconductor physics in the 1940s when there was no reason to believe that vacuum tubes would ever be displaced. But . . . who knows?

Further reading

Feynman, R. P. (1992). 'There's plenty of room at the bottom', *Journal of Microelectromechanical Systems*, **1**, 60.

This is the transcript of a talk given by Feynman in 1959, which explores the immense possibilities afforded by miniaturization. Although the progress in microelectronics since then has been truly impressive, only a fraction of the possibilities envisioned here have been realized so far.

Solutions to exercises

Chapter 1

E.1.1 $\mu = 27\,000\,\text{cm}^2/\text{Vs} = |e|\tau_\text{m}/m \rightarrow$ $\tau_\text{m} = 1.07\,\text{ps}$

$n_\text{s} = 1.6 \times 10^{11}/\text{cm}^2$ $\begin{array}{l} \rightarrow k_\text{f} = \sqrt{2\pi n_\text{s}} = 10^6/\text{cm} \\ \\ \rightarrow v_\text{f} = \hbar k_\text{f}/m = 1.67 \times 10^7\,\text{cm/s} \end{array}$ $\rightarrow L_\text{m} = v_\text{f}\tau_\text{m} = 0.18\,\mu\text{m}$

$D = v_\text{f}^2\tau_\text{m}/2 = 149.21\,\text{cm}^2/\text{s}, \ \tau_\varphi(1\,\text{K}) = 33.3\,\text{ps} \rightarrow L_\varphi = \sqrt{D\tau_\varphi} = 0.7\,\mu\text{m}$

E.1.2 (a) $V_x(B = 0) = 0.7$ mV, $I = 25.5\ \mu$A, $W = 0.38$ mm, $L = 1$ mm

$$\rho_{xx} = \frac{V_x}{I}\frac{W}{L} = \frac{0.7 \times 10^{-3}\,\text{V}}{25.5 \times 10^{-6}\,\text{A}} = 10.43\,\Omega \rightarrow \sigma_{xx} = 0.096\,\Omega^{-1}$$

$$V_\text{H}(B = 1\ \text{T}) = 33.3\ \text{mV}$$

$$\rho_{yx} = \frac{V_\text{H}}{I} = \frac{33.3 \times 10^{-3}\,\text{V}}{25.5 \times 10^{-6}\,\text{A}} = 1.306\,\text{K}\Omega = \frac{B}{en_\text{s}}$$

$$n_\text{s} = \frac{1\,\text{T}}{1.6 \times 10^{-19}\,\text{C} \times 1306\,\Omega} \approx 4.8 \times 10^{15}/\text{m}^2 \rightarrow 4.8 \times 10^{11}/\text{cm}^2$$

$$\mu = \frac{\sigma_{xx}}{en_\text{s}} = \frac{0.096\,\Omega^{-1}}{1.6 \times 10^{-19}\,\text{C} \times 4.8 \times 10^{15}/\text{m}^2} \approx 1.25 \times 10^6\,\text{cm}^2/\text{Vs}$$

(b)

peak number	*(m)*:	9	8	7	6	5	4
B	(T):	1.3	1.5	1.65	2.1	2.5	3.2
B^{-1}	(T^{-1}):	0.76	0.67	0.60	0.48	0.40	0.31

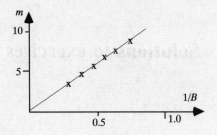

$$n_s = \frac{2e}{h}\frac{\Delta(m)}{\Delta(1/B)} = \frac{2\times1.6\times10^{-19}\,\text{C}}{6.63\times10^{-34}\,\text{J}\,\text{s}}\frac{1}{0.09\,\text{T}^{-1}} = 5.3\times10^{15}/\text{m}^2 \rightarrow 5.3\times10^{11}/\text{cm}^2$$

E.1.3 (a) For a hardwall potential,

$$E = E_s + E_i + \frac{\hbar^2 k^2}{2m} \quad \text{where} \quad E_i = i^2 E_1, \quad E_1 = \frac{\hbar^2\pi^2}{2mW^2} = 0.54\,\text{meV}$$

The number of electrons per unit length can be written as

$$n_{\text{L}} = \frac{2}{L}\ (\text{for spin}) \times \sum_i \frac{2k_{\text{f},i}}{2\pi/L} = \frac{2}{\pi\hbar}\sum_i \sqrt{2m(E_{\text{f}} - E_s - E_i)}$$

Normalizing the energies to E_1,

$$n_{\text{L}} = \frac{2}{W}\sum_i \sqrt{\frac{E_{\text{f}} - E_s}{E_1} - i^2}$$

(b) For a parabolic potential (see Eq.(1.6.5b)), $E_i = (i + 0.5)\hbar\omega_0$ so that

$$n_{\text{L}}W = 2\sum_i \sqrt{\frac{E_{\text{f}} - E_s}{E_1} - (i + 0.5)\frac{\hbar\omega_0}{E_1}}$$

Electron density versus Fermi energy for the two cases are shown in Fig. E.1.3. The question is what determines the Fermi energy in the narrow conductor. If we assume that it is equal to that in the wide conductor ($E_{\text{f}} - E_s = 17.2$ meV) then we obtain from the plot: (a) $n_{\text{L}} = 4.5 \times 10^6$/cm for the hardwall confinement and (b) $n_{\text{L}} = 3.2 \times 10^6$/cm for the parabolic confinement. In either case the electron density is smaller than what we would expect if we multiply the areal density by the width:

(a)

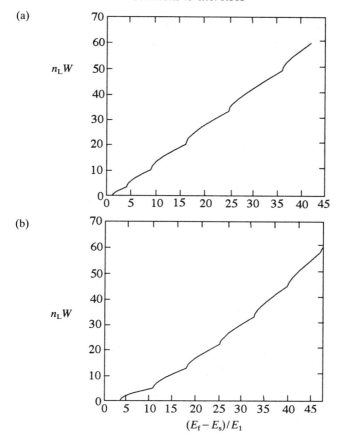

(b)

$(E_f - E_s)/E_1$

Fig. E.1.3. Plot of electron density versus Fermi energy for (a) hardwall potential and (b) parabolic potential (Courtesy of M. P. Samanta).

$$n_s W = 5 \times 10^{11}/\text{cm}^2 \times 1000\,\text{Å} = 5 \times 10^6/\text{cm}$$

This is what would happen if the narrow conductor were much shorter than the screening length. But in a long conductor we would expect the electron density to be equal to $n_s W$ in order to ensure charge neutrality.

From the plot we can see that this requires (a) $E_f - E_s = 19.5$ meV for hardwall confinement and (b) $E_f - E_s = 22.7$ meV for parabolic confinement. In either case $E_f - E_s$ is little larger than that in the wide conductor. The Fermi energy cannot be different in the wide conductor and in the narrow conductor, since they are in equilibrium. What happens is that charge transfer takes place so that a dipole layer builds up around each interface giving rise to a positive electrostatic potential in the narrow region relative to the wide regions. Consequently

$$E_s \text{ (wide)} - E_s \text{ (narrow)} = 2.3 \text{ meV for hard wall confinement}$$
$$= 5.5 \text{ meV for parabolic confinement}$$

E.1.4 We have seen (see Eq.(1.6.10)) that for parabolic confinement in a magnetic field

$$E(n,k) = E_s + (n + \tfrac{1}{2})\hbar\omega_{c0} + \frac{\hbar^2 k^2}{2m} \frac{\omega_0^2}{\omega_{c0}^2}$$

where

$$\omega_{c0}^2 = \omega_0^2 + (eB/m)^2$$

Proceeding as in E.1.3, the electron density can be written as

$$n_L W = \frac{2\omega_{c0}}{\omega_0} \sum_i \sqrt{\frac{E_f - E_s}{E_1} - (i + 0.5)\frac{\hbar\omega_{c0}}{E_1}}$$

We rewrite this in the form

$$\frac{n_L W}{(\omega_{c0}/\omega_0)^{3/2}} = 2\sum_i \sqrt{\frac{E_f - E_s}{(\omega_{c0}/\omega_0)E_1} - (i + 0.5)\frac{\hbar\omega_0}{E_1}}$$

Note that this is exactly the same as what we had in E.1.3 with the following replacements:

$$n_L W \rightarrow \frac{n_L W}{(\omega_{c0}/\omega_0)^{3/2}} \quad \text{and} \quad \frac{E_f - E_s}{E_1} \rightarrow \frac{E_f - E_s}{(\omega_{c0}/\omega_0)E_1}$$

Thus we can use the same plots as in E.1.3 (for the parabolic confining potential) if we scale the variables appropriately.

(a) If we assume that the Fermi energy remains constant (as is appropriate for a short conductor) then the number of modes changes whenever the cut-off energy for one of the subbands

$$\hbar\omega_{c0} = \left[(\hbar\omega_0)^2 + (\hbar eB/m)^2 \right]^{1/2}$$

crosses the Fermi energy. The number of modes is given by

$$M = \text{Int}\left[\frac{E_f - E_s}{\hbar\omega_{c0}} + \frac{1}{2} \right]$$

where Int(x) denotes the highest integer less than x. Noting that $\hbar\omega_0 = 3.9$ meV, $m = 0.067\, m_0$, we obtain,

$$M = 4 \quad \text{if} \quad 4.9\,\text{meV} > \hbar\omega_{c0} > 3.9\,\text{meV} \rightarrow 1.7\,\text{T} > B$$
$$M = 3 \quad \text{if} \quad 6.9\,\text{meV} > \hbar\omega_{c0} > 4.9\,\text{meV} \rightarrow 3.3\,\text{T} > B > 1.7\,\text{T}$$
$$M = 2 \quad \text{if} \quad 11.5\,\text{meV} > \hbar\omega_{c0} > 6.9\,\text{meV} \rightarrow 6.2\,\text{T} > B > 3.3\,\text{T}$$
$$M = 1 \quad \text{if} \quad 34.4\,\text{meV} > \hbar\omega_{c0} > 11.5\,\text{meV} \rightarrow 19.7\,\text{T} > B > 6.2\,\text{T}$$
$$M = 0 \quad \text{if} \quad \hbar\omega_{c0} > 34.4\,\text{meV} \rightarrow B > 19.7\,\text{T}$$

(b) If we assume that the electron density remains constant (as is appropriate for a long conductor) then we calculate the quantity

$$\nu = \frac{n_L W}{(\omega_{c0}/\omega_0)^{3/2}}$$

The number of modes can then be obtained making use of the plot in E.1.3b:

$$M = 1 \quad \text{if} \quad 0 < \nu < 5 \qquad M = 2 \quad \text{if} \quad 5 < \nu < 13 \qquad M = 3 \quad \text{if} \quad 13 < \nu < 22$$
$$M = 4 \quad \text{if} \quad 22 < \nu < 33 \qquad M = 5 \quad \text{if} \quad 33 < \nu < 45 \qquad M = 6 \quad \text{if} \quad 45 < \nu < 58$$

Hence,

$$M = 6 \quad \text{if} \quad \hbar\omega_{c0} < 4.2\,\text{meV} \rightarrow B < 0.9\,\text{T}$$
$$M = 5 \quad \text{if} \quad 4.2\,\text{meV} < \hbar\omega_{c0} < 5.1\,\text{meV} \rightarrow 0.9\,\text{T} < B < 1.9\,\text{T}$$
$$M = 4 \quad \text{if} \quad 5.1\,\text{meV} < \hbar\omega_{c0} < 6.7\,\text{meV} \rightarrow 1.9\,\text{T} < B < 3.1\,\text{T}$$
$$M = 3 \quad \text{if} \quad 6.7\,\text{meV} < \hbar\omega_{c0} < 9.6\,\text{meV} \rightarrow 3.1\,\text{T} < B < 5.0\,\text{T}$$
$$M = 2 \quad \text{if} \quad 9.6\,\text{meV} < \hbar\omega_{c0} < 18.1\,\text{meV} \rightarrow 5.0\,\text{T} < B < 10.2\,\text{T}$$
$$M = 1 \quad \text{if} \quad \hbar\omega_{c0} > 18.1\,\text{meV} \rightarrow B > 10.2\,\text{T}$$

Chapter 2

E.2.1

$$I = \frac{2e}{h} M \left[\mu_L - \mu_R \right] = \frac{2e}{h} N \left[\mu_L - \mu' \right] = \frac{2e}{h} N \left[\mu'' - \mu_R \right]$$

To simplify the algebra set $\mu_L = 1$, $\mu_R = 0$. This yields

$$1 - \mu' = \mu'' = \frac{M}{N}$$

Hence
$$eV_{\text{applied}} = \frac{1 + \mu'}{2} - \frac{\mu''}{2} = 1 - \frac{M}{N}, \quad I = \frac{2e}{h} M$$

Thus the contact resistance is reduced by the factor $(1 - M/N)$:

$$R_c = \frac{h}{2e^2} \frac{1}{M} \left[1 - \frac{M}{N} \right] = \frac{h}{2e^2} \left[\frac{1}{M} - \frac{1}{N} \right]$$

As we might expect, the contact resistance is zero when $N = M$.

E.2.2 The conductance of a ballistic conductor is given by $G = (2e^2/h)M$ where M is the number of modes. In problem E.1.4. we discussed the variation in M as a function of the magnetic field for a short conductor and for a long conductor. The former is appropriate in this case.

E.2.3 (a) Proceeding as described in Section 2.4 (see Eq.(2.4.9)) with $V_4 = 0$, we can write

$$\begin{Bmatrix} I_1 \\ I_2 \\ I_3 \end{Bmatrix} = \frac{2e^2}{h} \begin{bmatrix} T_0 & -T_L & -T_F \\ -T_R & T_0 & -T_L \\ -T_F & -T_R & T_0 \end{bmatrix} \begin{Bmatrix} V_1 \\ V_2 \\ V_3 \end{Bmatrix}$$

where
$$T_0 \equiv T_F + T_L + T_R$$

Inverting
$$\begin{Bmatrix} V_1 \\ V_2 \\ V_3 \end{Bmatrix} = \begin{bmatrix} R_{11} & R_{12} & R_{13} \\ R_{21} & R_{22} & R_{23} \\ R_{31} & R_{32} & R_{33} \end{bmatrix} \begin{Bmatrix} I_1 \\ I_2 \\ I_3 \end{Bmatrix}$$

where
$$R_{21} = \frac{h}{2e^2\Delta} \left[T_L T_0 + T_F T_R \right] \quad R_{23} = \frac{h}{2e^2\Delta} \left[T_R T_0 + T_F T_L \right]$$

and
$$\Delta = T_0^3 - 2T_0 T_L T_R - T_L^2 T_F - T_R^2 T_F - T_0 T_F^2$$
$$= \left[T_L + T_R \right] \left[T_L^2 + T_R^2 + 2T_F^2 + 2T_F T_L + 2T_F T_R \right]$$

Since $I_1 = -I_3$ and $I_2 = 0$, $V_2 = I_1[R_{21} - R_{23}]$ so that the Hall resistance is given by

$$R_H = \frac{V_2}{I_1} = \frac{h}{2e^2\Delta} \left[T_L - T_R \right] \left[T_L + T_R \right]$$
$$= \frac{h}{2e^2} \frac{T_L - T_R}{T_L^2 + T_R^2 + 2T_F^2 + 2T_F T_L + 2T_F T_R}$$

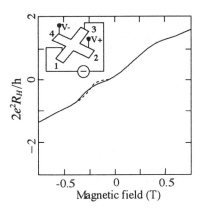

Fig. E.2.3. Hall resistance calculated using the measured values of T_F, T_L and T_R. The dashed line shows the directly measured Hall resistance. Adapted with permission from Fig. 9 of K. L. Shepard, M. L. Roukes and B. P. van der Gaag (1992). *Phys. Rev. B*, **46**, 9648.

E.2.4 Proceeding as described in Section 2.4 (see Eq.(2.4.9)) with $V_3 = 0$, we can write

$$\begin{Bmatrix} I_1 \\ I_2 \\ I_4 \end{Bmatrix} = \frac{2e^2}{h} \begin{bmatrix} T_0 & -T_L & -T_R \\ -T_R & T_0' & -T_F' \\ -T_L & -T_F' & T_0' \end{bmatrix} \begin{Bmatrix} V_1 \\ V_2 \\ V_4 \end{Bmatrix}$$

where $T_0 \equiv T_F + T_L + T_R = M$ and $T_0' \equiv T_F' + T_L + T_R$

Inverting
$$\begin{Bmatrix} V_1 \\ V_2 \\ V_4 \end{Bmatrix} = \begin{bmatrix} R_{11} & R_{12} & R_{14} \\ R_{21} & R_{22} & R_{24} \\ R_{41} & R_{42} & R_{44} \end{bmatrix} \begin{Bmatrix} I_1 \\ I_2 \\ I_4 \end{Bmatrix}$$

where $$R_{11} = \frac{h}{2e^2\Delta} \left[T_0'^2 - T_F'^2 \right]$$

and $$\Delta = T_0 \left[T_0'^2 - T_F'^2 \right] - T_F' \left[T_L^2 + T_R^2 \right] - 2T_0' T_L T_R$$

Since $I_2 = I_4 = 0$, $V_1 = I_1 R_{11}$ so that the conductance is given by

$$G = R_{11}^{-1} = \frac{2e^2}{h} \left[T_0 - \frac{T_F' \left[T_L^2 + T_R^2 \right] + 2T_0' T_L T_R}{T_0'^2 - T_F'^2} \right]$$

$$= \frac{2e^2}{h}\left[T_0 - \frac{T_F'\left[T_L^2 + T_R^2\right] + 2T_0'\, T_L T_R}{\left[T_L + T_R\right]\left[2T_F' + T_L + T_R\right]}\right]$$

$$= \frac{2e^2}{h}\left[T_0 - \frac{T_F'\left[T_L + T_R\right] + 2T_L T_R}{2T_F' + T_L + T_R}\right]$$

$$= \frac{e^2}{h}\left[T_0 + T_F + \frac{\left[T_L - T_R\right]^2}{2T_F' + T_L + T_R}\right]$$

If there is no magnetic field, then $T_L = T_R$, so that

$$G = \frac{e^2}{h}\left[T_0 + T_F\right] = \frac{e^2 M}{h} \quad \text{if} \quad T_F = 0$$

$$= \frac{2e^2 M}{h} \quad \text{if} \quad T_F = T_0$$

E.2.5 Proceeding as we did in Section 2.6 we can write the current in lead p at energy E_n due to a scattering state (q,k) as

$$i_{p,n}(q) = \frac{2e}{h}\left[\delta_{pq}\delta_{n,0} - T_{pq}(E_n,E)\right]$$

Assuming that a scattering state originating in lead q is occupied according to the Fermi function $f_q(E)$ we have,

$$I_p = \sum_{q,n}\int f_q(E)i_{p,n}(q)\mathrm{d}E = \frac{2e}{h}\int\left[f_p(E) - \sum_{q,n} T_{pq}(E_n,E)\,f_q(E)\right]\mathrm{d}E$$

E.2.6 We can write

$$\chi_{n,k'}(y)\left[E_s + \frac{(\hbar k + eBy)^2}{2m} + \frac{p_y^2}{2m} + U(y)\right]\chi_{m,k}(y) = E\,\chi_{m,k}(y)\,\chi_{n,k'}(y)$$

$$\chi_{m,k}(y)\left[E_s + \frac{(\hbar k' + eBy)^2}{2m} + \frac{p_y^2}{2m} + U(y)\right]\chi_{n,k'}(y) = E\,\chi_{n,k'}(y)\,\chi_{m,k}(y)$$

Subtracting we obtain

$$\frac{\hbar}{m}(k - k')\int\left[\chi_{m,k}\left(\frac{\hbar(k + k')}{2} + eBy\right)\chi_{n,k'}\right]\mathrm{d}y = 0$$

Hence $\displaystyle\int\left[\chi_{m,k}\left(\frac{\hbar(k+k')}{2}+eBy\right)\chi_{n,k'}\right]dy=0$ if $k\neq k'$

By normalizing the wavefunctions $\chi(y)$ appropriately we can ensure that

$$\int\left[\chi_{m,k}\left(\frac{\hbar(k+k')}{2}+eBy\right)\chi_{n,k'}\right]dy=\delta_{k,k'}$$

as stated.

E.2.7 We can write

$$f_1(E)-f_2(E)=\int_{\mu_2}^{\mu_1}\left[\frac{d}{dE'}\frac{1}{\exp[(E-E')/k_BT]+1}\right]dE'$$

$$=\int_{\mu_2}^{\mu_1}\left[-\frac{d}{d(E-E')}\frac{1}{\exp[(E-E')/k_BT]+1}\right]dE'$$

Defining $$F_T(E)\equiv-\frac{d}{dE}\left(\frac{1}{\exp(E/k_BT)+1}\right)$$

we can write $$f_1(E)-f_2(E)=\int_{\mu_2}^{\mu_1}F_T(E-E')dE'$$

From Eq.(2.5.1), $$I=\frac{2e}{h}\int\overline{T}(E)dE\int_{\mu_2}^{\mu_1}F_T(E-E')dE'$$

$$=\int_{\mu_2}^{\mu_1}\left[\frac{2e}{h}\int\overline{T}(E)F_T(E-E')dE\right]dE'$$

Eqs.(2.5.4)–(2.5.6) follow readily.

Chapter 3

E.3.1 (a)

$$I_P=0=T_{P1}\left[\mu_P-\mu_1\right]+T_{P2}\left[\mu_P-\mu_2\right]\quad\rightarrow\quad\mu_P=\frac{T_{P1}\mu_1+T_{P2}\mu_2}{T_{P1}+T_{P2}}$$

Setting $\mu_1=1$ and $\mu_2=0$: $\quad\mu_P=\dfrac{T_{P1}}{T_{P1}+T_{P2}}$

(b) T_{P1}

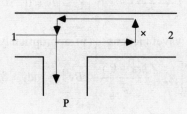

$$t_{P1} = \sqrt{\varepsilon}\, e^{ikd_1} \left[1 + \frac{ib\sqrt{1-T}\, e^{i2kd_2}}{1 - ia\sqrt{1-T}\, e^{i2kd_2}} \right] \qquad \text{(coherent)}$$

$$T_{P1} = \varepsilon \left[1 + \frac{b^2(1-T)}{1 - a^2(1-T)} \right] = \varepsilon\, \frac{1 + c(1-T)}{1 - a^2(1-T)} \qquad \text{(incoherent)}$$

T_{P2}

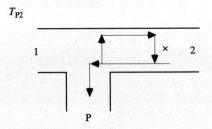

$$t_{P2} = \frac{\sqrt{\varepsilon}\,\sqrt{T}\, e^{ik(d_3+d_2)}}{1 - ia\sqrt{1-T}\, e^{i2kd_2}} \quad \text{and} \quad T_{P2} = \frac{\varepsilon T}{1 - a^2(1-T)}$$

(c) Assuming weak coupling, we have $c \sim 1$, $a \sim 0$ and $b \sim 1$, so that in the incoherent case

$$T_{P1} \approx \varepsilon(2-T) \quad T_{P2} = \varepsilon T$$

Hence $$\mu_P = \frac{2-T}{2} = 1 - \frac{T}{2} \qquad \text{(incoherent)}$$

This is simply the average potential of the $+k$ and the $-k$ states in Fig. 2.3.1. In the coherent case

$$t_{P1} \approx \sqrt{\varepsilon}\, e^{ikd_1} \left[1 + i\sqrt{1-T}\, e^{i2kd_2} \right] \to T_{P1} \approx \varepsilon \left[(2-T) - 2\sqrt{1-T}\, \sin(2kd_2) \right]$$

$$t_{P2} \approx \sqrt{\varepsilon}\,\sqrt{T}\, e^{ik(d_3+d_2)} \to T_{P2} \approx \varepsilon T$$

Hence $\qquad \mu_P \approx \dfrac{(1 - \frac{T}{2}) - \sqrt{1 - T}\,\sin(2kd_2)}{1 - \sqrt{1 - T}\,\sin(2kd_2)}$ $\qquad$ (coherent)

E.3.2 Consider an infinite wire with a transverse confining potential $U(y)$ and a zero vector potential (Fig. 3.3.3). The Hamiltonian is written as (see Eq.(3.3.2))

$$H_{\text{op}} = -\frac{\hbar^2}{2m}\frac{\partial^2}{\partial x^2} - \frac{\hbar^2}{2m}\frac{\partial^2}{\partial y^2} + U(y)$$

Its eigenfunctions are given by

$$\psi_{\beta,m}(x) = \frac{1}{\sqrt{L}}\chi_m(y)\exp\left[i\beta x\right] \qquad (\text{E.3.1})$$

where $\qquad \left[-\dfrac{\hbar^2}{2m}\dfrac{\partial^2}{\partial y^2} + U(y)\right]\chi_m(y) = \varepsilon_{m,0}\,\chi_m(y)$

and $\qquad\qquad \varepsilon_{m,\beta} = \varepsilon_{m,0} + \dfrac{\hbar^2\beta^2}{2m}$

Substituting into the eigenfunction expansion (Eq.(3.3.17)) we obtain

$$G^R(x,y;x',y') = \frac{1}{L}\sum_{m,\beta}\frac{\chi_m(y)\chi_m(y')\exp\left[i\beta(x-x')\right]}{E - \varepsilon_{m,0} - (\hbar^2\beta^2/2m) + i\eta}$$

where we have assumed that the transverse mode wavefunctions are $\chi_m(y)$ real. Replacing the summation by an integral according to the usual prescription

$$\sum_{\beta} \;\to\; \frac{L}{2\pi}\int d\beta$$

we obtain $\quad G^R(x,x') = -\dfrac{m}{\pi\hbar^2}\sum_{m}\chi_m(y)\chi_m(y')\displaystyle\int_{-\infty}^{+\infty}\dfrac{\exp\left[i\beta(x-x')\right]}{\beta^2 - k_m^2(1 + i\delta)}d\beta$

where $\qquad k_m^2 = \dfrac{2m(E - \varepsilon_{m,0})}{\hbar^2}$ $\quad$ and $\quad \delta = \dfrac{\eta}{E - U_0}$

The integral is best done using contour integration techniques. The integrand has two poles at

$$\beta = \pm k_m \sqrt{1 + i\delta} \approx \pm k_m \left(1 + \frac{i\delta}{2}\right)$$

The infinitesimal quantity δ serves to move the poles off the real axis as shown in Fig. E.3.2.

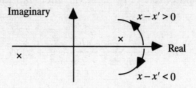

Fig. E.3.2. Contours in the complex β-plane used to do the integral.

For $(x - x') > 0$ we close the contour in the upper half plane (where the integrand is bounded) and thereby pick up the residue for the pole at positive k to obtain ($\delta \to 0$):

$$G^{\mathrm{R}}(x, y; x', y') = \sum_m -\frac{i}{\hbar v_m} \chi_m(y) \chi_m(y') \exp\left[+ik_m(x - x')\right]$$

For $(x - x') < 0$ we close the contour in the lower half plane and pick up the residue for the pole at negative k

$$G^{\mathrm{R}}(x, y; x', y') = \sum_m -\frac{i}{\hbar v_m} \chi_m(y) \chi_m(y') \exp\left[-ik_m(x - x')\right]$$

in agreement with the result obtained earlier using more elementary techniques (see Eq.(3.3.15)).

E.3.3

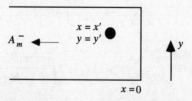

Fig. E.3.3. Retarded Green's function for a semi-infinite wire.

Let us first calculate the Green's function for a *continuous* semi-infinite wire that terminates on one side ($x = 0$) in an infinite potential (see Fig. E.3.3) where the wavefunction must go to zero. The eigenfunctions are given by (cf. Eq.(E.3.1))

$$\psi_{m,\beta}(x) = \sqrt{\frac{2}{L}}\chi_m(y)\sin[\beta x]; \quad \varepsilon_{m,\beta} = \varepsilon_{m,0} + \frac{\hbar^2\beta^2}{2m} \quad \text{(E.3.2)}$$

Substituting into the eigenfunction expansion (Eq.(3.3.17)) we obtain

$$G^R(x,y;x,y') = \frac{2}{L}\sum_m\sum_{\beta>0}\frac{\chi_m(y)\chi_m(y')\sin^2(\beta x)}{E - \varepsilon_{m,0} - (\hbar^2\beta^2/2m) + i\eta}$$

where we have set $x' = x$ since we will only need the Green's function between two points with the same x-coordinate. Converting the summation over β into an integral according to the prescription

$$\sum_\beta \to \frac{L}{\pi}\int d\beta$$

we have

$$G^R(x,y;x,y') = \frac{2}{\pi}\sum_m\chi_m(y)\chi_m(y')\int_0^\infty\frac{\sin^2(\beta x)}{E - \varepsilon_{m,0} - (\hbar^2\beta^2/2m) + i\eta}d\beta$$

Noting that $\qquad \sin^2(\beta x) = \dfrac{2 - \exp(2i\beta x) - \exp(-2i\beta x)}{4}$

we can write

$$G^R(x,y;x,y') = \frac{1}{2\pi}\sum_m\chi_m(y)\chi_m(y')\int_{-\infty}^{+\infty}\frac{[1 - \exp(i2\beta x)]}{E - \varepsilon_{m,0} - (\hbar^2\beta^2/2m) + i\eta}d\beta$$

and using contour integration as in the last example we obtain (cf. Eq.(3.3.15))

$$G^R(x,y;x,y') = -\sum_m\frac{2\sin(k_m x)}{\hbar v_m}\chi_m(y)\exp[ik_m x]\chi_m(y') \quad \text{(E.3.3)}$$

where $\qquad k_m \equiv \dfrac{\sqrt{2m(E - \varepsilon_{m,0})}}{\hbar} \quad$ and $\quad v_m \equiv \dfrac{\hbar k_m}{m}$

So far we have assumed both x- and y-coordinates to be continuous. What we need is the Green's function on a discrete lattice, having lattice points spaced by a, between two points along one edge of the lead (Fig. 3.5.2). We can obtain this from Eq.(E.3.3) simply by setting y and y' to the desired values and setting $x = a$:

$$g^R(p_i, p_j) = \left[G^R(x, y; x, y')\right]_{x=a; y=p_i; y'=p_j}$$

$$= -\sum_m \frac{2\sin(k_m a)}{\hbar v_m} \chi_m(p_i)\exp\left[ik_m a\right]\chi_m(p_j)$$

Using the relationship between the velocity and the wavenumber on a discrete lattice (see Eq.(3.5.8b))

$$\hbar v_m = 2at\sin(k_m a)$$

we obtain the desired result.

It is important to note that we set $x = a$ and not, say, equal to zero. The eigenfunctions in Eq.(E.3.2) satisfy the discretized Schrödinger equation for a semi-infinite lead only if the first point on the lead is located at $x = a$. To see this, consider the tight-binding equations for a semi-infinite lead:

$$(n > 1) \qquad (E - U_0 - 2t)\psi(n) + t\psi(n+1) + t\psi(n-1) = 0 \qquad \text{(E.3.4a)}$$

$$(n = 1) \qquad (E - U_0 - 2t)\psi(1) + t\psi(2) = 0 \qquad \text{(E.3.4b)}$$

Eq.(E.3.4a) is satisfied by a solution of the form

$$\psi(n) = A_1\sin\left[kna\right] + A_2\cos\left[kna\right]$$

for all values of A_1 and A_2, as long as $E = U_0 + 2t(1 - \cos(ka))$. But Eq.(E.3.4b) is satisfied only if $A_2 = 0$, showing that our choice of wavefunction in Eq.(E.3.1) is correct. But if the first point were somewhere other than $n = 1$, a different combination of A_1 and A_2 would be required.

Actually for a one-dimensional lead we can obtain the Green's function quite simply by solving the tight-binding equations directly:

$$(n > 1) \qquad (E - U_0 - 2t)g^R(n,1) + tg^R(n+1,1) + tg^R(n-1,1) = 0 \qquad \text{(E.3.5a)}$$

$$n = 1 \qquad (E - U_0 - 2t)g^R(1,1) + tg^R(2,1) = 1 \qquad \text{(E.3.5b)}$$

Eq.(E.3.5a) is satisfied by an outgoing wave solution of the form

$$g^R(n,1) = g^R(1,1)\exp\left[ik(n-1)a\right]$$

as long as $E = U_0 + 2t(1 - \cos(ka))$. Using this solution in Eq.(E.3.5b) we obtain

$$g^R(1,1) = \left[E - U_0 - 2t + t\exp(ika)\right]^{-1} = -\frac{1}{t}\exp\left[ika\right]$$

which is the single-moded version of Eq.(3.5.18).

E.3.4 From Eq.(3.4.6) we can write for $n \neq m$,

$$\left| s_{nm} \right|^2 = \frac{\hbar^2 v_n v_m}{a^2} \sum_{i,j,j',i'} \chi_n(q_j) G^R(j,i) \chi_m(p_i) \chi_n(q_{j'}) G^A(i',j') \chi_m(p_{i'})$$

where we have made use of the relation $G^A(i',j') = G^R(j',i')^*$. Hence we can write

$$T_{qp} = \sum_{n \in q} \sum_{m \in p} \left| s_{nm} \right|^2$$

$$= \sum_{i,j,j',i'} \Gamma_q(j',j) G^R(j,i) \Gamma_p(i,i') G^A(i',j') = \text{Tr}\left[\Gamma_q G^R \Gamma_p G^A \right]$$

where

$$\Gamma_p(i,i') \equiv \sum_{m \in p} \chi_m(p_i) \frac{\hbar v_m}{a} \chi_m(p_{i'})$$

E.3.5:

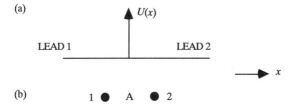

(a)

(b)   1 ●   A   ● 2

Fig. E.3.5 (*a*) A 1-D conductor with a delta function scatterer. (*b*) The conductor is represented by a single lattice point A and each mode in the leads is represented by a single node.

We can write the wavefunction as

$$\Psi(x) = e^{ikx} + s_{11} e^{-ikx} \quad x < 0 \quad \text{and} \quad \Psi(x) = s_{21} e^{ikx} \quad x > 0$$

In order to satisfy the Schrödinger equation

$$E\Psi + \frac{\hbar^2}{2m} \frac{\partial^2 \Psi}{\partial x^2} = U_0 \Psi \delta(x)$$

the first derivative of Ψ must be discontinuous at $x = 0$:

$$\left[\frac{\partial \Psi}{\partial x} \right]_{x=0^+} - \left[\frac{\partial \Psi}{\partial x} \right]_{x=0^-} = \frac{2mU_0}{\hbar^2} \Psi(x=0)$$

while Ψ itself must be continuous at $x = 0$. Hence

$$s_{21} = 1 + s_{11}$$

$$iks_{21} = ik(1 - s_{11}) + \frac{2mU_0}{\hbar^2} s_{21} \quad \rightarrow \quad s_{21}(1 - \frac{2U_0}{i\hbar v}) = 1 - s_{11}$$

so that, $\qquad s_{11} = \dfrac{U_0}{i\hbar v - U_0}$ and $s_{21} = \dfrac{i\hbar v}{i\hbar v - U_0}$

where the velocity $v \ (= \hbar k/m = (2mE)^{1/2}/\hbar)$ is the same in both leads, since the potential is the same. From the symmetry of the problem, $s_{22} = s_{11}$ and $s_{12} = s_{21}$.

(b) We choose a discrete lattice consisting of a single point 'A' so that all matrices are of dimensions (1×1), as in Fig. E.3.5.

From Eq.(3.5.9a): $\qquad EI - H_C = \left[E - (U_0/a) - 2t \right]$

From Eq.(3.5.18): $\qquad \Sigma_1^R = \Sigma_2^R = -t \exp[ika]$

From Eq.(3.5.17): $\quad G^R = \left[E - (U_0/a) - 2t + 2t \exp(ika) \right]^{-1}$

Making use of Eq.(3.5.8a,b), we can write

$$G^R = \left[-(U_0/a) + 2it \sin(ka) \right]^{-1} = \frac{a}{i\hbar v - U_0}$$

The S-matrix is now obtained from Eq.(3.4.3) (dropping the factor a from G^R since Eq.(3.4.3) applies to a continuous representation):

$$S_{qp} = -\delta_{qp} + \frac{i\hbar v}{i\hbar v - U_0} = \begin{bmatrix} \dfrac{U_0}{i\hbar v - U_0} & \dfrac{i\hbar v}{i\hbar v - U_0} \\[2ex] \dfrac{i\hbar v}{i\hbar v - U_0} & \dfrac{U_0}{i\hbar v - U_0} \end{bmatrix}$$

E.3.6 (a) The results follow readily from Eq.(3.2.4). *(b)* Plots are shown in Fig. E.3.6. with 1, 5 and 15 lattice points between the scatterers. The plots are truncated for $E > 4t$, where

$$t = \frac{\hbar^2}{2ma^2} = 23 \text{ meV} \times (\text{no. of points between the scatterers} + 1)^2$$

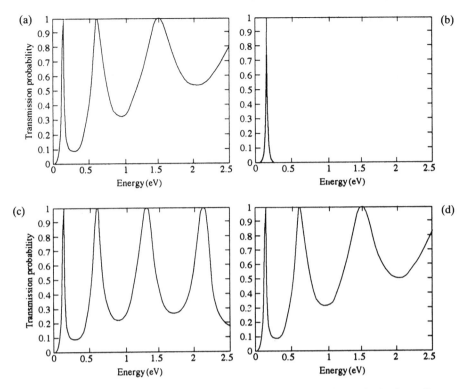

Fig. E.3.6. Transmission probability versus energy. (*a*) Exact, (*b*) 1 point in the well, (*c*) 5 points and (*d*) 15 points (courtesy of M. P. Samanta).

E.3.7.(a)

$$\begin{bmatrix} c & \sqrt{\varepsilon} & \sqrt{\varepsilon} \\ \sqrt{\varepsilon} & a & b \\ \sqrt{\varepsilon} & b & a \end{bmatrix}\begin{bmatrix} c & \sqrt{\varepsilon} & \sqrt{\varepsilon} \\ \sqrt{\varepsilon} & a & b \\ \sqrt{\varepsilon} & b & a \end{bmatrix} = \begin{bmatrix} c^2 + 2\varepsilon & \sqrt{\varepsilon}(a+b+c) & \sqrt{\varepsilon}(a+b+c) \\ \sqrt{\varepsilon}(a+b+c) & a^2+b^2+\varepsilon & 2ab+\varepsilon \\ \sqrt{\varepsilon}(a+b+c) & 2ab+\varepsilon & a^2+b^2+\varepsilon \end{bmatrix}$$

$$= \begin{bmatrix} 1 & 0 & 0 \\ 0 & 1 & 0 \\ 0 & 0 & 1 \end{bmatrix}$$

Hence we must have

$$a^2 + b^2 = 1 - \varepsilon, \quad c^2 = 1 - 2\varepsilon, \quad 2ab = -\varepsilon, \quad a + b + c = 0$$

These relations are satisfied if

$$a = (1-c)/2, \quad b = (1+c)/2, \quad c = \sqrt{1-2\varepsilon}$$

Other possible solutions are obtained by interchanging a and b, or by taking c as a negative quantity.

(b) Using Eq.(3.2.1), $(P = \exp(i\theta))$

$$t = \begin{bmatrix} \sqrt{\varepsilon} & \sqrt{\varepsilon} \end{bmatrix} \left[I - \begin{pmatrix} P & 0 \\ 0 & P \end{pmatrix} \begin{pmatrix} a & b \\ b & a \end{pmatrix} \begin{pmatrix} P & 0 \\ 0 & P \end{pmatrix} \begin{pmatrix} a & b \\ b & a \end{pmatrix} \right]^{-1} \begin{bmatrix} P & 0 \\ 0 & P \end{bmatrix} \begin{bmatrix} \sqrt{\varepsilon} \\ \sqrt{\varepsilon} \end{bmatrix}$$

$$= \begin{bmatrix} \sqrt{\varepsilon} & \sqrt{\varepsilon} \end{bmatrix} \left[I - \begin{pmatrix} (a^2+b^2)P^2 & 2abP^2 \\ 2abP^2 & (a^2+b^2)P^2 \end{pmatrix} \right]^{-1} \begin{bmatrix} P\sqrt{\varepsilon} \\ P\sqrt{\varepsilon} \end{bmatrix}$$

$$= \begin{bmatrix} \sqrt{\varepsilon} & \sqrt{\varepsilon} \end{bmatrix} \begin{bmatrix} 1-(1-\varepsilon)P^2 & \varepsilon P^2 \\ \varepsilon P^2 & 1-(1-\varepsilon)P^2 \end{bmatrix}^{-1} \begin{bmatrix} P\sqrt{\varepsilon} \\ P\sqrt{\varepsilon} \end{bmatrix}$$

$$= \frac{\varepsilon P}{(1-P^2)(1-c^2P^2)} \begin{bmatrix} 1 & 1 \end{bmatrix} \begin{bmatrix} 1-(1-\varepsilon)P^2 & -\varepsilon P^2 \\ -\varepsilon P^2 & 1-(1-\varepsilon)P^2 \end{bmatrix} \begin{bmatrix} 1 \\ 1 \end{bmatrix}$$

$$= \frac{2\varepsilon P}{(1-c^2P^2)}$$

Hence $$T = \frac{4\varepsilon^2}{\left| 1-c^2P^2 \right|^2} = \frac{4\varepsilon^2}{1-2c^2\cos(2\theta)+c^4} \tag{E.3.6}$$

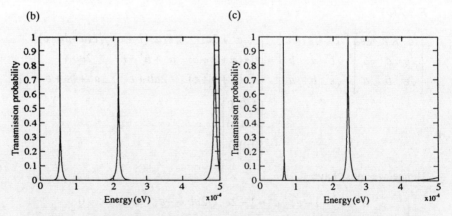

Fig. E.3.7. Transmission probability through a single-moded ring with radius $r = 1000$ Å calculated (*b*) from Eq.(E.3.6) with $\varepsilon = 0.025$ and (*c*) numerically from Eq.(3.5.20) using a lattice with 140 points along the ring (courtesy of M. P. Samanta).

(d)

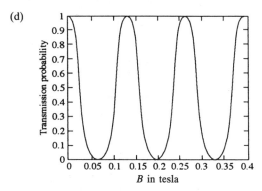

Fig. E.3.7.(*d*) Transmission probability through a single-moded ring with radius *r* = 1000 Å versus magnetic field calculated numerically from Eq.(3.5.20) at an energy corresponding to the second peak in Fig. E.3.7c, *E* = 0.244 meV (courtesy of M. P. Samanta).

E.3.8 For a ballistic conductor

$$G = \frac{2e^2}{h} \frac{\hbar^2}{L^2} \sum_m \sum_k v_m^2(k) \left| G^R(m,k;m,k) \right|^2$$

Neglecting the real part of the self-energy we can write from Eq.(3.7.4)

$$\Sigma_1^R(m,k;m,k) = \Sigma_2^R(m,k;m,k) = -\frac{i}{2} \frac{\hbar v_m(k)}{L}$$

Hence
$$G^R(m,k;m,k) = \frac{1}{(E - \varepsilon_{m,k}) + i(\hbar v_m(k)/L)}$$

so that
$$G = \frac{2e^2}{h} \sum_m \sum_k \frac{\hbar^2 v_m^2(k)/L^2}{(E - \varepsilon_{m,k})^2 + (\hbar^2 v_m^2(k)/L^2)}$$

$$\approx \frac{2e^2}{h} \sum_m \sum_k \frac{\hbar v_{m,k}}{L} \pi \delta(E - \varepsilon_{m,k}) = \frac{2e^2}{h} \sum_m \int \hbar v_{m,k} \delta(E - \varepsilon_{m,k}) dk$$

$$= \frac{2e^2}{h} \sum_m 1 = \frac{2e^2 M}{h}$$

E.3.9 We note that reversing the magnetic field is equivalent to taking the complex conjugate of the Hamiltonian (see Eq.(3.5.2)). Hence

$$\left[E - H - i\eta\right]_{+B} \left[G^A\right]_{+B} = I \quad \rightarrow \quad \left[E - H + i\eta\right]_{-B} \left[G^A\right]_{+B}^* = I$$

But
$$[E - H + i\eta]_{-B} [G^R]_{-B} = I$$

Hence
$$[G^R]_{-B} = [G^A]^*_{+B}$$

Since the advanced function is the conjugate transpose of the retarded function we obtain the desired result ('t' denotes transpose):

$$[G^R]_{-B} = [G^R]^t_{+B}$$

From Eq.(3.4.6) (assuming zero magnetic field in the leads)

$$[s_{nm}]_{-B} = -\delta_{nm} + i\hbar\sqrt{v_n v_m} \iint \chi_n(y_q) [G^R_{qp}(y_q; y_p)]_{-B} \chi_m(y_p) dy_q dy_p$$

$$= -\delta_{mn} + i\hbar\sqrt{v_m v_n} \iint \chi_m(y_p) [G^R_{pq}(y_p; y_q)]_{+B} \chi_n(y_q) dy_q dy_p$$

$$= [s_{mn}]_{+B}$$

Chapter 4

E.4.1 (a)

G_{pq}:	$q = 1$	$q = 2$	$q = 3$	$q = 4$
$p = 1$	0	pG_C	$(1-p)G_C$	0
$p = 2$	G_C	0	0	0
$p = 3$	0	0	pG_C	$(1-p)G_C$
$p = 4$	0	$(1-p)G_C$	0	pG_C

Note that we have written $G_{33} = pG_C$ in order to have all the sums and columns add up to the same number G_C, assuming all leads to have the same number of modes. However, the actual currents are unaffected by what we choose for the diagonal elements of G_{pq}.

(b)

$$I_2 = 0: \qquad\qquad\qquad V_2 = V_1$$
$$I_3 = 0: \qquad\qquad\qquad V_3 = V_4$$

$$I_1 = I = pG_C(V_1 - V_2) + (1-p)G_C(V_1 - V_3) = (1-p)G_C(V_2 - V_3)$$

Hence
$$R_H = \frac{V_2 - V_3}{I_1} = \frac{1}{G_C} \frac{1}{1-p}$$

E.4.2

G_{pq}:	$q = 1$	$q = 2$	$q = 3$	$q = 4$	$q = 5$
$p = 1$	0	0	$(1-p)G_C$	0	pG_C
$p = 2$	G_C	0	0	0	0
$p = 3$	0	0	pG_C	0	$(1-p)G_C$
$p = 4$	0	$(1-p)G_C$	0	pG_C	0
$p = 5$	0	pG_C	0	$(1-p)G_C$	0

Setting $V_4 = 0$,

$I_2 = 0$: $\qquad\qquad\qquad\qquad\qquad\qquad\qquad\qquad\qquad V_2 = V_1$

$I_3 = 0$: $\qquad\qquad\qquad\qquad\qquad\qquad\qquad\qquad\qquad V_3 = V_5$

$I_5 = 0 = p(V_5 - V_2) + (1 - p)(V_5 - V_4)$ $\qquad\qquad V_5 = pV_2$

$I_1 = I = pG_C(V_1 - V_5) + (1-p)G_C(V_1 - V_3) = G_C(V_1 - V_5)$

Hence
$$R_H = \frac{V_2 - V_3}{I_1} = \frac{1}{G_C}$$

Chapter 5

E.5.1 $\qquad\qquad \mathbf{J} = -\sigma\nabla\varphi, \quad \nabla.\mathbf{J} = 0 \quad \rightarrow \quad \nabla^2\varphi = 0$

In a circular geometry we can write

$$\varphi(\mathbf{r}) = \frac{-V\ln(r/L_{min})}{\ln(L_{max}/L_{min})} \quad \rightarrow \quad \mathbf{E}(\mathbf{r}) = \hat{r}\,\frac{V/r}{\ln(L_{max}/L_{min})}$$

Hence
$$\mathbf{J}(\mathbf{r}) = \hat{r}\,\frac{\sigma V/r}{\ln(L_{max}/L_{min})}$$

so that the net current is given by

$$I = \int \mathbf{J}.d\mathbf{S} = \frac{\pi\sigma V}{\ln(L_{max}/L_{min})} \quad \rightarrow \quad G = \frac{I}{V} = \frac{\pi\sigma}{\ln(L_{max}/L_{min})}$$

E.5.2 We can write Eq.(5.5.30) in the form

$$\frac{-\Delta\sigma}{e^2/\pi h} = \left[\Psi\left(\frac{1}{2} + \frac{B_m}{B}\right) - \Psi\left(\frac{1}{2} + \frac{B_\varphi}{B}\right)\right]$$

where
$$B_m = \frac{\hbar}{4|e|D\tau_m} \qquad B_\varphi = \frac{\hbar}{4|e|D\tau_\varphi}$$

$$\mu = 2.7 \text{ m}^2/\text{V s} = \frac{|e|\tau_m}{m} \quad \rightarrow \quad \tau_m = 1 \text{ ps}$$

$$\sigma = |e|n_s\mu = 1.6 \times 10^{-19} \text{ C} \times 1.6 \times 10^{15}/\text{m}^2 \times 2.7 \text{ m}^2/\text{V s} = 6.9 \times 10^{-4} \ \Omega^{-1}$$

$$\sigma = e^2 N_s D \rightarrow D = \frac{6.9 \times 10^{-4} \ \Omega^{-1}}{(1.6 \times 10^{-19} \text{ C})^2 \times 2.9 \times 10^{13}/\text{cm}^2 \text{ eV}} = 149 \text{ cm}^2/\text{sec}$$

Hence $B_m = 111$ G.

By adjusting τ_φ we can fit the experimental data. The data at T = 0.3 K can be fit if we assume $\tau_\varphi = 175$ ps ($B_\varphi = 0.64$ Gauss), so that

$$-\frac{\Delta R}{R} = \frac{\Delta \sigma}{\sigma} = 0.018 \left[\Psi\left(\frac{1}{2} + \frac{111\,G}{B}\right) - \Psi\left(\frac{1}{2} + \frac{0.64\,G}{B}\right) \right]$$

E.5.3 (a) We can estimate the period by noting that there are 19 oscillations in a range of 30 mT: $B_p = 1.58$ mT. Hence

$$S = \frac{h}{|e|B_p} = 2.62 \times 10^{-12} \text{ m}^2 \quad \rightarrow \quad d = 1.83 \ \mu\text{m}$$

(b) For a thick ring there is a continuous distribution of periods because the area S is not unique. The longest period corresponds to the smallest area which is that enclosed by the inner circumference

$$B_p(\text{max}) = \frac{h}{|e|S_{min}} = 5.29 \text{ mT}$$

while the shortest period corresponds to the largest area which is that enclosed by the outer circumference

$$B_p(\text{min}) = \frac{h}{|e|S_{max}} = 1.32 \text{ mT}$$

Due to this wide distribution of periods the conductance changes look more like noise than like oscillations.

E.5.4 We can define $\langle \cos\theta \rangle$ as

$$\langle \cos\theta \rangle \equiv \frac{\int\limits_{-\pi}^{+\pi} \cos\theta\, P(\theta)\mathrm{d}\theta}{\int\limits_{-\pi}^{+\pi} P(\theta)\mathrm{d}\theta} \quad \text{so that} \quad \frac{1}{\tau_\mathrm{m}} = \frac{1 - \langle\cos\theta\rangle}{\tau}$$

Now let us evaluate the conductivity correction due to the first of the ladder diagrams shown in Fig. 5.5.10:

$$\Gamma(\mathbf{k'},\mathbf{k}) = \frac{|U_{\mathbf{k-k'}}|^2}{L^2} = \frac{U_0^2}{L^2} P(\theta - \theta')$$

Inserting into Eq.(5.5.21) we obtain (we are ignoring spin and dividing by 2)

$$\Delta\sigma^{(1)} = \frac{e^2\hbar U_0^2}{2\pi L^4} \sum_\mathbf{k} v_x(\mathbf{k})|G(\mathbf{k})|^2 \sum_{\mathbf{k'}} v_x(\mathbf{k'})|G(\mathbf{k'})|^2 P(\theta - \theta')$$

$$= \frac{e^2\hbar U_0^2}{2\pi} \iint \frac{k}{4\pi^2} \frac{\hbar k \cos\theta}{m} \frac{1}{(E - \varepsilon_k)^2 + (\eta)^2} \mathrm{d}k\mathrm{d}\theta$$

$$\times \iint \frac{k'}{4\pi^2} \frac{\hbar k' \cos\theta'}{m} \frac{1}{(E - \varepsilon_{k'})^2 + (\eta)^2} P(\theta - \theta')\mathrm{d}k'\,\mathrm{d}\theta$$

$$= \frac{e^2\hbar U_0^2}{2\pi} \iint \frac{1}{(4\pi\hbar\eta)^2} k k'\, \delta(E - \varepsilon_k)\delta(E - \varepsilon_{k'})\mathrm{d}\varepsilon_k\mathrm{d}\varepsilon_{k'}$$

$$\times \iint \cos\theta \cos\theta'\, P(\theta - \theta')\mathrm{d}\theta'\,\mathrm{d}\theta$$

$$= \frac{e^2 U_0^2}{32\pi^3\hbar\eta^2} \frac{2mE}{\hbar^2} \pi \int \cos\theta\, P(\theta)\mathrm{d}\theta$$

Making use of the relations (note that we are ignoring spin)

$$n_\mathrm{s} = \frac{mE}{2\pi\hbar^2} \quad \text{and} \quad \eta = \frac{\hbar}{2\tau} = \frac{mU_0^2}{4\pi\hbar^2} \int P(\theta)\mathrm{d}\theta$$

we obtain

$$\Delta\sigma^{(1)} = \frac{e^2}{32\pi^3\hbar\eta^2} \frac{4\pi\hbar^2\eta}{m} 4\pi n_\mathrm{s}\pi\langle\cos\theta\rangle = \frac{e^2 n_\mathrm{s}}{m} \frac{\hbar}{2\eta}\langle\cos\theta\rangle$$

$$= \frac{e^2 n_\mathrm{s}}{m} \tau\langle\cos\theta\rangle$$

Similarly if we evaluate the next ladder diagram we obtain

$$\Delta\sigma^{(2)} = \frac{e^2 n_s}{m} \tau \langle \cos\theta \rangle^2$$

and so on. So the total conductivity is given by

$$\Delta\sigma = \frac{e^2 n_s}{m} \tau \left[1 + \langle \cos\theta \rangle + \langle \cos\theta \rangle^2 + \ldots \right]$$

$$= \frac{e^2 n_s}{m} \frac{\tau}{1 - \langle \cos\theta \rangle} = \frac{e^2 n_s \tau_m}{m}$$

Chapter 6

E.6.1

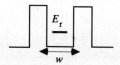

For a rough estimate we assume an effective well width equal to the actual well width plus (say) a fourth of the width of each barrier: $w = 50 + 12.5 + 12.5 = 75$ Å.

$$E_r \sim \frac{h^2}{8mw^2} = \frac{(6.63 \times 10^{-34} \text{ J s})^2}{8 \times 0.067 \times 9.1 \times 10^{-31} \text{ kg} \times (75 \times 10^{-10} \text{ m})^2} \sim 100 \text{ meV}$$

For a more accurate calculation, see for example, Chapter 1 of S. Datta (1989), *Quantum Phenomena*, Addison–Wesley.

$$v = \sqrt{\frac{2E_r}{m}} = \sqrt{\frac{2 \times 100 \times 10^{-3} \times 1.6 \times 10^{-19} \text{ J}}{0.067 \times 9.1 \times 10^{-31} \text{ kg}}} = 7.2 \times 10^7 \text{ cm/sec}$$

$$\nu = \frac{v}{2w} = \frac{7.2 \times 10^7 \text{ cm/sec}}{2 \times 75 \times 10^{-8} \text{ cm}} = 4.8 \times 10^{13}/\text{s} \quad \rightarrow \quad \hbar\nu = 31.8 \text{ meV}$$

To estimate the transmission probability through one of the barriers, we note that the decay constant is given by

$$\gamma = \sqrt{\frac{2m(U - E_r)}{\hbar^2}}$$

$$= \sqrt{\frac{2 \times 0.067 \times 9.1 \times 10^{-31} \text{ kg} \times 200 \times 1.6 \times 10^{-22} \text{ J}}{(1.06 \times 10^{-34} \text{ J s})^2}} = 5.9 \times 10^8 /\text{m}$$

Hence $\quad T \sim \exp(-\gamma d) = 5.3 \times 10^{-2} \rightarrow \Gamma_1 = \Gamma_2 = \hbar v T = 1.7 \text{ meV}$

$$\Gamma_\varphi = \hbar/\tau_\varphi = 0.7 \text{ meV}$$

Hence the fraction of the current that is coherent is given by

$$\frac{\Gamma_1 + \Gamma_2}{\Gamma_1 + \Gamma_2 + \Gamma_\varphi} = \frac{3.4}{4.1} = 83\%$$

E.6.2 $$I_P = \frac{2e}{\hbar} \frac{\Gamma_1 \Gamma_2}{\Gamma_1 + \Gamma_2} \sim 68 \text{ nA}$$

$$\frac{I_{\text{peak}}}{S} = I_P \frac{m}{\pi \hbar^2} \mu_1 \sim 68 \text{ nA} \times 2.9 \times 10^{10}/\text{cm}^2 \text{ meV} \times 10 \text{meV}$$

$$= 19.7 \text{ kA/cm}^2$$

E.6.3

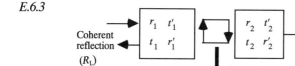

Fig. E.6.3. A resonant tunneling diode consists of two barriers with scattering matrices as shown in series. Scattering processes cause electrons to leak out of the coherent stream as shown.

It is easiest to calculate the coherent transmission probability T_L and the coherent reflection probability R_L and then obtain the scattering probability T_S from the relation

$$T_S + T_L + R_L = 1$$

To calculate T_L and R_L we sum the different paths as shown in Fig. 3.2.2, taking care to insert a factor of $\exp[-2\alpha W]$ every time we traverse the well. We obtain

$$T_L(E_L) = \left| \frac{t_1 t_2 \exp[-\alpha W]}{1 - r_1' r_2 \exp[-2\alpha W]} \right|^2$$

$$R_L(E_L) = \left| r_1 + \frac{t_1 r_2 t_1'\exp[-2\alpha W]}{1 - r_1' r_2 \exp[-2\alpha W]} \right|^2 = \left| \frac{r_1 - r_2 \exp[-2\alpha W]}{1 - r_1' r_2 \exp[-2\alpha W]} \right|^2$$

where we have made use of the relation: $r_1 r_1' - t_1 t_1' = 1$ (since the S-matrix is unitary, its determinant is one). Hence,

$$T_L(E_L) = \frac{T_1 T_2 \exp[-2\alpha W]}{1 + R_1 R_2 \exp[-4\alpha W] - 2\sqrt{R_1 R_2}\,\exp[-2\alpha W]\cos\theta(E_L)} \equiv \frac{N_T}{D}$$

$$R_L(E_L) = \frac{R_1 + R_2 \exp[-4\alpha W] - 2\sqrt{R_1 R_2}\,\exp[-2\alpha W]\cos\theta(E_L)}{1 + R_1 R_2 \exp[-4\alpha W] - 2\sqrt{R_1 R_2}\,\exp[-2\alpha W]\cos\theta(E_L)} \equiv \frac{N_R}{D}$$

We will now simplify the three quantities D, N_T and N_R one by one, making suitable approximations.

$$D = \left(1 - \sqrt{R_1 R_2}\,\exp[-2\alpha W]\right)^2 + 2\sqrt{R_1 R_2}\,\exp[-2\alpha W](1 - \cos\theta(E_L))$$

Approximate the cosine function by a quadratic function and assuming that T_1 ($= 1 - R_1$), T_2 ($= 1 - R_2$) and αW are all much less than one, we obtain

$$D \approx \left(\frac{T_1}{2} + \frac{T_2}{2} + 2\alpha W\right)^2 + \theta^2$$

Using the same approximations we can write

$$N_T \approx T_1 T_2$$

$$N_R = \left(\sqrt{R_1} - \sqrt{R_2}\,\exp[-2\alpha W]\right)^2 + 2\sqrt{R_1 R_2}\,\exp[-2\alpha W](1 - \cos\theta(E_L))$$

$$\approx \left(\frac{T_1}{2} - \frac{T_2}{2} - 2\alpha W\right)^2 + \theta^2$$

Hence,
$$T_L \approx \frac{N_T}{D} = \frac{T_1 T_2}{\theta^2 + \left(\dfrac{T_1 + T_2 + 4\alpha W}{2}\right)^2}$$

$$1 - R_L \approx 1 - \frac{N_R}{D} = \frac{T_1(T_2 + 4\alpha W)}{\left(\dfrac{T_1}{2} + \dfrac{T_2}{2} + 2\alpha W\right)^2 + \theta^2}$$

$$T_S = 1 - R_L - T_L \approx \frac{4\alpha W T_1}{\left(\dfrac{T_1}{2} + \dfrac{T_2}{2} + 2\alpha W\right)^2 + \theta^2}$$

Writing

$$\theta \approx \frac{d\theta}{dE_L}(E_L - E_r) = \frac{E_L - E_r}{\hbar v}$$

and defining $\Gamma_1 \equiv \hbar v T_1$, $\Gamma_2 \equiv \hbar v T_2$, $\Gamma_\varphi \equiv \hbar v(4\alpha W)$
we obtain

$$T_L(E_L) \approx \frac{\Gamma_1 \Gamma_2}{\left(E_L - E_r\right)^2 + \left(\dfrac{\Gamma_1 + \Gamma_2 + \Gamma_\varphi}{2}\right)^2} = \frac{\Gamma_1 \Gamma_2}{\Gamma_1 + \Gamma_2 + \Gamma_\varphi} A_\varphi(E_L - E_r)$$

$$T_S(E_L) \approx \frac{\Gamma_1 \Gamma_\varphi}{\left(E_L - E_r\right)^2 + \left(\dfrac{\Gamma_1 + \Gamma_2 + \Gamma_\varphi}{2}\right)^2} = \frac{\Gamma_1 \Gamma_\varphi}{\Gamma_1 + \Gamma_2 + \Gamma_\varphi} A_\varphi(E_L - E_r)$$

where $A_\varphi(\varepsilon)$ is a Lorentzian function with a linewidth of $\Gamma \equiv \Gamma_1 + \Gamma_2 + \Gamma_\varphi$.

$$A_\varphi(\varepsilon) \approx \frac{\Gamma}{\varepsilon^2 + \left(\Gamma/2\right)^2}$$

E.6.4

	ENERGY (E)	NUMBER OF ELECTRONS (N)
IV ————	$E_1 + E_2 + 2U_0$	2
III ————	$E_2 + 0.5U_0$	1
II ————	$E_1 + 0.5U_0$	1
I ————	0	0

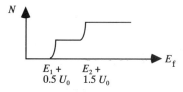

Fig. E.6.4 (*a*) Many-particle states in a structure with two levels E_1 and E_2. (*b*) Number of electrons in the structure as a function of the Fermi energy.

Equilibrium statistical mechanics tells us that at zero temperature any system goes to the state having the smallest value of $(E - NE_f)$. It is easy to see from Fig. E.6.4 that the states of the system will change as follows if we gradually increase the Fermi energy:

$$E_f < E_1 + 0.5U_0 \qquad\qquad\qquad \text{State I}$$
$$E_1 + 0.5U_0 < E_f < E_2 + 1.5U_0 \qquad \text{State II}$$
$$E_f > E_2 + 1.5U_0 \qquad\qquad\qquad \text{State IV}$$

As long as $E_2 > E_1$, state III is never reached. The number of electrons in the structure changes as shown with Fermi energy. It can be shown from Eq.(6.3.4) that the conductance spectrum consists of two peaks located at the energies where the particle numbers change:

$$G(E_f) = \frac{e^2}{h}\left(\frac{\Gamma_1^1\Gamma_2^1}{\Gamma_1^1+\Gamma_2^1}L[E_f - E_1 - 0.5U_0] + \frac{\Gamma_1^2\Gamma_2^2}{\Gamma_1^2+\Gamma_2^2}L[E_f - E_2 - 1.5U_0]\right)$$

Chapter 7

E.7.1

(a)
$$\frac{\sin\theta_1}{\sin\theta_2} = \frac{k_t/k_1}{k_t/k_2} = \frac{k_2}{k_1}$$

where k_t is the transverse wavevector (the same in both media), while k_1 and k_2 are the magnitudes of the wavevector in the two media:

$$E = \frac{\hbar^2 k_1^2}{2m} = U + \frac{\hbar^2 k_2^2}{2m} \quad\rightarrow\quad \frac{k_1}{k_2} = \sqrt{\frac{E}{E-U}}$$

hence
$$\frac{\sin\theta_1}{\sin\theta_2} = \sqrt{\frac{E-U}{E}}$$

(b) In this case

$$E = \frac{\hbar^2 k_1^2}{2m_1} = \frac{\hbar^2 k_2^2}{2m_2} \quad\rightarrow\quad \frac{k_1}{k_2} = \sqrt{\frac{m_1}{m_2}}$$

$$\frac{\sin\theta_1}{\sin\theta_2} = \sqrt{\frac{m_2}{m_1}}$$

Chapter 8

E.8.1

$$\nabla . \mathbf{J} = \frac{i\hbar}{2m}\left[\Psi(\nabla^2\Psi)^* - \Psi^*(\nabla^2\Psi)\right] - \frac{e\mathbf{A}}{m}\cdot\left[\Psi^*\nabla\Psi + \Psi\nabla\Psi^*\right] - \frac{e\nabla.\mathbf{A}}{m}\left[\Psi^*\Psi\right]$$

$$\left[\mathbf{p} - e\mathbf{A}\right]^2\Psi = \left[-i\hbar\nabla - e\mathbf{A}\right]\cdot\left[-i\hbar\nabla - e\mathbf{A}\right]\Psi$$
$$= -\hbar^2\nabla^2\Psi + e^2\mathbf{A}^2\Psi + 2ie\hbar\mathbf{A}.\nabla\Psi + ie\hbar\Psi(\nabla.\mathbf{A})$$

$$\Psi\left[\mathbf{p} - e\mathbf{A}\right]^2\Psi^* - \Psi^*\left[\mathbf{p} - e\mathbf{A}\right]^2\Psi$$
$$= -\hbar^2\left[\Psi\nabla^2\Psi^* - \Psi^*\nabla^2\Psi\right] - 2ie\hbar\mathbf{A}.\left[\Psi\nabla\Psi^* + \Psi^*\nabla\Psi\right] - 2ie\hbar\left[\nabla.\mathbf{A}\right]\Psi^*\Psi$$

Hence

$$i\hbar\nabla.\mathbf{J} = \frac{\Psi\left[\mathbf{p} - e\mathbf{A}\right]^2\Psi^* - \Psi^*\left[\mathbf{p} - e\mathbf{A}\right]^2\Psi}{2m} = \left[H_C\Psi\right]^*\Psi - \Psi^*\left[H_C\Psi\right]$$

since
$$H_C = \frac{\left[\mathbf{p} - e\mathbf{A}\right]^2}{2m} + U$$

E.8.2 Using Eq.(8.5.3) we can write

$$H_C G^n - G^n H_C = H_C G^R \Sigma^{in} G^A - G^R \Sigma^{in} G^A H_C \qquad (E.8.1)$$

From Eq.(8.5.2), $H_C G^R = EG^R - \Sigma^R G^R - I$ (E.8.2a)

Taking the Hermitian conjugate $G^A H_C = EG^A - G^A \Sigma^A - I$ (E.8.2b)

Substituting (E.8.2a,b) into (E.8.1),

$$H_C G^n - G^n H_C = \left[EG^R - \Sigma^R G^R - I\right]\Sigma^{in}G^A - G^R\Sigma^{in}\left[EG^A - G^A\Sigma^A - I\right]$$
$$= G^R\Sigma^{in} - \Sigma^{in}G^A - \Sigma^R G^n + G^n\Sigma^A$$

where we have made use of Eq.(8.5.3) again.

E.8.3 We can write

$$\text{Tr}\left[\Sigma^{in}G^p\right] = \iiint D(r,r';\hbar\omega)G^n(r,r';E - \hbar\omega)G^p(r',r;E)drdr'\,d(\hbar\omega)$$

$$\text{Tr}\left[\Sigma^{\text{out}}G^{\text{n}}\right] = \iiint D(r,r';\hbar\omega)G^{\text{p}}(r,r';E+\hbar\omega)G^{\text{n}}(r',r;E)drdr'\,d(\hbar\omega)$$

Hence $\quad \int \text{Tr}\left[\Sigma^{\text{out}}G^{\text{n}}\right]dE$

$$= \iiiint D(r,r';\hbar\omega)G^{\text{p}}(r,r';E+\hbar\omega)G^{\text{n}}(r',r;E)dEdrdr'\,d(\hbar\omega)$$

$$= \iiiint D(r',r;\hbar\omega)G^{\text{p}}(r',r;E)G^{\text{n}}(r,r';E-\hbar\omega)dEdrdr'\,d(\hbar\omega)$$

$$= \int \text{Tr}\left[\Sigma^{\text{in}}G^{\text{p}}\right]dE \quad \text{if} \quad D(r',r) = D(r,r')$$

E.8.4 Using Eq.(8.1.5) we can write

$$\begin{bmatrix} E - H_{\text{C}} - \Sigma^{\text{T}} & +i\Sigma^{\text{in}} \\ +i\Sigma^{\text{out}} & E - H_{\text{C}} + \Sigma^{\overline{\text{T}}} \end{bmatrix} \begin{bmatrix} G^{\text{T}} & -iG^{\text{n}} \\ -iG^{\text{p}} & -G^{\overline{\text{T}}} \end{bmatrix}$$

$$= \begin{bmatrix} E - H_{\text{C}} - \Sigma^{\text{R}} - i\Sigma^{\text{in}} & +i\Sigma^{\text{in}} \\ +i\Sigma^{\text{out}} & E - H_{\text{C}} - \Sigma^{\text{R}} - i\Sigma^{\text{out}} \end{bmatrix} \begin{bmatrix} G^{\text{R}} + iG^{\text{n}} & -iG^{\text{n}} \\ -iG^{\text{p}} & G^{\text{R}} + iG^{\text{p}} \end{bmatrix}$$

Consider the (11) term of the matrix obtained after multiplication. Making use of Eqs.(8.5.3), (8.5.2a) and (8.2.3) we obtain

$$[11] \rightarrow \left[E - H_{\text{C}} - \Sigma^{\text{R}} - i\Sigma^{\text{in}}\right]\left[G^{\text{R}} + iG^{\text{n}}\right] + \Sigma^{\text{in}}G^{\text{p}}$$

$$= I + i\left[E - H_{\text{C}} - \Sigma^{\text{R}}\right]G^{\text{n}} + \Sigma^{\text{in}}\left[G^{\text{n}} + G^{\text{p}} - iG^{\text{R}}\right]$$

$$= I + \Sigma^{\text{in}}\left[iG^{\text{A}} + G^{\text{n}} + G^{\text{p}} - iG^{\text{R}}\right] = I$$

Similarly we can work out the other elements of the matrix.

$$[12] \rightarrow \Sigma^{\text{in}}\left[-G^{\text{n}} - iG^{\text{A}} + iG^{\text{R}} - G^{\text{p}}\right] = 0 \qquad \text{etc., etc.}$$

Index